Degrowth and Tourism

The sustainability of tourism is increasingly under question given the challenges of overtourism, COVID-19 and the contribution of tourism to climate and environmental change. *Degrowth and Tourism* provides an original response to the central problem of growth in tourism, an imperative that has been intrinsic within tourism practice, and directs the reader to rethink the impacts of tourism and possible alternatives beyond the sustainable growth discourse.

Using a multi-scaled approach to investigate degrowth's macro effects and micro indications in tourism, this book frames degrowth in tourism in terms of business, destination and policy initiatives. It uses a combination of empirical research, case studies and theory to offer new perspectives and approaches to analyse issues related to overtourism, COVID-19, small-scale tourism operations and entrepreneurship, mobility and climate change in tourism. Interdisciplinary chapters provide studies on animal-based tourism, nature-based tourism, domestic tourism, developing community-centric tourism and many other areas, within the paradigm of degrowth.

This book offers significant insight on both the implications of degrowth paradigm in tourism studies and practices, as well as tourism's potential contributions to the degrowth paradigm, and will be essential reading for all those interested in sustainable tourism and transformations through tourism.

C. Michael Hall is a Professor at the University of Canterbury, New Zealand; Docent in Geography, University of Oulu, Finland; Visiting Professor in Tourism at Linnaeus University, Kalmar, Sweden; and a Guest Professor in the Department of Service Management and Service Studies, Lund University, Helsingborg, Sweden. He has written widely on tourism, regional development, heritage, food and global environmental change.

Linda Lundmark is an Associate Professor at the Department of Geography at Umeå University, Sweden. Her research interests among others are tourism, mobility, climate change and natural resources as part of contemporary and future development prospects in rural and sparsely populated areas in the far North. She is currently the chair of the Centre of Regional Science scientific advisory board and has received funding for several large research projects from Swedish research councils. She is also part of several research networks on tourism and tourism in polar areas.

Jundan Jasmine Zhang is a Postdoc at the Department of Ecology, Swedish University of Agricultural Sciences. She has a PhD in Tourism from University of Otago, New Zealand. Her main research interest lies in understanding the relationships between human and 'nature' in the context of global tourism. Jasmine has published in journals such as *Annals of Tourism Research*, *Journal of Sustainable Tourism* and *Tourism Geographies*, on subjects ranging from political ecology of tourism to tourism methodologies. She is currently doing research under the project *SVALUR – Understanding Resilience and Long-Term Environmental Change in the High Arctic: Narrative-Based Analyses from Svalbard*, where she is dedicated to bringing forward the multiplicity of knowledge on environmental changes through environmental humanitarian approaches.

Contemporary Geographies of Leisure, Tourism and Mobility
Series Editor: C. Michael Hall
Professor at the Department of Management, Marketing and Entrepreneurship, School of Business, University of Canterbury, Christchurch, New Zealand

The aim of this series is to explore and communicate the intersections and relationships between leisure, tourism and human mobility within the social sciences.

It will incorporate both traditional and new perspectives on leisure and tourism from contemporary geography, e.g. notions of identity, representation and culture, while also providing for perspectives from cognate areas such as anthropology, cultural studies, gastronomy and food studies, marketing, policy studies and political economy, regional and urban planning, and sociology, within the development of an integrated field of leisure and tourism studies.

Also, increasingly, tourism and leisure are regarded as steps in a continuum of human mobility. Inclusion of mobility in the series offers the prospect to examine the relationship between tourism and migration, the sojourner, educational travel, and second home and retirement travel phenomena.

The series comprises two strands:

Contemporary Geographies of Leisure, Tourism and Mobility aims to address the needs of students and academics, and the titles will be published simultaneously in hardback and paperback.

Routledge Studies in Contemporary Geographies of Leisure, Tourism and Mobility is a forum for innovative new research intended for research students and academics, and the titles will initially be available in hardback only. Titles include:

Tourism in Asian Cities
Edited by Saurabh Kumar Dixit

Degrowth and Tourism
New Perspectives on Tourism Entrepreneurship, Destinations and Policy
Edited by C. Michael Hall, Linda Lundmark and Jundan Jasmine Zhang

For more information about this series, please visit: www.routledge.com/Contemporary-Geographies-of-Leisure-Tourism-and-Mobility/book-series/SE0522

Degrowth and Tourism

New Perspectives on Tourism Entrepreneurship, Destinations and Policy

Edited by C. Michael Hall, Linda Lundmark and Jundan Jasmine Zhang

Routledge
Taylor & Francis Group
LONDON AND NEW YORK

First published 2021
by Routledge
2 Park Square, Milton Park, Abingdon, Oxon OX14 4RN

and by Routledge
52 Vanderbilt Avenue, New York, NY 10017

Routledge is an imprint of the Taylor & Francis Group, an informa business

British Library Cataloguing-in-Publication Data
A catalogue record for this book is available from the British Library

Library of Congress Cataloging-in-Publication Data
Names: Hall, Colin Michael, 1961– editor. | Lundmark, Linda, editor. | Zhang, Jundan Jasmine, editor.
Title: Degrowth and tourism : new perspectives on tourism entrepreneurship, destinations and policy / edited by C. Michael Hall, Linda Lundmark and Jundan Jasmine Zhang.
Description: Abingdon, Oxon ; New York, NY: Routledge, 2021. | Series: Contemporary geographies of leisure, tourism and mobility | Includes bibliographical references and index.
Identifiers: LCCN 2020039891 (print) | LCCN 2020039892 (ebook) | ISBN 9780367335656 (hardback) | ISBN 9780429320590 (ebook)
Subjects: LCSH: Sustainable tourism. | Overtourism.
Classification: LCC G156.5.S87 D44 2021 (print) | LCC G156.5.S87 (ebook) | DDC 338.4/79104—dc23
LC record available at https://lccn.loc.gov/2020039891
LC ebook record available at https://lccn.loc.gov/2020039892

ISBN: 978-0-367-33565-6 (hbk)
ISBN: 978-0-367-70034-8 (pbk)
ISBN: 978-0-429-32059-0 (ebk)

Typeset in Times New Roman
by Apex CoVantage, LLC

Contents

Figures

Tables

Contributors

Bailey Ashton Adie, School of Business, Law and Communications, Solent University, E Park Terrace, Southampton SO14 0YN, UK

Alberto Amore, School of Business, Law and Communications, Solent University, E Park Terrace, Southampton SO14 0YN, UK

Paul W. Ballantine, UC Business School, University of Canterbury, Private Bag 4800, Christchurch 8140, New Zealand

Karla A. Boluk, Department of Recreation and Leisure Studies, University of Waterloo, 200 University Ave W, Waterloo, Ontario, N2L 3G1, Canada

Sandro Carnicelli, School of Business and Creative Industries, Marketing, Events and Tourism, University of the West of Scotland, Blantyre, Glasgow, G72 0LH, Scotland, UK

Dean B. Carson, Department of Epidemiology and Global Health, Norrlands universitetssjukhus Epidemiologi och global hälsa, Umeå Universitet, 90185 Umeå, Sweden

Doris A. Carson, Department of Geography, Samhällsvetarhuset, Umeå universitet, Hörsalstorget 4, 901 87 Umeå, Sweden

Erika Andersson Cederholm, Department of Service Management and Service Studies, Lunds universitet, Campus Helsingborg, Box 882, 251 08 Helsingborg, Sweden

O. Cenk Demiroglu, Department of Geography, Samhällsvetarhuset, Umeå universitet, Hörsalstorget 4, 901 87 Umeå, Sweden

Marco Eimermann, Department of Geography, Samhällsvetarhuset, Umeå universitet, Hörsalstorget 4, 901 87 Umeå, Sweden

Peter Fredman, Faculty of Environmental Sciences and Natural Resource Management, Norwegian University of Life Sciences (NMBU) and Mid Sweden University, Department of Economics, Geography, Law and Tourism (EJT), SE-831 25 Östersund, Sweden

C. Michael Hall, Department of Management, Marketing and Entrepreneurship, University of Canterbury, Christchurch, New Zealand; Linnaeus University School of Business and Economics, Kalmar, Sweden; Department of Service Management and Service Studies, Lunds University, Helsingborg, Sweden; Geography, Oulu University, Finland

Freya Higgins-Desbiolles, School of Management, UniSA Business School, University of South Australia, Adelaide, South Australia, 5001, Australia

Chris Krolikowski, Centre for Tourism and Leisure Management, UniSA Business School, University of South Australia, Adelaide, South Australia, 5001, Australia

Outi Kulusjärvi, Geography Research Unit, PO Box 3000, 90014 University of Oulu, Oulu, Finland

Urban Lindgren, Department of Geography, Samhällsvetarhuset, Umeå universitet, Hörsalstorget 4, 901 87 Umeå, Sweden

Linda Lundmark, Department of Geography, Samhällsvetarhuset, Umeå universitet, Hörsalstorget 4, 901 87 Umeå, Sweden

Lusine Margaryan, Mid Sweden University, Department of Economics, Geography, Law and Tourism (EJT) SE-831 25 Östersund, Sweden

Anja Pabel, Centre for Tourism and Regional Opportunities, School of Business and Law, Central Queensland University, Cairns Campus, Cairns, Queensland, 4970, Australia

Bruce Prideaux, Centre for Tourism and Regional Opportunities, School of Business and Law, Central Queensland University, Cairns Campus, Cairns, Queensland, 4870, Australia

Esteban Ruiz-Ballesteros, Department of Social Anthropology, Psychology and Public Health, Universidad Pablo de Olavide, 41013 Sevilla, Spain

Jarkko Saarinen, Geography Research Unit, PO Box 3000, 90014 University of Oulu, Oulu, Finland; and School of Tourism and Hospitality, College of Business & Economics, University of Johannesburg, Bunting Road Campus, Auckland Park, Johannesburg, South Africa

Siamak Seyfi, Geography Research Unit, PO Box 3000, 90014 University of Oulu, Oulu, Finland

Carina Sjöholm, Department of Service Management and Service Studies, Lunds universitet, Campus Helsingborg, Box 882, 251 08 Helsingborg, Sweden

Stian Stensland, Faculty of Environmental Sciences and Natural Resource Management, Norwegian University of Life Sciences (NMBU), P.O. Box 5003, NO-1432, Ås, Norway

Ethemcan Turhan, Department of Spatial Planning and Environment, Faculty of Spatial Sciences, University of Groningen, Postbus 800, 9700 AV Groningen, the Netherlands

Gayathri Wijesinghe, UniSA Business School, University of South Australia, Adelaide, South Australia, 5001, Australia

Jundan Jasmine Zhang, Department of Ecology, Swedish University of Agricultural Science, Ulls väg 16, 75651 Uppsala, Sweden

Preface and acknowledgments

The idea of this book resulted from a conference session at the 27th Nordic Symposium for Tourism and Hospitality in Alta, Norway. We would like to thank the conference organizers for allowing us to hold the session, and the delegates who attended the session, as well as those who joined the lively discussions during and after the conference. Thanks also go to the reviewers who provided insightful comments on some early ideas of this book.

This research would not have been possible without the support of the Swedish research council for sustainable development FORMAS, which funded the research projects "Mobilising the rural: post-productivism and the new economy", "Mobilities, micro-urbanisation and changing settlement patterns in the sparsely populated North", "Rethinking urban tourism development: dealing with sustainability in the age of over-tourism" and "Climate change and the double amplification of Arctic tourism: challenges and potential solutions for tourism and sustainable development in an Arctic context".

Jasmine would also like to specifically acknowledge colleagues who have shared their enthusiasm on the ideas of degrowth and other related topics in the Department of Geography at Umeå University, for instance, Charlotta Hedberg, Desirée Enlund, Sabina Bergstén and Madeleine Eriksson. Other colleagues who have inspired her to continue this line of enquiries include Maxim Vlasov, Xaviera Sánchez de la Barquera Estrada and those who are actively involved in the Transition Network (Omställningsnätverket). Finally, her deepest appreciation goes to her family and friends in Umeå, not least Anton, Linus and Nova. Thank you for keeping me grounded and helping me to change perspectives every now and then, and for reminding me of the joy and lightness of life.

Michael would like to specifically thank a number of colleagues with whom he has undertaken related conversations and research over the years. In particular, thanks go to Bailey Adie, Alberto Amore, Paul Ballantine, Dorothee Bohn, Chris Chen, Tim Coles, Hervé Corvellec, David Duval, Stefan Gössling, Martin Gren, Johan Hultman, Maria Juschten, Joya Kemper, Myung Ja Kim, Tessa McKegg, Dieter Müller, Jan-Henrik Nilsson, Girish Prayag, Yael Ram, Anna Laura Raschke, Jarkko Saarinen, Anna Dóra Sæþórsdóttir, Dan Scott, Allan Williams, Kimberley Wood, and Maria José Zapata Campos for their thoughts, as well as for the stimulation of Agnes Obel, Ann Brun, Beirut, Paul Buchanan, Bill Callahan,

Nick Cave, Bruce Cockburn, Ebba Fosberg, Mark Hollis, Margaret Glaspy, Aimee Mann, Larkin Poe, Vinnie Reilly, Henry Rollins, Matthew Sweet, Emma Swift, Henry Wagon and The Guardian, BBC6, JJ and KCRW – for making the world much less confining. Special mention must also be given to the Malmö Saluhall; Balck, Packhuset, and Postgarten in Kalmar; and Nicole Aignier and the Hotel Grüner Baum in Merzhausen. Finally, and most importantly, Michael would like to thank the Js and the Cs who stay at home and mind the farm.

Following on previous acknowledgements, Linda's first thought goes to her husband Peter and their beloved children Lovena, Eino, Elvie and Viljo, to whom she owes tremendous gratitude and who are the joy of her life. Secondly, thoughts go to her mother and father and to her sister and her family, who have contributed to the discussions by sharing experiences from their lives. Last but not least, she expresses her gratitude to colleagues at the Department of Geography, Umeå University, as well as within her research networks and other networks for challenging ideas and for adding to her knowledge on the themes presented in this volume.

We also wish to gratefully acknowledge the help and support of Jody Cowper for proofreading and editing. Finally, we would both like to thank Emma Travis, Lydia Kessell and all at Routledge for their continuing support.

C. Michael Hall, Linda Lundmark, Jundan Jasmine Zhang

Acronyms

BECCS	bio-energy with carbon capture and storage
CBT	community-based tourism
CIPRA	International Commission for the Protection of the Alps
DMO	destination marketing organisation
GBR	Great Barrier Reef
GBRMPA	Great Barrier Reef Marine Park Authority
GDP	gross domestic product
GHG	greenhouse gas
IATA	International Air Transport Association
ILO	International Labour Organisation
IMF	International Monetary Fund
IPCC	Intergovernmental Panel on Climate Change
KPI	key performance indicator
LED	low energy demand
LTG	limits to growth
MRE	material/resource/energy
NBT	nature-based tourism
OECD	Organisation for Economic Co-operation and Development
SDGs	Sustainable Development Goals
SSP	shared socio-economic pathway
TCI	Tourism Climate Index
TNQ	Tropical North Queensland
TNQDTP	Tropical North Queensland Destination Tourism Plan
TTNQ	Tropical Tourism North Queensland
UNDP	United Nations Development Programme
UNEP	United Nations Environment Programme
UNWTO	United Nations World Tourism Organization
WCED	World Commission on Environment and Development
WEF	World Economic Forum
WHA	World Heritage Area
WTMA	Wet Tropics Management Authority
WTR	wet tropics rainforests
WTTC	World Travel and Tourism Council

1 Introduction

Degrowth and tourism: implications and challenges

Linda Lundmark, Jundan Jasmine Zhang and C. Michael Hall

Introduction

As an industry and a social practice that is based on people's ability to move from place to place, tourism requires continual rethinking and reconceptualising. Commodified cultural souvenirs; local residents displaced from neighbourhoods that have become tourism precincts; overcrowded tourists' destinations even at remote destinations like the High Arctic and the Antarctic; and not least constant flights, cruise ships and other tourism traffic emitting greenhouse gases into the atmosphere, are all signs of 'mass tourism' or, at worse, 'overtourism' (Phi, 2019; Adie, Falk, & Savioli, 2020; Sæþórsdóttir, Hall, & Wendt, 2020; Veríssimo, Moraes, Breda, Guizi, & Costa, 2020). This abruptly came to a halt in early 2020 when the novel coronavirus COVID-19 put many mobilities, economies and societies on hold (Gössling, Scott, & Hall, 2020). As a result, a number of researchers have argued that this pandemic provides an opportunity to rethink tourism practices, particularly the more unhealthy and unsustainable aspects of tourism (Cheer, 2020; Hall, Scott, & Gössling, 2020; Everingham & Chassagne, 2020; Renaud, 2020). Importantly, the impacts of COVID-19 on tourism throughout the world and its social, economic and environmental repercussions actually provide even more of a justification to move beyond tourism business as usual to suggest that a radical reform is both necessary and possible to bring to the global tourism agenda. As a result, it is therefore not surprising that a growing number of commentators argue that a degrowth perspective is more relevant than ever today for tourism businesses and destinations, and that tourism can also contribute to a more economically and environmentally equitable society (Hall, 2019; Higgins-Desbiolles, Carnicelli, Krolikowski, Wijesinghe, & Boluk, 2019), especially in the post COVID-19 era (Everingham & Chassagne, 2020).

Although this book was put into production prior to COVID-19 – stirring up the many messy and unsettling discussions on economy, environmental changes and healthcare systems that have been witnessed – the contributors to this book did come together from the common concern for the future of our planet and society. This anthology aims to provide new perspectives on tourism practice and theory from the paradigm of degrowth. Drawing on conceptual and empirical studies, this edited book crosses disciplinary boundaries and attempts to further integrate degrowth paradigms into tourism studies. At the same time, we also hope that the stories and voices coming through the chapters may contribute to the wider discussions among degrowth-interested groups and scholars.

The development of the degrowth concept is situated in a number of discourses of transition and shares similarities with many transition initiatives around the world (Escobar, 2015). Arguably, the contemporary degrowth concept is unique because it criticises the 'growth imperative' or growthism which is embedded within the dominant notion of development and modernity, including sustainable development (Demaria, Schneider, Sekulova, & Martinez-Alier, 2013; Hall, 2019). In many ways, degrowth's promise and proposals speak to UNDP's Agenda 2030 and the Sustainable Development Goals (SDGs) (see also Chapter 14, this volume). For instance, it advocates for the reduction of working hours, co-housing, voluntary simplicity and downshifting to promote the transition from a materialistic lifestyle towards a more participatory and environmental friendly society, while proposals on enhancing public services and redistributive taxation and including cooperative and circulative economy aim to bring more equity and socio-economic opportunities to more people (Cosme, Santos, & O'Neill, 2017). However, it is important to stress that degrowth is not a single voice, as diverse actors and movements are concerned with subjects ranging from ecological economics, technologies, and political ecology and economy to indigenous knowledge and green theology (Paulson, 2017). This diversity is also noticeable in the various chapters in the present work. Furthermore, degrowth is approached at different scales, for example, at that of individual consumers and entrepreneurs, through to businesses and communities and, in a tourism context, destinations and business networks. These different scales also provide insights into some of the ideological tensions that exist with respect to perspectives on degrowth, especially in a tourism context, and the extent to which the market and proposals for green growth are able to meet degrowth goals.

This introduction is divided into three main sections. The first part provides a broad overview of the degrowth literature via a scoping review of how the notion of degrowth has grown as a research subject in general and in tourism studies in particular. The second section then examines some of the main implications and challenges of employing degrowth in tourism research and the potential link to practice. The final section introduces the reader to the various chapters in this book and how they resonate with some of the identified implications and challenges.

Degrowth in tourism research

The topic of degrowth has substantially expanded in terms of the number of publications in just over a decade. A comprehensive review of all of those is not within the scope of this introductory chapter, but a brief bibliometric review of degrowth literature can help provide an overview of knowledge production derived from the notion of degrowth, as well as how issues of degrowth within tourism studies relate to the overall development of the degrowth paradigm. Following the PRISMA (Preferred Reporting Items for Systematic Reviews and Meta-Analyses) guidelines and applying search terms such as *degrowth*, *de-growth* and *décroissance* on Scopus and Web of Science (WoS), a list of 671 eligible peer-reviewed records was identified, of which 36 records fall within the field of tourism. Figure 1.1 shows how the peer-reviewed publications including the aforementioned keywords in their titles, abstracts and keywords have grown since 2007. It also clearly shows that the period 2018–2020 witnessed a major growth of degrowth-related publications within tourism studies.

As a term originally used by scholars and activists to critique the ideology and costs of growth-based development (Kallis et al., 2018), degrowth has been increasingly and extensively applied in different domains and with different foci. Using the software VOSviewer 1.6.14 for sourcing and analysing information on important themes, scholarly relationships, and research networks and communities (van Eck & Waltman, 2020), Figure 1.2 illustrates the diversity of keywords used in the degrowth literature and how the focus can change over time.

As shown in Table 1.1, the most frequently used keywords in the general degrowth literature (used 28 times or more) are mainly concerned with sustainability and economic (growth) matters. Climate change appears in tenth place. Mirroring the roots of the degrowth paradigm, *capitalism* occurs 49 times. In the smaller selection of degrowth and tourism literature, one interesting difference is that *sustainable development* and *sustainability* did not make it into the ten most frequently used keywords, although *sustainable tourism* is in ninth place. The issue of *overtourism*, although appearing rather recently, is in fifth place followed by *Spain*, showing the focus that has been on problems created by tourism in the Barcelona region (e.g. Ramos & Mundet, 2020).

Visualising the most frequently occurring keywords and the links between them illustrates the difference between tourism-specific research and general research on degrowth. The overall research is closely linked to the degrowth key term, and

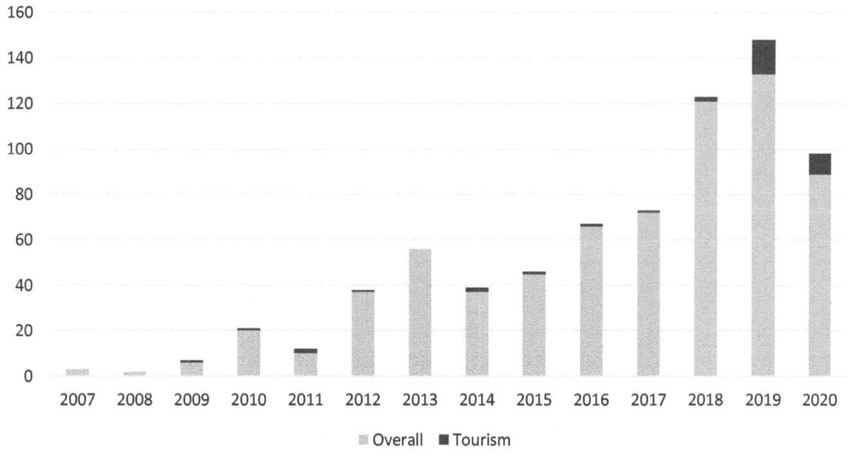

Figure 1.1 Number of degrowth publications overall and in tourism every year in Web of Science and Scopus

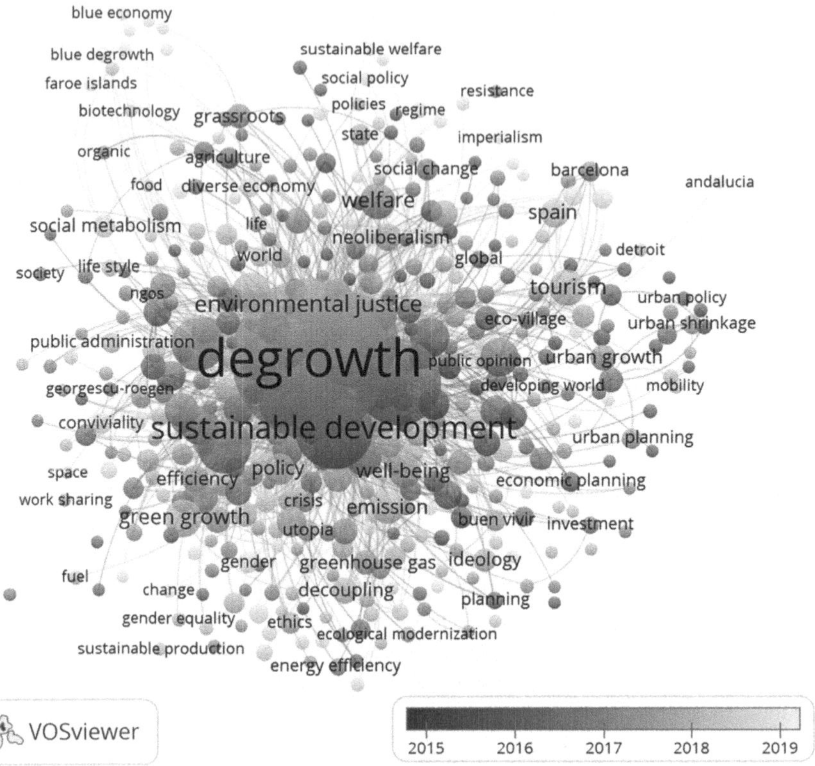

Figure 1.2 The diversity of keywords used in the degrowth literature and change in focus over time

degrowth is in the centre of the network. For tourism research, there is a less obvious core interest in degrowth, and the network is thus more spread out towards one side (see Figures 1.3a and Figure 1.3b, respectively).

In order to see how the topics in tourism degrowth literature relate to the overall degrowth literature, keywords in each of these two bodies of literature were ordered into three broader clusters in VOSviewer. As shown in Table 1.2, the topics in the overall literature appear to concentrate on a) the ideals and political aspects of degrowth; b) economic issues and sustainable development; and c) climate change and environmental justice. In the tourism literature, two clusters emerge with respect to research on tourism and degrowth: a) economic development, tourism development and issues of overtourism; and b) issues regarding sustainable tourism development and management. Although the second cluster within the tourism literature is more integrated into the core network of overall degrowth literature, indicating an explicit influence of the degrowth concept and

Table 1.1 The top 12 most frequently occurring keywords used in the identified records in Scopus and Web of Science

Keywords overall	*Number of occurrences*	*Total link strength*	*Keywords tourism*	*Number of occurrences*	*Total link strength*
Degrowth	432	534	Degrowth	23	53
Sustainability	150	290	Tourism development	9	36
Economic growth	128	248	Ecotourism	10	35
Sustainable development	106	207	Economic growth	6	24
Economics	52	104	Overtourism	6	24
Growth	48	101	Spain	5	21
Ecological economics	43	100	Tourism management	6	20
Capitalism	49	94	Tourist destination	6	20
Transition	36	87	Sustainable tourism	8	19
Climate change	44	80	Tourism market	5	19
Post-growth	29	75	Sustainability	6	18
Environmental justice	28	61	Sustainable development	5	17

Note: Analysed in VOSviewer 1.6.14

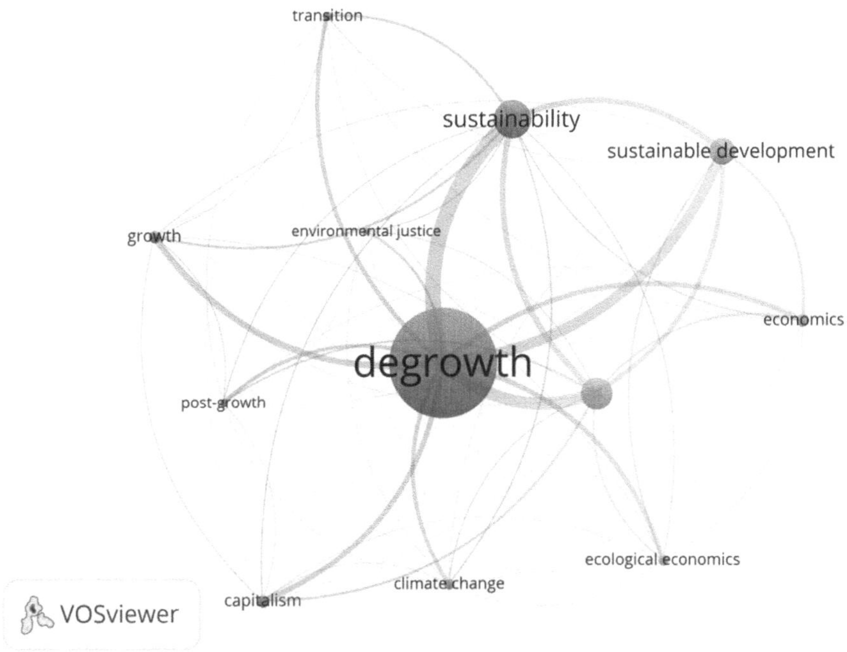

Figure 1.3a The 12 overall keywords occurring most frequently and the links between them

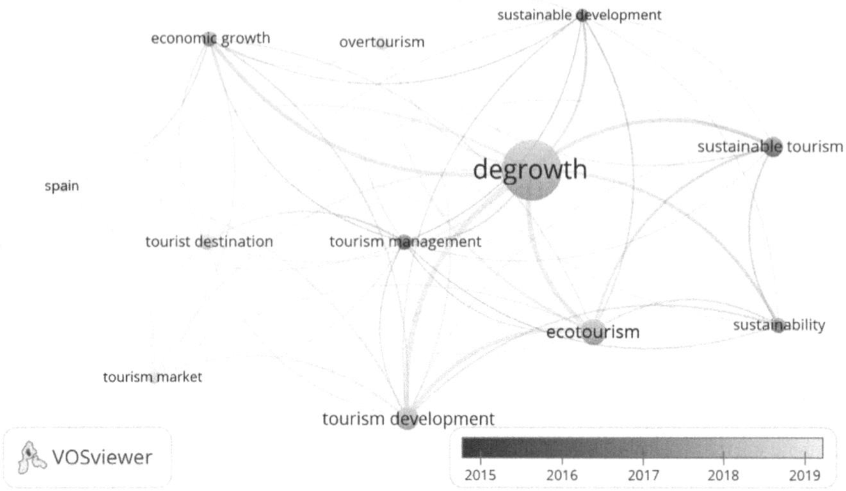

Figure 1.3b The 12 tourism keywords occurring most frequently and the links between them

Table 1.2 Clusters of keywords

Overall keywords clustering			*Tourism keywords clustering*	
First cluster	*Second cluster*	*Third cluster*	*First cluster*	*Second cluster*
Capitalism	Ecological economics	Climate change	Economic growth	Degrowth
Degrowth	Economic growth	Environmental justice	Overtourism	Ecotourism
Growth	Economics		Spain	Sustainability
Post-growth	Sustainable development		Tourism development	Sustainable development
Sustainability			Tourism market	Sustainable tourism
Transition			Tourist destination	Tourism management

research on tourism scholars, the first cluster of tourism literature indicates a rather applied approach to employ the degrowth notion to understand social, economic and environmental issues within tourism contexts.

A closer look at the tourism and degrowth literature provides us with further information on the two identified themes. The references included in this review are inserted under the keywords and clusters they belong to in Table 1.3.

The geographic scope and scale most prominent in the tourism and degrowth literature so far is urban (Blanco-Romero et al., 2019; Blázquez-Salom et al.,

Table 1.3 Major themes and corresponding authors

Economic development, tourism development and issues of overtourism	*Sustainable tourism development and management*
Economic growth Hall, 2009, 2010, 2011; Buhr, Isaksson, & Hagbert, 2018; Blázquez-Salom, Blanco-Romero, Vera-Rebollo, & Ivars-Baidal, 2019; Milano, Novelli, & Cheer, 2019; Valdivielso & Moranta, 2019; Andreoni, 2020; Everingham & Chassagne, 2020; Menton et al., 2020; Said & MacMillan, 2020	**Degrowth** Hall, 2009; Panzer-Krause, 2017; Buhr et al., 2018; Littig, 2018; Blanco-Romero, Blázquez-Salom, Morell, & Fletcher, 2019; Blázquez-Salom et al., 2019; Boluk, Cavaliere, & Higgins-Desbiolles, 2019; Chassagne & Everingham, 2019; Cheung & Li, 2019; Fletcher, Murray Mas, Blanco-Romero, & Blázquez-Salom, 2019; Gascón, 2019; Higgins-Desbiolles et al., 2019; Milano et al., 2019; Navarro-Jurado et al., 2019; Panzer-Krause, 2019; Perkumiene & Pranskuniene, 2019; Renkert, 2019; Valdivielso & Moranta, 2019; Andreoni, 2020; Carver, 2020; Everingham & Chassagne, 2020; Menton et al., 2020; Renaud, 2020; Said & MacMillan, 2020
Overtourism Cheung & Li, 2019; Fletcher et al., 2019; Higgins-Desbiolles et al., 2019; Milano et al., 2019; Navarro-Jurado et al., 2019; Perkumiene & Pranskuniene, 2019; Romero-Padilla, Cerezo-Medina, Navarro-Jurado, Romero-Martínez, & Guevara-Plaza, 2019; Valdivielso & Moranta, 2019	**Ecotourism** Hall, 2011; Canavan, 2014; Meana Acevedo, 2016; Chassagne & Everingham, 2019; Cheung & Li, 2019; Fletcher et al., 2019; Higgins-Desbiolles et al., 2019; Panzer-Krause, 2019; Perkumiene & Pranskuniene, 2019; Valdivielso & Moranta, 2019; Leposa, 2020
Spain Blázquez-Salom et al., 2019; Milano et al., 2019; Navarro-Jurado et al., 2019; Romero-Padilla et al., 2019; Valdivielso & Moranta, 2019	**Sustainability** Hall, 2009, 2010; Canavan, 2014; Panzer-Krause, 2017; Buhr et al., 2018; Blázquez-Salom et al., 2019; Boluk et al., 2019; Panzer-Krause, 2019; Perkumiene & Pranskuniene, 2019; Renkert, 2019; Leposa, 2020; Menton et al., 2020; Said & MacMillan, 2020
Tourism development Hall, 2010; Canavan, 2014; Blanco-Romero et al., 2019; Boluk et al. 2019; Fletcher et al., 2019; Gascón, 2019	**Sustainable development**Hall, 2009, 2011; Meana Acevedo, 2016; Littig, 2018; Boluk et al., 2019; Fletcher et al., 2019; Andreoni, 2020; Menton et al., 2020; Said & MacMillan, 2020
Tourist market Sun, Lin, & Higham, 2020	**Sustainable tourism** Hall, 2009, 2010, 2011; Canavan, 2014; Hollenhorst, Houge-Mackenzie, & Ostergren, 2014; Panzer-Krause, 2017; Boluk et al., 2019; Chassagne & Everingham, 2019; Cheung & Li, 2019; Fletcher et al., 2019; Higgins-Desbiolles et al., 2019; Panzer-Krause, 2019; Perkumiene & Pranskuniene, 2019; Valdivielso & Moranta, 2019; Ramos & Mundet, 2020
Tourism destination Hernández-Martín, Álvarez-Albelo, & Padrón-Fumero, 2015; Panzer-Krause, 2017; Sun et al., 2020	**Tourism management** Hall, 2009; Canavan, 2014; Hernández-Martín et al., 2015; Milano et al., 2019; Sun et al., 2020

2019; Navarro-Jurado et al., 2019) or focusing on the city (Romero-Padilla et al., 2019). There is also a strong focus on Europe and especially the Mediterranean region, concentrating on Barcelona (Ramos & Mundet, 2020), Costa del sol (Navarro-Jurado et al., 2019), Malaga and Marbella (Romero-Padilla et al., 2019), as well as coastal tourism (Romero-Padilla et al., 2019). This is partly due to the mature state of the destinations and the popularity and accessibility of the region among tourists, causing what has been commonly addressed as overtourism (Milano et al., 2019), as well as the strong European focus on notions of degrowth as opposed to the more North American emphasis on the notion of a steady-state economy (Kerschner, 2010; Buch-Hansen, 2014; Koch, 2015; O'Neill, 2015). The island of Rugen (Panzer-Krause, 2017) and the example of rural Ireland (Panzer-Krause, 2019) are atypical in this context. There are also examples from parts of the world other than Europe such as Namibia (Carver, 2020) and Hong Kong (Cheung & Li, 2019).

Relating degrowth to tourism

The bibliometric review demonstrates how a body of tourism knowledge has grown from the expanding intellectual and academic works of degrowth paradigm. While components in the degrowth movement and paradigm are not completely new for tourism researchers, such as critiques on consumerism and commodification, as well as stresses on responsibility, well-being and equality, the explicit recognition of degrowth and its implications in tourism is still rather recent (Hall, 2009, 2010; Andriotis, 2014). The early usage of the term 'degrowth' was inspired by the first international degrowth conference in Paris in 2008 (e.g. Hall, 2009) as well as Daly's (1972, 1974, 2014) work on a steady-state economy, yet it was only following the emergence of interest in overtourism that the concept gained substantial attention in tourism studies, leading to a special issue dedicated on this topic (Fletcher et al., 2019) and a monograph discussing the potential for degrowth within tourism (Andriotis, 2018). This increase of attention on degrowth can also be partly explained by the metanarratives and discourses on climate change and Anthropocene, which may have contributed to growing awareness of the damages caused by growth imperatives (Gren & Huijbens, 2014, 2015; Hall, Baird, James, & Ram, 2016). These have, in turn, led to anxieties towards political, social, economic and environmental crises, resulting in an even more urgent search for alternatives and feasible solutions for a sustainable future.

Similarly, while tourism has been portrayed in many contexts as the 'alternative' and more 'environmentally friendly' industry, the general public have become increasingly aware that this is not necessarily the case, leading to new contested geographies of destinations (Gyimóthy, Morales Pérez, Widtfeldt Meged, & Wilson, 2020; Morales-Pérez, Garay, & Wilson, 2020). While social, environmental and cultural dimensions have been said to be important in a sustainable tourism operation, the continuing pursuit of efficiency and sufficiency in a tourism context has made tourism scholars increasingly question if tourism is contributing to a more sustainable future or rather is a 'tragedy of the commons'

(Hall, 2009; Hall, Gössling, & Scott, 2015; Büscher & Fletcher, 2017), because economic growth is limited by both physical and social boundaries if future possibilities are to remain for future generations. Degrowth therefore re-emerged in a context that influences tourism scholars as both social theory and social movement (Demaria, Schneider, Sekulova, & Martinez-Alier, 2013; Schneider, Kallis, & Martinez-Alier, 2010; Martínez-Alier, Pascual, Vivien, & Zaccai, 2010). Kallis et al. (2018) map out how different disciplines and fields can potentially contribute to a more coherent degrowth knowledge system, and among others, scholars from anthropology and social sciences have shed light on cases and insights from peripheral niches around the world where societies can and are willing to live with low-throughput and steady-state socio-economics (p. 302). Some of the recent works done in the tourism field using degrowth can also be said to contribute to such potentials.

Fletcher et al. (2019) identify 'overtourism' as a manifestation of the permeating growth imperative that has evoked or renewed tourism scholars' interests to engage with the degrowth and limits to growth literature (Sæþórsdóttir et al., 2020), where a more radical and determined approach is promoted than the previous more 'business as usual' approaches (UNWTO & UNDP, 2017; WTTC and McKinsey & Company, 2017) that only treats overtourism as a management problem (Hall, 2019). Indeed, Fletcher et al. (2019) argue that the tourism industry itself is a form of global economic expansion which demands serious rethinking and re-politicization, if ideas on degrowth are to aid the tourism industry and its capacity to contribute to a more sustainable society. In keeping with this approach, several authors discuss the relationships between degrowth and tourism in terms such as oversaturation (Blanco-Romero et al., 2019), tourism saturation (Blázquez-Salom et al., 2019) and excess tourism (Romero-Padilla et al., 2019), defining it as social discontent towards tourists (Fletcher et al., 2019; Higgins-Desbiolles et al., 2019), while Sæþórsdóttir et al. (2020), focusing more on the environmental dimension and visitor perceptions, return to long-standing notions such as carrying capacity. Some of the early scholars on the topic were talking about steady-state tourism (Hall, 2009, 2010), mainly referring to the unsustainable trajectories of volume-oriented perpetual growth in tourism at destination and global scales, and also critiquing notions of green growth (Hall, 2013, 2015).

Within the tourism literature there is also a substantial interest in local residents in tourism destinations and their responses to tourism development, especially Airbnb and tourism-related gentrification (Amore, 2019; Amore, de Bernardi, & Arvanitis, 2020). The critique towards the sustainability of tourism lies in the fact of the right to be a tourist, and a tourism business often appears to be given more weight than the rights of local residents (Higgins-Desbiolles et al., 2019). There is also a focus on community-based tourism (Chassagne & Everingham, 2019), social movements that resist environmental impacts (Navarro-Jurado et al., 2019) and housing issues (Romero-Padilla et al., 2019; Amore et al., 2020). One such example is the debate surrounding the short-term rental of properties to tourists via Internet platforms, such as Airbnb. Although short-term rental housing has long been available to visitors in tourism destinations, new technology has increased the visibility in a global market and in cities with urban-cultural tourism, and housing is a new form of accommodation that because of its effect on the availability and cost of housing to permanent residents in some destinations generates new conflicts (Müller & Hall, 2018; Gössling & Hall, 2019; Romero-Padilla et al., 2019; Kuhzady, Seyfi, & Béal, 2020).

There is also a strong critical tradition in the tourism and degrowth literature. Among others these include the critique of the capitalist accumulation model (Blázquez-Salom et al., 2019) and recognition of tourism as a key facet of capitalism (Hall, 2015; Higgins-Desbiolles et al., 2019). Renkert (2019) and Chassagne and Everingham (2019) have argued for a shift away from Eurocentric ideas of tourism development and growth as well as tackling more exploitative models of tourism (Everingham & Chassagne, 2020). Such research also often crosses over into studies seeking more inclusive forms of tourism; although such a focus deals with issues of equity, wealth distribution, and economic, environmental and social justice in a tourism context, it does not necessarily mean that tourist growth will be limited by destinations, tourist firms or travel companies (Zapata Campos, Hall, & Backlund, 2018). Sometimes the focus on restricting tourism growth also contains elements of activism (Chassagne & Everingham, 2019; Hall, 2016; Higgins-Desbiolles, 2010), although such activist interventionist positions are generally limited with respect to tourism studies.

Given the long-standing relationship between debates on growth and sustainability (Saarinen, 2015, 2018), there is also a substantial engagement from a degrowth perspective with the sustainability issues of tourism development and the critique of aggregate quantitative growth in a tourism context (Hall, 2009, 2010, 2011), including when described as green growth (Hall, 2015; Panzer-Krause, 2017). As a result of the interface between sustainable tourism and degrowth thinking, some proponents of degrowth strategies in tourism, such as Panzer-Krause (2017), are advocating for a stronger consideration of ecological and social-cultural aspects of development in economic geography, whereas others argue for issues of social justice and equity (Blanco-Romero et al., 2019) to be brought to the forefront of research. This latter perspective is also significant because of degrowth being proposed as a way of improving the quality of everyday life (Romero-Padilla et al., 2019). As a topic this connects to several research contributions: the connection between tourism, Buen Vivir and degrowth (Chassagne & Everingham, 2019) or quality of life (Navarro-Jurado et al., 2019), where Buen Vivir is seen as positive for transformative change and degrowth is a consequence of another way of life (Everingham & Chassagne, 2020). The interest in Buen Vivir – "the good life" that comes from the collective well-being derived from being part of a community (Salazar, 2015) – highlights the challenges of implementing degrowth thinking. The notion of Buen Vivir was incorporated into the Ecuadorian constitution in 2008 and the Bolivian constitution in 2009, but the transfer from idea to reality has been met with resistance. As Mercado (2017) observes under the heading of 'a struggle to incorporate into modern politics and economics':

> Not surprisingly, the principles of Buen Vivir are difficult to carry out in today's economic and political environment. The governments of Ecuador

and Bolivia are struggling to reconcile their goals to lead social change with the day-to-day financial needs of the country and the demands for development from large corporations.

Implications and challenges of degrowth in tourism practice

The difficulties of implementing notions of Buen Vivir in policy practice are reflected in the wider difficulties of achieving degrowth in practice at destination, community and business levels. Although Fletcher et al. (2019) argue that "the search for post-growth, post-capitalist, post-development and/or degrowth alternatives has become a social and intellectual imperative" (p. 1758), its incorporation into tourism practices is somewhat limited.

Andriotis (2018) suggests seeing degrowth as useful in the context of alternative forms of tourism rather than mass tourism, pointing to the need for developing ways of 'degrowth-inspired traveling'. But tourists as subjects are largely ignored thus far in studies on degrowth and tourism. Moreover, the articulations of alternative forms of tourism have tended to remain just that – alternative – and have failed to shift the dominant growth orientation of tourism.

Degrowth does have linkages to considerations of the role of demarketing and social marketing in tourism, which seeks to shift tourist and business behaviours towards more sustainable forms of travel and tourism consumption (Armstrong & Kern, 2011; Hall, 2014; Armstrong Soule & Reich, 2015). Although implementation of such measures tended to be applied in limited situations, often connected to what would now be regarded as issues of overtourism, rather than being part of a comprehensive rethinking of the nature of tourism growth and development. Similarly, although clearly an important theme and integral to the nature of the global travel system (Hall, 2010), the relationship between mobility and degrowth has only received limited consideration (Higgins-Desbiolles et al., 2019; Renaud, 2020). Therefore, travel is also not very visible as part of the degrowth and tourism research with few exceptions, for example, with respect to the right to travel (Perkumiene & Pranskuniene, 2019) and travel emissions (Sun et al., 2020). In the case of the latter, a range of interventions have been proposed to reduce emissions and potentially reduce the emissions to and in specific destinations (Gössling, Scott, & Hall, 2015; Oklevik et al., 2019; Sun et al., 2020), but to achieve such measures at a system-wide level is regarded as extremely problematic (Gössling, Hall, Peeters, & Scott, 2010; Scott, Hall, & Gössling, 2016; Scott, Gössling, Hall, & Peeters, 2016).

Another key point that is missing from much of the discussion on degrowth is that, as with concerns regarding overtourism, there are often very different contexts for tourism at different geographic locations (Sæþórsdóttir et al., 2020). Similar to existing policies and development strategies that influence the growth and decline of different places, concepts such as degrowth are potentially also open to differentiated interpretation and implementation strategies in response to local and regional contexts, effects, challenges depending and governance arrangements (Rasoolimanesh, Ramakrishna, Hall, Esfandiar, & Seyfi, 2020). One apparent example is the urban-rural divide and how that is often constructed by different development policy and social norms (Sæþórsdóttir et al., 2020). How degrowth can be used in understanding and challenging the geographies of growth and decline, in order to build a sustainable society, as well as the role of governance and institutional arrangements in the applicability and selection of degrowth strategies, is clearly an important area of research that requires further investigation.

In the tourism and degrowth literature so far, the concept has tended to be focused on dealing with issues in places that have grown or 'overgrown', rather than necessarily dealing with the global system. In his seminal paper in a tourism context, Hall (2009) drew on the sustainable consumption literature as well as the concept of degrowth to highlight the differences between a green growth strategy that focused on efficiency as well as retaining a commitment to growth and a consumer-oriented sufficiency strategy that sought to slow down the rate of consumption and suggested that both were needed in order to achieve degrowth in tourism. Significantly, he also suggested that focusing on reductions in consumption in isolation could lead to economic recession, something that is extremely relevant to the impacts of COVID-19 on the tourism industry (Hall, 2009, 2015; see also Chapter 14, this volume). Similarly, Paulson (2017) points out that the difference between voluntary and involuntary degrowth lies in how 'development' has been historically perceived and carried out in different contexts, and the boundaries can be blurred. Historical processes of economic recession and resource depletion, or the impacts of pandemics (Hall et al., 2020), are often not regarded as degrowth and actually may have little long-term impact on the negative aspects of growthism (Forster et al., 2020), yet historical trajectory sometimes can be messed up with contemporary (mis)representations. For instance, it is a popular myth that Scandinavian countries have a high potential to put a degrowth paradigm into practice due to their relatively high GDPs and high levels of environmental awareness and socio-economic stability. However, in reality, many rural areas in Scandinavia as well as other European countries and elsewhere in the so-called developed world are struggling with economic recession and depopulation, and degrowth has yet to enter their political discourse (Buhr et al., 2018). Degrowth in different economic-geographic contexts may therefore need to be treated more specifically so that regional nuances and complexities can be better captured in research.

A further challenge to the practice of degrowth is the positionality of researchers in relation to degrowth theory. The question of degrowth is highly related to one's philosophical and ideological positioning – in other words, one's views on how the world and society should be operating and how one's own being and becoming are situated in the world. Demmer and Hummel (2017) argue that all scientific practice, not least the ones where researchers work with the degrowth notion, involves an 'ontological politics'. Indeed, in reflecting on the creation of "new structures and systems over time as part of a sustainable transition process" (Hall, 2016, p. 368), Hall suggested that:

> effective behavioural change is likely to also require changes in research practices. Science is not the same as journalism, but there is a lesson to be

> learnt with respect to communication and its political nature. Peer-reviewed research remains vital; however, although potentially unpalatable to many, perhaps the model that may best be followed, if sustainable tourism and mobility are to be achieved is not the path of so-called "objective science", but a more value-driven activist/advocacy-based interventionist model of upstream social marketing that seeks to change socio-technical regimes.
>
> (Hall, 2016, p. 369)

Nevertheless, such ontological and activist positioning raises significant questions as to what kind of knowledge is created through the engagement of degrowth ideas in tourism studies and the implications of such knowledge in policy-making for tourism destinations' marketing and management.

Entrepreneurship, destinations and policy: the contributions of this book

The existing literature on degrowth and tourism indicates that overall, degrowth perspectives link tourism scholars to a number of other fields, including consumer behaviour, ecological economics, geography, political science, social marketing, sustainability and studies of socio-technical transitions. While acknowledging that the previous works undertaken on degrowth and tourism have been invaluable, the contributions to this volume seek a) to broaden the scope of the current discussion of degrowth and tourism; b) to make links between other conceptual frameworks with degrowth ideas, other than the currently most recognized frames such as political economy and political ecology; and c) to shed light on other areas of potentially relevant research and of interest for tourism scholars' knowledge making.

A multi-level approach is adapted as different chapters investigate degrowth's macro effects and micro indications in tourism at different scales, and the book is divided into three main sections organized under the themes of entrepreneurship, destinations and policies that provide theoretically situated and empirically grounded illustrations of the relevance of degrowth in tourism studies. The first section focuses as the way that individual tourism entrepreneurs have sought to embrace notions of degrowth in their activities, with such research building on the substantial body of research in tourism on so-called lifestyle and values-based entrepreneurship and sustainable tourism business operations in particular (Shaw & Williams, 2004; Andersson Cederholm & Hultman, 2010; Natale, Arcidiacacono, & Martino, 2013; Font, Garay, & Jones, 2016; Tomassini, Font, & Thomas, 2020).

While the definition of entrepreneurship is not uncontroversial, it has increasingly come to be seen as an essential social innovation for achieving sustainability goals (Hörisch, 2015). The ideas brought by sustainable entrepreneurs into existing markets interact and influence with consumers' motivation and behaviours and, to some extent, create space for changes that contribute to sustainable consumption. Meanwhile, Næss (2020) calls for a radical innovation that explores pathways for transitioning to sustainable degrowth, where sustainable entrepreneurship such as social and ecological entrepreneurship have been recognized as important elements in a degrowth context (Hörisch, 2015). In tourism studies, Panzer-Krause (2019) shows how tourism entrepreneurs in rural Ireland form sustainability networks in the wake of social and economic transitions if rural areas and degrowth strategies are used as a part of social innovations.

Three chapters in the section *Degrowth and Tourism Entrepreneurship* share a geographic focus on Scandinavian countries and highlight the relationship between lifestyle entrepreneurship and alternative business philosophies that can be aligned with degrowth ideas. In Chapter 2, Andersson Cederholm and Sjöholm present two examples of lifestyle-oriented animal-based tourism in Sweden: horse-based tourism and hunting tourism, illustrating enterprises engaging with decommodification practices as a way to resist commodification. Drawing on ideas of gift-economic exchange, these authors argue the non-growth orientation shown in these tourism lifestyle entrepreneurial ventures is related to a range of economic and non-economic values that are embedded in a social and cultural context. Margaryan and colleagues in Chapter 3 look at the confluence of degrowth and lifestyle entrepreneurship. Drawing on interview studies and national surveys, they focus on the case of small-scale, nature-based tourism entrepreneurs in Norway and Sweden, and demonstrate the prerogative of lifestyle values in this business, such as the importance of working outdoors, being self-reliant, working with people with similar interests, contributing to sustainable development, utilising local resources or educating people about nature. These values, they argue, enable the lifestyle entrepreneurs to become viable and vigorous actors and 'agents of degrowth'. Looking at the context of rural areas in Sweden, a contrasting perspective is provided in Chapter 4 in Eimermann et al.'s study of the phenomenon of "holistic simplifiers" and understanding its mobility aspects with the help of georeferenced statistic data. Importantly, the chapter questions some of the myths that surround sea and tree change migration such as that movement contributes to more sustainable lifestyles and practices. The authors also point out some methodological issues that may be considered for future research in tourism degrowth.

Significantly, one of the clearest potential contributions of tourism to research on degrowth is the importance of place-based cases that relate to the destination focus of much tourism research. Section 2, *Degrowth and Tourism Destinations*, collects four chapters from four different destinations. In Chapter 5 Kulusjärvi brings us to northern Finland. Through ethnographic studies and employing a poststructural political economy approach, Kulusjärvi argues that it is important for tourism destinations to diverge from the growth-focused tourism strategies and make space within the current political structures for alternative economic pathways, agency and subjectivities, in order to enable a network of degrowth economy in tourism destinations. In comparing two Italian World Heritage destinations at different stages of their development of heritage tourism in Chapter 6, Amore and Adie critically examine how the notion of degrowth can be understood in destination policies and in practice. Chapter 7 shifts focus to peripheral destinations in Australia. Similar to the heritage tourism destinations, tourism in these remote contexts often followed a boosterist growth paradigm while the urban-rural dynamic continues to create a developmental gap. Looking at the issues from a degrowth perspective, Carson and Carson (Chapter 7) discuss the possibilities and barriers for using degrowth to reposition and redistribute resources between tourism destinations and markets. Chapter 8 takes up the challenge of climate change for nature-based tourism destinations. Advocating degrowth-oriented strategies and ideas for changing the track of the ever-increasing environmental damages caused by the tourism industry, Prideaux and Pabel point out that a suitable economic model or models would be needed in order to make the transition from the current global production system to a new one that supports degrowth principles. Hindrances exist, but the authors see it as feasible to implement degrowth strategies at the destination level and therefore propose a framework to assist tourism destinations to achieve degrowth goals and to cope with declining ecosystems.

Six chapters are included in the last section, *Degrowth and Tourism Policy*, examining how a degrowth paradigm may impose on tourism-related policies, and what role policies may take in facilitating a degrowth-inspired tourism industry. In Chapter 9 Saarinen takes a critical overview of various recent policy documents and examines the critical issues in sustainable tourism and its connections to the neoliberalisation of tourism development thinking, in particularly in terms of governance. The chapter also provides some guidance and directions on structures and strategies for the tourism industry shifting towards a degrowth or post-growth economy. Boluk et al. in Chapter 10 continue this discussion of neoliberalism and growth imperative in sustainability discourse and tourism development, highlighting the importance of rethinking tourism through placing the rights of local communities at the centre of tourism practice. Ruiz-Ballesteros in Chapter 11 also considers communitarian logics and philosophies such as reciprocity, commons and conviviality in community-based tourism practices in Ecuador that can be important factors for degrowth. Chapter 12 by Ballantine looks at the way in which domestic tourism campaigns may be used as a degrowth strategy by persuading consumers to travel domestically rather than internationally. A bottom-up approach is also pursued in Chapter 13, where Demiroglu and Turhan use web content analysis to examine how climate activists and lobbyist groups within the Nordic ski tourism community have challenged mainstream tourism consumption and consequently the possibilities of establishing new norms that may influence new travel models on a mass scale. Chapter 14 by Hall and Seyfi looks at the way in which the impacts of COVID-19 on tourism may be understood from a degrowth perspective. The authors highlight that the effects of the pandemic illustrate the recessionary potential for degrowth strategies unless accompanied by appropriate policy measures. The final chapter by the editors, Chapter 15, posits some potential future research directions with respect to tourism and degrowth at various scales.

Material degrowth is relatively easy to grasp but ideologically and politically difficult to implement. However, the sustainability of tourism is increasingly under question given problems of overtourism in a wide range of destinations prior to COVID-19, the contribution of tourism to climate and environmental change, and the potential for a return to growthism in response to the economic impacts of COVID-19 (Hall et al., 2020). Nevertheless, COVID-19 has been seen by a number of commentators as a transformative opportunity to reset tourism's growth focus and its accompanying economic, socio-cultural and environmental effects (e.g. Galvani, Lew, & Perez, 2020; Ioannides & Gyimóthy, 2020; see also the special issue of *Tourism Geographies*, *22*(3): Visions of travel and tourism after the global COVID-19 transformation of 2020 (Lew et al., 2020)). Although the various chapters in this book necessarily embrace a transformative vision, they do provide original responses to the central problem of growth in tourism at multiple scales and how a degrowth perspective may contribute to a more sustainable vision of tourism and its broader contribution to economic, social and environmental well-being.

References

Adie, B. A., Falk, M., & Savioli, M. (2020). Overtourism as a perceived threat to cultural heritage in Europe. *Current Issues in Tourism*, *23*(14), 1737–1741.

Amore, A. (2019). *Tourism and urban regeneration: Processes compressed in time and space*. Abingdon: Routledge.

Amore, A., de Bernardi, C., & Arvanitis, P. (2020). The impacts of Airbnb in Athens, Lisbon and Milan: A rent gap theory perspective. *Current Issues in Tourism*. doi:10.1080/13683500.2020.1742674

Andersson Cederholm, E., & Hultman, J. (2010). The value of intimacy – Negotiating commercial relationships in lifestyle entrepreneurship. *Scandinavian Journal of Hospitality and Tourism*, *10*(1), 16–32.

Andreoni, V. (2020). The trap of success: A paradox of scale for sharing economy and degrowth. *Sustainability*, *12*(8). doi:10.3390/SU12083153

Andriotis, D. K. (2014). Tourism development and the degrowth paradigm. *Turističko poslovanje*, *2014*(13), 37–45.

Andriotis, D. K. (2018). *Degrowth in tourism: Conceptual, theoretical and philosophical issues*. Wallingford: CABI.

Armstrong, E. K., & Kern, C. L. (2011). Demarketing manages visitor demand in the blue mountains national park. *Journal of Ecotourism*, *10*(1), 21–37.

Armstrong Soule, C. A., & Reich, B. J. (2015). Less is more: Is a green demarketing strategy sustainable? *Journal of Marketing Management*, *31*(13–14), 1403–1427.

Blanco-Romero, A., Blázquez-Salom, M., Morell, M., & Fletcher, R. (2019). Not tourismphobia but urban-philia: Understanding stakeholders' perceptions of urban touristification. *Boletin De La Asociacion De Geografos Espanoles,* (83), 30. doi:10.21138/bage.2834

Blázquez-Salom, M., Blanco-Romero, A., Vera-Rebollo, F., & Ivars-Baidal, J. (2019). Territorial tourism planning in Spain: From boosterism to tourism degrowth? *Journal of Sustainable Tourism*, *27*(12), 1764–1785.

Boluk, K. A., Cavaliere, C. T., & Higgins-Desbiolles, F. (2019). A critical framework for interrogating the united nations sustainable development goals 2030 agenda in tourism. *Journal of Sustainable Tourism*, *27*(7), 847–864.

Buch-Hansen, H. (2014). Capitalist diversity and de-growth trajectories to steady-state economies. *Ecological Economics*, *106*, 167–173.

Buhr, K., Isaksson, K., & Hagbert, P. (2018). Local interpretations of degrowth: Actors, arenas and attempts to influence policy. *Sustainability*, *10*(6), 14. doi:10.3390/su10061899

Büscher, B., & Fletcher, R. (2017). Destructive creation: Capital accumulation and the structural violence of tourism. *Journal of Sustainable Tourism*, *25*(5), 651–667.

Canavan, B. (2014). Sustainable tourism: Development, decline and de-growth. Management issues from the Isle of man. *Journal of Sustainable Tourism*, *22*(1), 127–147.

Carver, R. (2020). Lessons for blue degrowth from Namibia's emerging blue economy. *Sustainability Science*, *15*(1), 131–143.

Chassagne, N., & Everingham, P. (2019). Buen Vivir: Degrowing extractivism and growing wellbeing through tourism. *Journal of Sustainable Tourism*, *27*(12), 1909–1925.

Cheer, J. M. (2020). Human flourishing, tourism transformation and COVID-19: A conceptual touchstone. *Tourism Geographies*. doi:10.1080/14616688.2020.1765016.

Cheung, K. S., & Li, L. H. (2019). Understanding visitor – Resident relations in overtourism: Developing resilience for sustainable tourism. *Journal of Sustainable Tourism*, *27*(8), 1197–1216.

Cosme, I., Santos, R., & O'Neill, D. W. (2017). Assessing the degrowth discourse: A review and analysis of academic degrowth policy proposals. *Journal of Cleaner Production*, *149*, 321–334.

Daly, H. E. (1972). In defense of a steady-state economy. *American Journal of Agricultural Economics*, *54*(5), 945–954.

Daly, H. E. (1974). The economics of the steady state. *The American Economic Review*, *64*(2), 15–21.

Daly, H. E. (2014). *From uneconomic growth to a steady-state economy*. Cheltenham: Edward Elgar.

Demaria, F., Schneider, F., Sekulova, F., & Martinez-Alier, J. (2013). What is degrowth? From an activist slogan to a social movement. *Environmental Values*, *22*(2), 191–215.

Demmer, U., & Hummel, A. (2017). Degrowth, anthropology, and activist research: The ontological politics of science. *Journal of Political Ecology*, *24*(1), 610–622.

Escobar, A. (2015). Degrowth, post development, and transitions: A preliminary conversation. *Sustainability Science*, *10*(3), 451–462.

Everingham, P., & Chassagne, N. (2020). Post COVID-19 ecological and social reset: Moving away from capitalist growth models towards tourism as Buen Vivir. *Tourism Geographies*. doi:10.1080/14616688.2020.1762119

Fletcher, R., Murray Mas, I., Blanco-Romero, A., & Blázquez-Salom, M. (2019). Tourism and degrowth: An emerging agenda for research and praxis. *Journal of Sustainable Tourism*, *27*(12), 1745–1763.

Font, X., Garay, L., & Jones, S. (2016). Sustainability motivations and practices in small tourism enterprises in European protected areas. *Journal of Cleaner Production*, *137*, 1439–1448.

Forster, P. M., Forster, H. I., Evans, M. J., Gidden, M. J., Jones, C. D., Keller, C. A., . . . Turnock, S. T. (2020). Current and future global climate impacts resulting from COVID-19. *Nature Climate Change*. doi:10.1038/s41558-020-0883-0

Galvani, A., Lew, A. A., & Perez, M. S. (2020). COVID-19 is expanding global consciousness and the sustainability of travel and tourism. *Tourism Geographies*, *22*(3), 567–576.

Gascón, J. (2019). Tourism as a right: A "frivolous claim" against degrowth? *Journal of Sustainable Tourism*, *27*(12), 1825–1838.

Gössling, S., & Hall, C. M. (2019). Sharing versus collaborative economy: How to align ICT developments and the SDGs in tourism? *Journal of Sustainable Tourism*, *27*(1), 74–96.

Gössling, S., Hall, C. M., Peeters, P., & Scott, D. (2010). The future of tourism: Can tourism growth and climate policy be reconciled? A mitigation perspective. *Tourism Recreation Research*, *35*(2), 119–130.

Gössling, S., Scott, D., & Hall, C. M. (2015). Inter-market variability in CO_2 emission-intensities in tourism: Implications for destination marketing and carbon management. *Tourism Management*, *46*, 203–212.

Gössling, S., Scott, D., & Hall, C. M. (2020). Pandemics, tourism and global change: A rapid assessment of COVID-19. *Journal of Sustainable Tourism*. doi:10.1080/09669582.2020.1758708

Gren, M., & Huijbens, E. H. (2014). Tourism and the Anthropocene. *Scandinavian Journal of Hospitality and Tourism*, *14*(1), 6–22.

Gren, M., & Huijbens, E. H. (Eds.). (2015). *Tourism and the Anthropocene*. Abingdon: Routledge.

Gyimóthy, S., Morales Pérez, S., Widtfeldt Meged, J., & Wilson, J. (2020). Editorial: Contested spaces in the sharing economy. *Scandinavian Journal of Hospitality and Tourism*, *20*(3), 205–211.

Hall, C. M. (2009). Degrowing tourism: Décroissance, sustainable consumption and steady-state tourism. *Anatolia*, *20*(1), 46–61.

Hall, C. M. (2010). Changing paradigms and global change: From sustainable to steady-state tourism. *Tourism Recreation Research*, *35*(2), 131–143.

Hall, C. M. (2011). Policy learning and policy failure in sustainable tourism governance: From first- and second-order to third-order change? *Journal of Sustainable Tourism*, *19*(4–5), 649–671.

Hall, C. M. (2013). Framing behavioural approaches to understanding and governing sustainable tourism consumption: Beyond neoliberalism, "nudging" and "green growth"? *Journal of Sustainable Tourism*, *21*(7), 1091–1109.

Hall, C. M. (2014). *Tourism and social marketing*. Abingdon: Routledge.

Hall, C. M. (2015). Economic greenwash: On the absurdity of tourism and green growth. In V. Reddy & K. Wilkes (Eds.), *Tourism in the green economy* (pp. 339–358). London: Earthscan.

Hall, C. M. (2016). Intervening in academic interventions: Framing social marketing's potential for successful sustainable tourism behavioural change. *Journal of Sustainable Tourism*, *24*(3), 350–375.

Hall, C. M. (2019). Constructing sustainable tourism development: The 2030 agenda and the managerial ecology of sustainable tourism. *Journal of Sustainable Tourism*, *27*(7), 1044–1060.

Hall, C. M., Baird, T., James, M., & Ram, Y. (2016). Climate change and cultural heritage: Conservation and heritage tourism in the Anthropocene. *Journal of Heritage Tourism*, *11*(1), 10–24.

Hall, C. M., Gössling, S., & Scott, D. (Eds.). (2015). *The Routledge handbook of tourism and sustainability*. Abingdon: Routledge.

Hall, C. M., Scott, D., & Gössling, S. (2020). Pandemics, transformations and tourism: Be careful what you wish for. *Tourism Geographies*, *22*(3), 577–598.

Hernández-Martín, R., Álvarez-Albelo, C. D., & Padrón-Fumero, N. (2015). The economics and implications of moratoria on tourism accommodation development as a rejuvenation tool in mature tourism destinations. *Journal of Sustainable Tourism*, *23*(6), 881–899.

Higgins-Desbiolles, F. (2010). In the eye of the beholder? Tourism and the activist academic. In P. M. Burns, C. A. Palmer, & J. M. Lester (Eds.), *Tourism and visual culture* (pp. 98–106). Wallingford: CABI.

Higgins-Desbiolles, F., Carnicelli, S., Krolikowski, C., Wijesinghe, G., & Boluk, K. A. (2019). Degrowing tourism: Rethinking tourism. *Journal of Sustainable Tourism*, *27*(12), 1926–1944.

Hollenhorst, S. J., Houge-Mackenzie, S., & Ostergren, D. M. (2014). The trouble with tourism. *Tourism Recreation Research*, *39*(3), 305–319.

Hörisch, J. (2015). The role of sustainable entrepreneurship in sustainability transitions: A conceptual synthesis against the background of the multi-level perspective. *Administrative Sciences*, *5*(4), 286–300.

Ioannides, D., & Gyimóthy, S. (2020). The COVID-19 crisis as an opportunity for escaping the unsustainable global tourism path. *Tourism Geographies*, *22*(3), 624–632.

Kallis, G., Kostakis, V., Lange, S., Muraca, B., Paulson, S., & Schmelzer, M. (2018). Research on degrowth. *Annual Review of Environment and Resources*, *43*(1), 291–316.

Kerschner, C. (2010). Economic de-growth vs. Steady-state economy. *Journal of Cleaner Production*, *18*(6), 544–551.

Koch, M. (2015). Climate change, capitalism and degrowth trajectories to a global steady-state economy. *International Critical Thought*, *5*(4), 439–452.

Kuhzady, S., Seyfi, S., & Béal, L. (2020). Peer-to-peer (P2P) accommodation in the sharing economy: A review. *Current Issues in Tourism*. https://doi.org/10.1080/13683500.2020.1786505

Leposa, N. (2020). Problematic blue growth: A thematic synthesis of social sustainability problems related to growth in the marine and coastal tourism. *Sustainability Science*, *15*(4), 1233–1244.

Lew, A., Cheer, J., Brouder, P., Teoh, S., Balslev Clausen, H., Hall, M., . . . Salazar, N. (Eds.). (2020). Visions of travel and tourism after the global COVID-19 transformation of 2020. *Tourism Geographies*, *22*(3), 455–747.

Littig, B. (2018). Good work? Sustainable work and sustainable development: A critical gender perspective from the Global North. *Globalizations*, *15*(4), 565–579.

Martínez-Alier, J., Pascual, U., Vivien, F.-D., & Zaccai, E. (2010). Sustainable de-growth: Mapping the context, criticisms and future prospects of an emergent paradigm. *Ecological Economics*, *69*(9), 1741–1747.

Meana Acevedo, R. (2016, January). Degrowth and tourism: The role of the tourism sector in planetary overreach. The need for an updated model change. In *Monografies de la Societat d'Historia Natural de les Balears* (pp. 79–90). Majorca: Societat d'Historia Natural de les Balears.

Menton, M., Larrea, C., Latorre, S., Martinez-Alier, J., Peck, M., Temper, L., & Walter, M. (2020). Environmental justice and the SDGs: From synergies to gaps and contradictions. *Sustainability Science*. doi:10.1007/s11625-020-00789-8

Mercado, C. (2017, December 25). Buen Vivir: A new era of great social change. *Pachamama Alliance*. Retrieved from https://blog.pachamama.org/buen-vivir-new-era-great-social-change

Milano, C., Novelli, M., & Cheer, J. M. (2019). Overtourism and degrowth: A social movements perspective. *Journal of Sustainable Tourism*, *27*(12), 1857–1875.

Morales-Pérez, S., Garay, L., & Wilson, J. (2020). Airbnb's contribution to socio-spatial inequalities and geographies of resistance in Barcelona. *Tourism Geographies*. doi:10.1080/14616688.2020.1795712

Müller, D., & Hall, C. M. (2018). From common ground to elite and commercial landscape. In C. M. Hall & D. Müller (Eds.), *The Routledge handbook of second home tourism and mobilities* (pp. 115–121). Abingdon: Routledge.

Natale, A., Arcidiacacono, C., & Martino, S. D. (2013). From "Gomorrah domain" to "don peppe diana lands". A southern Italian experience of work-based liberation, community networking, and well being. *Universitas Psychologica*, *12*(4), 1037–1047.

Navarro-Jurado, E., Romero-Padilla, Y., Romero-Martínez, J. M., Serrano-Muñoz, E., Habegger, S., & Mora-Esteban, R. (2019). Growth machines and social movements in mature tourist destinations Costa del Sol-Málaga. *Journal of Sustainable Tourism*, *27*(12), 1786–1803.

Næss, P. (2020). Sustainable development: A question of 'modernization'or 'degrowth'?. In A. Hagen & U. Higdem (Eds.), *Innovation in public planning* (pp. 91–109). Cham: Palgrave Macmillan.

Oklevik, O., Gössling, S., Hall, C. M., Steen Jacobsen, J. K., Grøtte, I. P., & McCabe, S. (2019). Overtourism, optimisation, and destination performance indicators: A case study of activities in Fjord Norway. *Journal of Sustainable Tourism*, *27*(12), 1804–1824.

O'Neill, D. W. (2015). What should be held steady in a steady-state economy? Interpreting Daly's definition at the national level. *Journal of Industrial Ecology*, *19*(4), 552–563.

Panzer-Krause, S. (2017). Un-locking unsustainable tourism destination paths: The role of voluntary compliance of tourism businesses with sustainability certification on the island of Rugen. *Zeitschrift Fur Wirtschaftsgeographie*, *61*(3–4), 174–190.

Panzer-Krause, S. (2019). Networking towards sustainable tourism: Innovations between green growth and degrowth strategies. *Regional Studies*, *53*(7), 927–938.

Paulson, S. (2017). Degrowth: Culture, power and change. *Journal of Political Ecology*, *24*(1), 425–448.

Perkumiene, D., & Pranskuniene, R. (2019). Overtourism: Between the right to travel and residents' rights. *Sustainability*, *11*(7), 2138.

Phi, G. T. (2019). Framing overtourism: A critical news media analysis. *Current Issues in Tourism*. doi:10.1080/13683500.2019.1618249

Ramos, S. P., & Mundet, L. (2020). Tourism-phobia in Barcelona: Dismantling discursive strategies and power games in the construction of a sustainable tourist city. *Journal of Tourism and Cultural Change*. doi:10.1080/14766825.2020.1752224

Rasoolimanesh, S. M., Ramakrishna, S., Hall, C. M., Esfandiar, K., & Seyfi, S. (2020). A systematic scoping review of sustainable tourism indicators in relation to the sustainable development goals. *Journal of Sustainable Tourism*. doi:10.1080/09669582.2020.1775621

Renaud, L. (2020). Reconsidering global mobility – Distancing from mass cruise tourism in the aftermath of COVID-19. *Tourism Geographies*. doi:10.1080/14616688.2020.1762116

Renkert, S. R. (2019). Community-owned tourism and degrowth: A case study in the Kichwa Anangu community. *Journal of Sustainable Tourism*, *27*(12), 1893–1908.

Romero-Padilla, Y., Cerezo-Medina, A., Navarro-Jurado, E., Romero-Martínez, J. M., & Guevara-Plaza, A. (2019). Conflicts in the tourist city from the perspective of local social movements. *Boletin De La Asociacion De Geografos Espanoles*, (83), 35. doi:10.21138/bage.2837

Saarinen, J. (2015). Conflicting limits to growth in sustainable tourism. *Current Issues in Tourism*, *18*(10), 903–907.

Saarinen, J. (2018). Beyond growth thinking: The need to revisit sustainable development in tourism. *Tourism Geographies*, *20*(2), 337–340.

Sæþórsdóttir, A. D., Hall, C. M., & Wendt, W. (2020). From boiling to frozen? The rise and fall of international tourism to Iceland in the era of overtourism. *Environments*, *7*(8), 59. doi:10.3390/environments7080059

Said, A., & MacMillan, D. (2020). 'Re-grabbing' marine resources: A blue degrowth agenda for the resurgence of small-scale fisheries in Malta. *Sustainability Science*, *15*(1), 91–102.

Salazar, J. F. (2015, July 24). Buen Vivir: South America's rethinking of the future we want. *The Conversation*. Retrieved from https://theconversation.com/buen-vivir-south-americas-rethinking-of-the-future-we-want-44507

Schneider, F., Kallis, G., & Martinez-Alier, J. (2010). Crisis or opportunity? Economic degrowth for social equity and ecological sustainability. Introduction to this special issue. *Journal of Cleaner Production*, *18*(6), 511–518.

Scott, D., Gössling, S., Hall, C. M., & Peeters, P. (2016). Can tourism be part of the decarbonized global economy? The costs and risks of alternate carbon reduction policy pathways. *Journal of Sustainable Tourism*, *24*(1), 52–72.

Scott, D., Hall, C. M., & Gössling, S. (2016). A review of the IPCC Fifth Assessment and implications for tourism sector climate resilience and decarbonization. *Journal of Sustainable Tourism*, *24*(1), 8–30.

Shaw, G., & Williams, A. M. (2004). From lifestyle consumption to lifestyle production: Changing patterns of tourism entrepreneurship. In R. Thomas (Ed.), *Small firms in tourism* (pp. 99–113). Oxford: Elsevier.

Sun, Y. Y., Lin, P. C., & Higham, J. (2020). Managing tourism emissions through optimizing the tourism demand mix: Concept and analysis. *Tourism Management*, *81*. doi:10.1016/j.tourman.2020.104161

Tomassini, L., Font, X., & Thomas, R. (2020). Narrating values-based entrepreneurs in tourism. *Journal of Tourism and Cultural Change*. doi:10.1080/14766825.2020.1793991

UNWTO & UNDP. (2017). *Tourism and the sustainable development goals – Journey to 2030*. Madrid: UNWTO.

Valdivielso, J., & Moranta, J. (2019). The social construction of the tourism degrowth discourse in the Balearic Islands. *Journal of Sustainable Tourism*, *27*(12), 1876–1892.

van Eck, N. J., & Waltman, L. (2020). VOSviewer Manual. In *Manual for VOSviewer version 1.6.14*. Leiden: Universiteit Leiden CWTS.

Veríssimo, M., Moraes, M., Breda, Z., Guizi, A., & Costa, C. (2020). Overtourism and tourismphobia: A systematic literature review. *Tourism*, *68*(2), 156–169.

WTTC and McKinsey & Company. (2017). *Coping with success: Managing overcrowding in tourism destinations*. London: WTTC.

Zapata Campos, J., Hall, C. M., & Backlund, S. (2018). Can MNC promote more inclusive tourism? Apollo tour operator's sustainability work. *Tourism Geographies*, *20*(4), 630–652.

Part 1

Degrowth and tourism entrepreneurship

2 Decommodification as a socially embedded practice

The example of lifestyle enterprise in animal-based tourism

Erika Andersson Cederholm and Carina Sjöholm

Introduction

> No, well . . . it depends on who you are as a person, but I am absolutely not . . . I am unfortunately, not interested in money and particularly not my own money. I have to say that . . . I also think money is pretty uninteresting. But as long as I can make a living, I am quite pleased.

The prior quote is from an interview with Emily, a lifestyle entrepreneur running a combined B&B and horse farm. She is one of several business owners taking part in a study on lifestyle entrepreneurship in Sweden. From this study we have learnt that Emily manages to make a living from her business, albeit with relatively small margins. This means that her non-pecuniary interest is probably not the result of economic abundance. Her business is not so prosperous that she can afford to be completely negligent about monetary matters. Hence, the quote does not seem to be about economic matters. Instead, it seems to be an expression of a specific form of self-presentation.

The growing literature on lifestyle entrepreneurship has often emphasized the non-growth and non-profit orientation among these business owners, and generally labels them as atypical entrepreneurs (Ioannides & Petersen, 2003). Some studies demonstrate how these business operators emphasize values other than merely economic values. These could be values related to sustainability (Ateljevic & Doorne, 2000; Wang, Li, & Xu, 2019), family values (Getz, Carlsen, & Morrison, 2004), quality of life in general (Marcketti, Niehm, & Fuloria, 2006) or an interest in nature and animals (Andersson Cederholm, 2015; Andersson Cederholm & Åkerström, 2016; Helgadóttir & Sigurdardóttir, 2008). Some studies have highlighted the simultaneous consumption and production of lifestyle in this form of enterprising, which can also be viewed as a means to emphasize personal interest and passion for the business while downplaying the economic side (Andersson Cederholm, 2018; Andersson Cederholm & Hultman, 2010; de Wit Sandström, 2018; Di Domenico, 2005; Sweeney, Docherty-Hughes, & Lynch, 2018). Emily's expression in the opening quote could thus be interpreted as a mode of emphasizing specific values, while downplaying others, thereby demonstrating a specific form of work ethos or way of doing business.

This chapter focuses on the non-growth orientations and practices in lifestyle entrepreneurship in tourism, discussed within the framework of degrowth. Degrowth can be described as an ongoing critical debate and emerging research area in academia, as well as a social movement. Degrowth has been defined as 'equitable downscaling of production and consumption that increases human well-being and enhances ecological conditions at the local and global level, in the short and long term' (Schneider, Kallis, & Martínes-Alier, 2010, p. 512). Degrowth is primarily associated with an ideological position originating in the Western world, expressed through practices such as voluntary simplicity, sharing resources in organized forms, or other active positioning against economization and consumerism (see also D'Alisa, Demaria, & Kallis, 2015; Martínes-Alier, Pascual, Vivien, & Zaccai, 2010). This position is visible in tourism as well, as some lifestyle entrepreneurs are ideologically oriented not only in their overall approach but also in what they are selling. Through their products, tangible or intangible, and through the entrepreneur's own lifestyle they express a critique of the consumerist society, creating an 'alternative' habitus and promoting 'alternative' markets.

However, the notion of 'lifestyle' in business, as well as the notion of degrowth, may not only be associated with an 'alternative' habitus and ideological positioning. Nor does it always encompass individual decision-making, but it could be embedded in various cultural and economic systems (Paulson, 2017). These could be different forms of gift-economic exchanges, often co-existing and intertwined in market relationships. These systems may restrict growth-oriented practices, although it is not always visible at the level of individual symbolic production and consumption. Following this line of thought, this chapter will focus on forms of non-growth practices that are not immediately associated with downsizing or alternative lifestyles. Consequently, these types of practices often fall under the radar of analyses searching for illustrative empirical examples of alternative markets. We will illustrate these practices using two forms of rural enterprises in Sweden: horse-based tourism and hunting tourism.

Our analysis will demonstrate how a non-growth orientation emanates from the way economic exchange is socially organized and culturally understood within the enterprise and the operators' social network. The aim of this chapter is to demonstrate and discuss how a non-growth approach may be embedded in specific forms of economic and social exchange. In particular, we focus on how gift-economic relationships, in combination with traditions of caring for animals and nature, are intertwined with market relations, and discuss how these may restrict or reduce a growth approach. Although gift-economic exchanges in modern societies are often deeply intertwined with market relationships, the exchange is based on a logic different to that of conventional market logic. Furthermore, these businesses are not necessarily based and formed around a political or ideological non-growth approach nor as a mode of alternative identity. We argue that their non-growth orientation is related more to the character of the exchange and the way both economic and non-economic values are embedded in a social and cultural context.

We start with a description and analysis of two cases of lifestyle-oriented tourism in Sweden. The first is horse-based tourism, based on a study conducted between 2009 and 2013 on horse farming, and comprising ethnographic interviews with 25 business owners. The second case draws on an ongoing research project that started in 2019 on hunting tourism entrepreneurship. The analysis presented here is based on in-depth interviews with eight hunting tourism operators as well as observations and informal conversations with the operators and their customers at hunting events. The interviews lasted two to three hours, with additional informal conversations and observations. All interviews are transcribed verbatim and anonymized. We have adopted a narrative practice perspective by focusing on patterns in which the story of work, life, animals and the community unfold in the interview context and in the observations, assuming that how the entrepreneurs talk about their business and present themselves is also a means of *doing* lifestyle-oriented business (see Gubrium & Holstein, 1998; Holstein & Gubrium, 1997). Hence, what we refer to in the analysis as 'practices' includes how they tell a story of themselves and their enterprise, how they perform this enterprise by telling that story, and how they interact with animals, co-workers and customers.

Following an account of the horse-based and hunting enterprises, a synthesizing analysis is provided of the type of social and economic exchange that characterizes these businesses. This is informed by the theoretical framework of Zelizer's concept of 'relational work' (Zelizer, 1985, 2013) and literature on gift-economic exchanges (Cheal, 1988; Godbout, 1998). As a result, three forms of decommodification practices are discussed, as well as how they can be understood in the context of degrowth.

The horse-based enterprise

Horse-based enterprises comprise relatively small-scale horse farms in Sweden, offering services such as tour riding, accommodation (for horses and/or humans) such as a farm stay or bed-and-breakfast, breeding, training and farm shops. All horse farms included in the study are owned or operated by women, mirroring the female domination in this market. Some of the farms are operated as a family business, where (primarily) daughters are working on the farm, as family employees or partners with their mothers (Andersson Cederholm, 2015). Although the study does not have the intention to focus on the actual financial situation of the business, it highlights how the owners talk about and experience the financial situation. A narrative of economic harshness is very common and is typically told in a self-ironic tone: 'you will not be a millionaire', or 'if you want a proper salary, you need another job'. This type of narrative is typically followed by an emphasis on the *other* values that the business provides.

Passion and practical knowledge

In stories of 'how it all started', the owners typically emphasize their own passionate interest for horses. Several of the horse farmers are lifestyle migrants, who have left employment as a teacher, physiotherapist, journalist or marketing manager and relocated. Some have worked in farm enterprises and have a farming background. A majority of the horse farmers in this study have a university education, which adds an extra dimension to their story of a working career and lifestyle spurred by free will. Their work and the enterprise are clearly framed in the narrative as a voluntary lifestyle choice that has not been chosen out of economic necessity or family tradition.

'Passion' is a recurring narrative in the interviews and appears to consist of several dimensions. Foremost, it is a mode of explaining, perhaps legitimating, their specific life choice. What may seem, at least on the surface, as economically irrational is often followed by a lengthy explanation. Several of the women have left successful and, compared with the horse enterprise, economically more viable careers in order to pursue a personal interest and try to make a living out of it. The 'passion' dimension also seems to indicate a specific kind of knowledge. It is a type of knowledge or perhaps professionalism that is based on an idea of a practical, emotional, or even intuitive knowledge. The expression of having 'an eye for horses' is often used, indicating a type of knowledge earned through genuine interest and practical experience.

SUSAN: This eye. We often talk about that . . . you need to have an eye for horses. And you need to want to be with them. Caring for them but at the same time be willing to sacrifice them . . . for what you want to achieve. Some horses will lose their abilities as riding horses when you put them in tour riding but, on the other hand, they will be safer riding horses. . . . Well, you need to accept that this is not only a hobby, and the horses need to be sustainable, they need to last. And then you have to be damn sociable as well.

Susan's trade-off reasoning may seem instrumental, but it is embedded in a narrative of care, particularly for the horses but also for people – 'you have to be damn sociable as well'. Furthermore, the caring aspect is part of a professionality – 'having an eye for horses' – and embedded in a narrative of sustainability. You need to care for them not only for altruistic reasons, but you need them as working comrades as well.

Money-talk and resistance to commodification

A recurrent theme in our analysis of both horse and hunting enterprises is the apparent engagement in talking about money and prices. Among the horse farmers this is often expressed through a form of 'negative' or 'dissociating' talk about money, such as the *non-interest* in pecuniary matters, as in the introductory quote by Emily. Among the horse farmers, it could also take the form of a conspicuous absence of any talk about prices. At first glance, this may seem paradoxical, perhaps irrational, if you are running a business. Nevertheless, the 'dissociating' money talk can be a means of highlighting values other than 'merely' economic ones. Valuing horses in quantitative, monetary terms is probably something they have to do often in daily business. However, economic valuation is relatively absent from the horse farmers' narrative when they talk about the overall values of the enterprise. Maggie puts it like this:

MAGGIE: To have this feedback, to meet people and see that they really like it here. That is actually much more important than money. . . . And the horses, they have such an incredible value. Both value and value, if you know what I mean. When it becomes too commercial, no . . . that is not how I like it.

INTERVIEWER: No?

MAGGIE: When commercialism takes over in a company like this, then it can go really bad. I have seen examples of that. It could go really wrong, for horses as well as people. So, you have to reach a level where it feels good. I want my horses to have a good life. It is a responsibility. Some of these horses I will never ever sell, never in my entire life, not even if I close down the business.

'Both value and value, if you know what I mean'. This type of narrative illustrates the ongoing negotiation and balancing work undertaken in the intersection between a commercial and a non-commercial sphere, typical of lifestyle businesses. The non-money interest can be seen as a type of building block for a work ethos and a work identity for the lifestyle entrepreneur. Although this can be interpreted as a symbolic demarcation, it does not necessarily entail an ideological positioning as a non-growth-oriented entrepreneur. Running a horse farm entails a caring ethos – for animals and people – and if you fail in your caring ethos, the outcome will probably be immediate. 'It can go really wrong', as Maggie warns.

Favour exchanges in friendship networks

What is quite significant for horse farming is that it is labour intensive. Horses need to be fed at regular hours, they need to be walked to and from the stable and the paddock, their boxes need to be cleaned and they need daily exercise. The owners are thus dependent on people helping out. Stables that offer accommodation for other people's horses often have the horse owners visiting on a daily basis to take care of their horses, but there is still a lot of work to be performed by the stable owner. The customers, who own the horses, often perform some form of stable duty, sometimes according to a schedule with a 'taking turn' system. Family members help out as well, and they may or may not be formally employed. There are also voluntary workers such as 'stable girls' who do not own a horse themselves but are willing to help out and receive free riding in return. Finally, there are also helpers who are labelled 'friends', 'neighbours' or simply part of the 'network'.

In total, there is a range of different types of 'helpers' who work for free, often based on a system of favour exchanges. The helpers who have a specific interest in horses, such as 'stable girls', may be a challenge to the balance of reciprocity, since they sometimes help out more than the stable owner may give in return. As Susan explains: 'Many people in this business want to do things for free, but you cannot accept too much of that'. However, some of these voluntary helpers may develop their skills with horses, and that provides the opportunity for a more balanced reciprocity. Susan continues: 'you can often give them what they want, which is knowledge. Knowledge is a reasonable salary . . . and you get help in return'.

However, if the voluntary workers become too skilled, the position becomes reversed and they are giving favours to the stable owners, not the opposite. Hence, the type of payback varies depending on who you are, if you are a close friend or more distant, and what type of work you can perform (Andersson Cederholm & Åkerström, 2016). Although favour exchange is a form of informal economic exchange, it is sometimes formalized bluntly, as one of the horse farmers said: 'If I scratch your back, you scratch mine'. However, among closer friends, reciprocity is often taken for granted and not articulated as a formal requirement. These are also the situations in which it may become ambiguous, and misunderstandings may become rather emotional (Andersson Cederholm & Åkerström, 2016). Some people are both friends and customers, and the roles may therefore shift according to the situation.

This demonstrates the socially embedded practice of running this type of lifestyle business, particularly the type of business that requires many helping hands. A large proportion of labour is performed in the form of favour exchanges which, although intertwined in market relationships, constrain market relations. If a customer is a potential friend, and vice versa, as is common in horse farming, the relationship differs from that in businesses where there is greater separation of the personal and commercial spheres.

The hunting enterprise

Hunting tourism, the sale of recreational hunting and hunting tours to visitors, is both similar to and different from horse-based enterprises. Both forms of tourism can be labelled special interest tourism. Hunting tourism is, however, even more 'special' in the sense that it requires a hunting licence. This means that you can expect the visitors to have at least a minimum level of knowledge, although this may vary among groups of visitors. Both hunting and horse-based recreation are forms of nature tourism. They also involve animals. In hunting, two types of animals are primarily in focus: the wild animals and the dogs. The dogs and the horses in both of these types of enterprises take similar roles: as working comrades, friends, family and sometimes as goods (if they are for sale). It is quite noticeable that the involvement of animals in these enterprises evokes an ethical aspect that clearly affects the way the operators talk about their business. Care for animals, animal populations and nature are common elements in the entrepreneurs' accounts and narratives.

In the hunting community in Sweden, ethical issues are widely debated, and hunting in general is a morally contested space (von Essen, 2018; von Essen, Allen, & Hansen, 2017). The issues of animal ethics and animal welfare in hunting are thus more outspoken and debated compared to in horse farming. For instance, a legitimizing discourse is discernible in public debates, in hunting magazines and in social media, in which hunters emphasize the tradition of stewardship in hunting, whereby local hunters perform free labour for the government in order to keep populations of wild animals at a sustainable level and to reduce the risk of wildlife traffic accidents. Although recreational hunting in Sweden is widely accepted among the general public, provided it has a utilitarian dimension and the game meat is properly cared for (Kagervall, 2014), it is nevertheless a contested moral space in which hunters debate ethical issues both within the hunting community and in a wider social space.

Hunting tourism is a multifaceted area in Sweden and mirrors the historical background of hunting practices in Sweden. On the one hand, there is a long tradition of 'folk hunting' emanating from subsistence hunting, and which provides the cultural base of most recreational hunting in Sweden. On the other hand, there is the hunting tradition among wealthy landowners with large domains and mansions, who hunt and invite friends and business associates to large hunting events. Hence, a traditional division between landowner aristocracy and hunting for subsistence among the rural working class is still discernible in Nordic as well as other European countries (Mischi, 2012; von Essen et al., 2017). Moreover, new forms of rural middle classes in European countries make the picture more nuanced (Heley, 2010). Due to these different traditions, the class dimension is complex and is similarly reflected in the hunting tourism arena.

Narratives of passion and practical knowledge

In interviews, hunting entrepreneurs often come back to the notion of a 'serious entrepreneur', often by dissociating themselves from a type of entrepreneur that they regard as non-serious or less serious. Hence, an image of what is ascribed to be a 'good' or 'serious' entrepreneur emerges through negative identification. A passion for hunting, often grounded in childhood or youth, is a recurrent topic, e.g. 'I used to hunt with my father'. Similar to the horse farmers, the passion for the activity, in this case hunting, is combined with a more general passion for nature and animals. The narrative is clearly embedded in a moral framework, and a non-serious hunting entrepreneur is often said to be someone who does not take ethics seriously. The moral ethos of the hunter is often described in terms of *jägarmässighet* (in Swedish), or 'sportsmanlike hunting' (von Essen, 2016), which implies specific codes of conduct.

Being a good entrepreneur in these cases equals being a good hunter. Similarly to the horse entrepreneurs, practical knowledge gained through experience and learning to 'read' animals and nature is highly valued. A passionate hunting tourism operator is also someone who can read people and knows their level of competence in regard to hunting skills. The entrepreneurs emphasize, often in a very detailed way, how they 'set the scene' for a specific hunting area, read signs to assess the customers' shooting skills and place the customers in shooting areas that match their skills. Hence, pedagogic skills and 'an eye for people' are highly regarded among the operators.

MICHAEL: Those who are serious operators try to build a whole package, providing accommodation, renovating the farms and so on, to make more money. So instead of selling two days hunting only, it is better in that way because you can arrange shooting practices the day before . . . and if you are a professional hunter then you are able to see 'aha, he knows how to shoot, and she knows . . .', then you will know that it will neither be too much nor too little (shooting) . . . so it is all about being professional, to try and find . . . to know your own land, to know that 'here you can shoot, but here it is not so good . . .' and the customer who cannot shoot will be put in this spot.

Money-talk and resistance to commodification

Starting a hunting enterprise in order to 'make money' is mentioned as a possibly plausible motive. However, to be in it '*just* for the money' is not quite regarded as acceptable. In Sweden, as well as in the other Nordic countries, one can see a scepticism or ambivalence towards commercial hunting tourism (Dahl & Sjöberg, 2010; Nygård & Uthardt, 2011; Oian & Skogen, 2016). Assuming that the entrepreneurs are aware of such an attitude among local hunting teams, they probably relate to that discourse and critique towards commercialism in hunting. However, they appear to emphasize 'right' commercialization, and often emphasize that commercial hunting is a means to maintain a 'living countryside' as it may offer job opportunities. Furthermore, being too commercial, or 'just in it for the money', is regarded as both economically and ecologically unsustainable. In hunting, there is often a direct link between the two forms of sustainability, simply expressed as 'if you shoot everything, there won't be any next year'. One tourist operator in one of the largest private hunting domains in Sweden told us that he would not mind expanding the hunting business, but then they would need more land. That would in fact be physically impossible because their land borders the sea at one end and is cut off by the highway at the other end. This entrepreneur contended that they cannot hunt more than they already do, if they are to preserve the stock. Growth would simply not be possible.

Being 'too commercial' further implies having the wrong attitude towards pricing. This is partly related to an ideal of the fair and honest businessperson who does not promise more than s/he can deliver. In hunting, one controversial issue is whether the hunting operator, or outfitter, can guarantee a killing or not. Hunting, as with most wildlife tourism, is an unpredictable business. It may even be difficult to guarantee that you may spot an animal of interest. The expectations among customers vary, which is one of the challenges identified by the entrepreneurs. Some of them talk about 'lowering expectations' among customers, referring to those customers who expect a certain number of shots or a particularly large trophy. One way to lower expectations is through strategic pricing strategies.

MICHAEL: You have to lower the expectations. But if you are dishonest and only look at the money, then you will have a customer who says 'well we have ordered fifteen . . .' and you can never say it is going to be fifteen animals on one day. And then maybe you are lucky that day, there are lots of them and the customer shoots fifteen at the first . . . and then you have that son of a bitch (the dishonest entrepreneur) saying that 'if you want to continue then it costs . . .' and then it goes on . . . it is so damn important to find that balance.

Money-talk in hunting enterprises seems to be about setting the right price, economically and symbolically. Hunters have a slightly different approach to the horse farmers. The hunting operators do not avoid talking about prices. On the contrary, having the 'right' pricing strategy seems to be a topic of much concern. This might be related to the direct link between ecological and economic sustainability, which is also an expression of being a serious, ethical entrepreneur. It may also be related to specific characteristics of hunting enterprises, such as the prevalence of price tags on certain types of wildlife. Particular species, as well as individual animals, such as a male red deer with large antlers, are costly to shoot.

As mentioned previously, the folk hunting tradition has a long history in Sweden. The democratic ideal of wildlife being accessible to all is prevalent, and some hunters are highly critical towards the high prices for trophy animals. Hence, many hunting operators express a non-growth orientation. It does not necessarily mean that they are not interested in earning money from their hunting business, or even letting their own business grow, but there is a strong and historically established tradition of adopting a stewardship role in hunting, in combination with a tradition of folk hunting. The commodification of wildlife, expressed in high price tags for trophy animals, becomes a symbol of a marketized logic that may seem contradictory to other values that many Swedish hunters place in high regard. We would like to argue that the ambivalence, or expressed reluctance, toward either the high prices for some animals or the chase for a large number of shots is a form of socially embedded decommodification practice.

Favour exchanges in friendship networks

In Sweden, one form of reciprocal domestic hunting tourism is established in which hunting in domains other than one's own home environment takes the form of a friendship or hunting team exchange. This informal exchange of hunting opportunities in other places is often not regarded as 'tourism' by the hunters or the providers of such arrangements, even though it involves consumption of services such as transport and accommodation. This form of tourism exists both in parallel and intertwined with a more marketized form of commercial hunting tourism. 'Exchange hunting' is an established concept among hunters in Sweden, and hunters explain that 'this is how hunting works'.

MICHAEL: if we for a moment ignore the tourism part and the money, this is how hunting works. My friend John arranges a hunt in Småland where he comes from, and then he invites me for a wild boar hunt, and then I invite him here. And then I invite someone else . . . this is how the network grows. And then you become friends.

Commercial enterprising and exchange hunting are two types of exchange that are both separate and intertwined. For the entrepreneurs, this entails a form of boundary or relational work (Zelizer, 2013), where the boundaries are negotiated, drawn and redrawn. Some entrepreneurs have elaborate narratives of how they draw the line legally within the company, to differentiate between friendship reciprocity and economic transaction. Michael, for instance, is describing in detail how he would have earned more money if he had not enjoyed hunting and therefore engaged in a friendship reciprocity:

> If we look at it from a purely legal aspect . . . well, if I didn't want to hunt myself, and sold all the hunting rights on my land, then I could have earned money. But since I have so many friends and I enjoy being invited to other places, then you need to have those hunts privately. So, if you look at the hunting business on a purely plus and minus basis, it will be minus.

Michael contends that he enjoys hunting, and hunting entails friendship reciprocity. He also invites both paying guests and friends to his own land to hunt. This means that he cannot run a purely business-like business.

Decommodifying practices in a mixed market/gift economy

The two types of businesses described here are different in many respects. Nevertheless, they have similarities that make it analytically relevant to analyze them side-by-side. They are rural based – the resources are wild nature or farmland, or a combination of both, and these set the scene for the business. Humans and animals are the main actors here, although they have different roles. As businesses, they offer services in the sport, recreation and tourism markets.

These enterprises could be labelled lifestyle enterprises, as they share many of the overall characteristics that are often said to be typical for these businesses, such as downplaying economic motives whilst emphasizing values of personal relationships, sociability and quality of life in general (Andersson Cederholm, 2018; Ateljevic & Doorne, 2000; Di Domenico, 2005; Getz et al., 2004; Marcketti et al., 2006; Sweeney et al., 2018). In this analysis, we have identified three characteristics or practices that detail the complex relationship between a commercial and a non-commercial social sphere. The first characteristic presented here relates to the overall values of running the business and how these values are expressed. 'Passion' for animals, nature and for the core activity of hunting or riding is emphasized. But 'passion' indicates more than just a personal interest. It seems to entail a passion to share with other people. An interest in servicing people seems equally important for these enterprises. Having an eye for nature, animals as well as an eye for people, a service skill, is often highlighted as a virtue as well as a condition for running the business. The emphasis on passion is narrated in relation to an opposite – being in business merely for profit. We would like to argue that the passion narrative is one of the decommodifying practices that are significant for these lifestyle enterprises. The passion imperative is not only about having a passion for running a business as such, although they do emphasize the freedom that comes with the status and value of self-employment. It is about the type of business they are running. To these entrepreneurs, the passion narrative in combination with a high regard for practical knowledge makes this type of business stand out from just 'any type of business'. Practical knowledge, with many years of training and learning to read nature, horses, dogs and wild animals, is an expression of being a 'serious' entrepreneur. This implies a certain ethical standard, partly in relation to moral standards when doing business, but also in relation to the third aspect – the animals or 'nature' in general.

What we have called 'money-talk' in the analysis consists primarily of narratives around the symbolic role of money, implying that talk about money is used as a means of expressing who you are as an entrepreneur or business operator, as well as a person more generally. In both types of enterprises, we can see a distancing from a commodification of values related to nature and animals. Although making money based on animals (domesticated or wild) as a resource is part of their business and does not in itself need to be a problem, it becomes a problem when animals are treated as *primarily* economic units, or obviously capitalized, such as having specific price tags. It is in these stories that we can discern resistance.

The third dimension we have discussed is friendship reciprocity and favour exchanges. Friendship reciprocity is an institutionalized form of hunting tourism, because hunters exchange hunting opportunities, and favour exchanges are common. For instance, a hunter may have a dog that is particularly skilful, and thus offer a hunter/dog service for a specific hunting arrangement, a service that is later reciprocated. Friendship reciprocity is also an important ingredient in the social-economic exchange of horse-based enterprises. On a continuum between at the one end personal friendships, towards more formalized network relations, to more clearly market-based and formalized relationships at the other, actors reciprocate services that are beyond formalized and monetary arrangements (Andersson Cederholm & Åkerström, 2016).

In both of these types of businesses, there is a constant and ongoing negotiation of when and how to draw a line between gift exchanges and market exchanges. One of the forerunners of interactionist economic sociology, Viviana Zelizer, has used the concept of relational work in order to understand how and by what means (money, favours or other forms of repayment) people draw boundaries between various life spheres, such as business, work, home or other intimate spheres, and different types of social relationships (Zelizer, 2005, 2013). These boundaries are culturally constituted (such as the taboo of paying someone to be your friend) but also negotiated in situ. Hence, they are never given, and always more or less open for social negotiation. For instance, in certain situations, it is considered appropriate to pay a friend for a service, but in other situations it would be out of the question and deemed as endangering or even corrupting a close friendship relationship. The relational work that exists among horse farmers and hunting operators is clearly apparent, and sometimes friendship and market relations are kept strictly separate, and sometimes they are entangled or blurred.

What is particularly relevant for the argument in this chapter is that these entanglements seem to restrain or hold back a market-based logic. The exchange of favours is an important dimension of an informal gift economy. Such exchanges exist outside the requirements of paid work and formalized work roles, although they can be more or less institutionalized. Previous studies of gift economies within modern market economies have highlighted the moral economy of interpersonal exchanges in relation to more formalized market exchanges (Cheal, 1988; Godbout, 1998). Cheal (1988, p. 19) defines the gift economy as 'a system of redundant transactions within a moral economy, which makes possible the extended reproduction of social relations', suggesting that it is not necessarily a separate sphere outside a formal economy or work relations, but is rather integrated into all spheres of life. Although intertwined with formal work relationships, a gift economy operates with a different logic than an economy that valorizes the exchange in either monetary or other quantifiable and measurable terms. A gift exchange aims at strengthening social relationships, and reciprocity can never be openly articulated as an obligation or, as Godbout (1998, p. 5) states: 'The magic of the gift can only operate as long as the underlying rules are not formulated'. This means that gift exchanges are inherently informal and non-articulated. They exist entangled, or in tandem, with market-based relationships.

Although horse-based enterprises or hunting businesses are commercial actors, they sustain many different forms of market- and non-market-based social relationships. Horse farming, for instance, is dependent on gift-economic relationships in order to make the business work (Andersson Cederholm & Åkerström, 2016). Regardless of whether the business is actually dependent on gift-economic relationships or not, friendship and favour exchanges are part of an established tradition among people sharing the same recreational interest. Both horse-based enterprises and hunting tourism are socially grounded in a market where leisure relationships and work-oriented relationships are intertwined. In addition, the work-oriented relationships are a mixture between gift-economic exchanges and market-based exchange. This provides the social conditions for specific economic orientations and practices. It also provides the backdrop to the situated negotiations that we have illustrated in our analysis. We argue that these structural and situated conditions encourage practices of decommodification. Offering riding or hunting in exchange for non-monetary repayment, or a mix of monetary and non-monetary exchange based on the specific situation and context, can thus be regarded as an enactment of decommodification.

As we have aimed to demonstrate in this chapter, non-marketized forms of exchange are not necessarily an expression of political and ideological positioning but are embedded in specific social and cultural practices that confine economic growth. Although some forms of lifestyle enterprising in tourism may promote values related to sustainability and degrowth (Ateljevic & Doorne, 2000; Wang et al., 2019), others represent a form of enterprising that may even seem counter-intuitive to degrowth. Recreational hunting, for instance, displays a multifaceted class structure where some forms of hunting are framed as exclusive and high-market, often with operators connected to the large landowners and mansions, whilst, as we mentioned previously, a long tradition of 'folk hunting' also prevails. This complex social arena, in combination with an increasing number of new hunters from an urban middle class, creates a multifaceted area of consumption related to new communities of practice among hunters (von Essen, 2019). These may display various forms of conspicuous consumption, such as expensive equipment and clothing. In addition, as hunting implies the killing of animals and the consumption of game meat, it brings ethical notions of animal welfare and the consumption of meat into the spotlight. Hence, consumption related to hunting as well as the game meat is a contested social space. This implies that it may not be easily associated with the degrowth movement, although the rise of care ethics is visible in hunting consumption (von Essen, 2019), and the consumption of game meat is often promoted as a sustainable form of meat consumption in line with a growing interest in locally produced food (Tidball, Tidball, & Curtis, 2013).

Some previous studies on degrowth have shown how non-marketized, locally embedded forms of exchange may even be criticized by non-consumerist social movements. This is clearly illustrated in, for instance, Bogadóttir and Olsen's (2017) study of the traditional whale hunt in the Faroe Islands, which is deeply embedded in the Faroese social life. The authors show how the practice of the whale hunt – the *grindadráp* – is ingrained with values that are often associated with degrowth, such as the socially embedded practice of cooperation where no monetary transactions are involved. However, a forceful anti-whaling campaign and activist groups have turned the tradition into a contested space, evoking a discursive struggle among the Faroese to defend the right to the whale hunt, portraying it as a sustainable, environmentally friendly practice. Other examples of social and economic practices that may be counter-intuitive to the 'equitable downscaling of production and consumption' (Schneider et al., 2010, p. 512) are local traditions of conspicuous consumption. An illustrative example is Gezon's (2017) analysis of the khat economy in Northern Madagascar, which demonstrates a culturally embedded consumerism through the accumulation of material goods as status objects. The growing and selling of the drug crop khat occupies a space on the edge of the formal economy and provides a source of income for many people. Apart from providing cash, the growth, distribution and consumption of khat is part of a lifestyle where the acquisition of status objects is highly valued. Gezon demonstrates how the khat economy is embedded in a complex relational economy which – despite people's preoccupation with the acquisition of material goods – does not primarily evolve around a capitalist growth motive. These types of examples are not so visible in the ongoing discussion on degrowth. As Gezon (2017, p. 583) puts it:

> A path to sustainability is therefore not only in *changing* social imaginaries but also in valorizing and leveraging ones that exist, but that may fall below the radar of traditional economic analysis and the cultural and political ideology in which it is embedded.

These examples from the literature, as well as our examples of lifestyle enterprises in Sweden, point to the complexities involved in social and economic practices. The study shows how market logic and gift-economic exchanges may be both intertwined and collide. It also demonstrates that degrowth may be encouraged and sustained by practices that are not visible at the level of individual production and consumption.

Conclusion

The entrepreneurs in our study are lifestyle-oriented, implying that the entrepreneurs' personal interests and lifestyle provide the foundation for the business. Furthermore, in the narrative of their choice of work and everyday business, they emphasize non-economic values as overarching goals. These could be values related to caring for animals, nature, people or, more generally, to live 'the good life' by being able to combine work with a passion for a hobby or lifestyle. Some downplay profit-oriented values as a means of expressing a form of alternative business identity; others do not and, instead, openly welcome opportunities for economic viability. The place or location – the stable, surrounding areas, untamed nature – where they perform their business and co-produce experiences with visitors is also a fundamental ingredient in the experiences they are offering. Therefore, they are quite similar to many other forms of lifestyle-oriented businesses in tourism. However, in contrast to what are sometimes depicted as illustrative examples of enterprising activity in line with a degrowth orientation, the enterprises in our analysis do not in general market and sell services or products with 'green' or 'alternative' labels, nor do they market their lifestyle as especially niche-oriented or alternative to a conventional market. Neither horse-based enterprising nor hunting tourism is generally associated with 'alternative' lifestyles, nor are they particularly associated with the degrowth movement. The examples discussed in this chapter become illustrative cases for our argument that decommodification practices may be grounded in forms of social and economic exchange which are not only, and sometimes not even, ideologically and politically infused. In this chapter we have identified three decommodifying practices common to the two types of enterprises in focus: *narratives of passion, money-talk* and *favour exchanges*. We argue that such structural, as well as interactional and situated social conditions constituting such practices, may serve to restrain a market-based logic and market-based dynamics. This is in line with previous studies that emphasize the social and cultural embeddedness of economic practices in relation to degrowth (Bogadóttir & Olsen, 2017; Gezon, 2017; Paulson, 2017). Our study, as well as some of the previous studies, have demonstrated that the enactment of a caring ethos which embraces humans as well as non-humans, the constant balancing of moral and economic values, and the sustainment of social relationship and communities are intertwined in the practices and enterprises in focus. We would like to conclude that, through theoretical frameworks and analytical approaches that shed light on the structural and interactive dynamics of social and economic exchange, and which demonstrate how moral and economic values are intertwined as well as constantly negotiated, a more nuanced picture of the type of economic practices and enterprises that may facilitate or constrain degrowth will be discernible.

References

Andersson Cederholm, E. (2015). Lifestyle enterprising: The 'ambiguity work' of Swedish horse-farmers. *Community, Work & Family*, *18*(3), 317–333.

Andersson Cederholm, E. (2018). Relational work in lifestyle enterprising: Sustaining the tension between the personal and the commercial. *Kultura i Spoleczenstwo*, *62*(4), 3–17.

Andersson Cederholm, E., & Åkerström, M. (2016). With a little help from my friends – Relational work in leisure-related enterprises. *The Sociological Review*, *64*(4), 748–765.

Andersson Cederholm, E., & Hultman, J. (2010). The value of intimacy – Negotiating commercial relationships in lifestyle entrepreneurship. *Scandinavian Journal of Hospitality and Tourism*, *10*(1), 16–32.

Ateljevic, I., & Doorne, S. (2000). 'Staying within the fence': Lifestyle entrepreneurship in tourism. *Journal of Sustainable Tourism*, *8*(5), 378–392.

Bogadóttir, R., & Olsen, E. S. (2017). Making degrowth locally meaningful: The case of the Faroese grindadráp. *Journal of Political Ecology*, *24*(1), 504–518.

Cheal, D. (1988). *The gift economy*. London and New York, NY: Routledge.

Dahl, F., & Sjöberg, G. (2010). Social sustainability of hunting tourism in Sweden. In A. Matilainen & S. Keskinarkaus (Eds.), *The social sustainability of hunting tourism in Northern Europe* (pp. 57–73). Helsinki: Ruralia Institute, University of Helsinki.

D'Alisa, G., Demaria, F., & Kallis, G. (Eds.). (2015). *Degrowth: A vocabulary for a new era*. London: Routledge.

de Wit Sandström, I. (2018). *Kärleksaffären: Kvinnor och köpenskap i kustens kommers*. Göteborg and Stockholm: Makadam förlag.

Di Domenico, M. (2005). Producing hospitality, consuming lifestyles: Lifestyle entrepreneurship in urban Scotland. In E. Jones & C. Haven-Tang (Eds.), *Tourism SMEs, service quality and destination competitiveness* (pp. 109–122). Wallingford: CABI Publishing.

Getz, D., Carlsen, J., & Morrison, A. (2004). *The family business in tourism and hospitality*. Wallingford: CABI Publishing.

Gezon, L. L. (2017). Beyond (anti)utilitarianism: Khat and alternatives to growth in northern Madagascar. *Journal of Political Ecology*, *24*(1), 582–594.

Godbout, J. T. (1998). *The world of the gift*. Montreal and Kingston: McGill-Queen's University Press.

Gubrium, J. F., & Holstein, J. A. (1998). Narrative practice and the coherence of personal stories. *The Sociological Quarterly*, *39*(1), 163–187.

Heley, J. (2010). The new squirearchy and emergent cultures of the new middle classes in rural areas. *Journal of Rural Studies*, *26*, 321–331.

Helgadóttir, G., & Sigurdardóttir, I. (2008). Horse-based tourism: Community, quality and disinterests in economic value. *Scandinavian Journal of Hospitality and Tourism*, *8*(2), 105–121.

Holstein, J. A., & Gubrium, J. F. (1997). Active interviewing. In D. Silverman (Ed.), *Qualitative research. Theory, method and practice* (pp. 113–129). London: Sage.

Ioannides, D., & Petersen, T. (2003). Tourism 'non-entrepreneurship' in peripheral destinations: A case study of small and medium tourism enterprises on Bornholm, Denmark. *Tourism Geographies*, *5*(4), 408–435.

Kagervall, A. (2014). *On the conditions for developing hunting and fishing tourism in Sweden* (Unpublished doctoral thesis). Swedish University of Agricultural Sciences, Umeå.

Marcketti, S. B., Niehm, L. S., & Fuloria, R. (2006). An exploratory study of lifestyle entrepreneurship and its relationship to life quality. *Family and Consumer Sciences Research Journal*, *34*(3), 241–259.

Martínes-Alier, J., Pascual, U., Vivien, F.-D., & Zaccai, E. (2010). Sustainable de-growth: Mapping the context, criticisms and future prospects of an emergent paradigm. *Ecological Economics*, *69*(9), 1741–1747.

Mischi, J. (2012). Contested rural activities: Class, politics and shooting in the French countryside. *Ethnography*, *14*(1), 64–84.

Nygård, M., & Uthardt, L. (2011). Opportunity or threat? Finnish hunters' attitudes to hunting tourism. *Journal of Sustainable Tourism*, *19*(3), 383–401.

Oian, H., & Skogen, K. (2016). Property and possession: Hunting tourism and the morality of landownership in rural Norway. *Society & Natural Resources*, *29*(1), 104–118.

Paulson, S. (2017). Degrowth: Culture, power and change. *Journal of Political Ecology*, *24*(1), 425–448.

Schneider, F., Kallis, G., & Martínes-Alier, J. (2010). Crisis or opportunity? Economic degrowth for social equity and ecological sustainability. Introduction to this special issue. *Journal of Cleaner Production*, *18*(6), 511–518.

Sweeney, M., Docherty-Hughes, J., & Lynch, P. (2018). Lifestyling entrepreneurs' sociological expressionism. *Annals of Tourism Research*, *69*, 90–100.

Tidball, K. G., Tidball, M. M., & Curtis, P. (2013). Extending the locavore movement to wild fish and game: Questions and implications. *Natural Sciences Education*, *42*, 185–189.

von Essen, E. (2016). *In the gap between legality and legitimacy: Illegal hunting in Sweden as a crime of dissent* (Unpublished doctoral thesis). Swedish University of Agricultural Sciences, Uppsala.

von Essen, E. (2018). The impact of modernization on hunting ethics: Emerging taboos among contemporary Swedish hunters. *Human Dimensions of Wildlife*, *23*(1), 21–38.

von Essen, E. (2019). How wild boar hunting is becoming a battleground. *Leisure Sciences*. doi:10.1080/01490400.2018.1550456

von Essen, E., Allen, M., & Hansen, H. P. (2017). Hunters, crown, nobles, and conversation elites: Class antagonism over the ownership of common fauna. *International Journal of Cultural Property*, *24*(2), 161–186.

Wang, C., Li, G., & Xu, H. (2019). Impact of lifestyle-oriented motivation on small tourism enterprises' social responsibility and performance. *Journal of Travel Research*, *58*(7), 1146–1160.

Zelizer, V. A. R. (1985). *Pricing the priceless child: The changing social value of children*. New York, NY: Basic Books.

Zelizer, V. A. R. (2005). *The purchase of intimacy*. Princeton, NJ: Princeton University Press.

Zelizer, V. A. R. (2013). *Economic lives: How culture shapes the economy*. Princeton, NJ: Princeton University Press.

3 Lifestyle entrepreneurs as agents of degrowth

The case of nature-based tourism businesses in Scandinavia

Lusine Margaryan, Peter Fredman, and Stian Stensland

Introduction

Concerns regarding the long-term destructiveness of uncurbed tourism growth have become more visible in recent years in light of the acute social and environmental pressure caused by overtourism, i.e. excessive concentration of tourists and tourism-related businesses in a number of urban as well as nature-based destinations (Fletcher, Murray Mas, Blanco-Romero, & Blázquez-Salom, 2019; Higgins-Desbiolles, Carnicelli, Krolikowski, Wijesinghe, & Boluk, 2019; Milano, Cheer, & Novelli, 2019). Furthermore, discontent with uncontrolled tourist flows happens within the larger context of growing public concern with global environmental degradation and climate change, expressed in vocal social protests in multiple countries.

A recent global assessment report on biodiversity and ecosystem services documented unprecedented and accelerating rates of nature transformation by humans, including profound alteration of 75% of terrestrial ecosystems and one million species under immediate threat of extinction (IPBES, 2019). Given these alarming symptoms, it is likely that the optimistic era of sustainable development, green growth and ecological modernization, dominating scientific and public discourses since the 1980s, is approaching its twilight (Gómez-Baggethun, 2020; Parrique, 2019). In other words, the high hopes for the prolific 'marriage' of conventional growth-oriented economics with environmental agenda remain unmet. The neoliberal promise of growth as the panacea for global ills, including the just-in-time technological fix, has not delivered results when it comes to securing a safe and high-quality environment for future generations (Parrique, 2019). This worry has already permeated the minds of future generations quite literally, exemplified by active participation of children in environmental protests and school strikes in the summer of 2019 in many countries. In this context, tourism emerges as a global power, which directly contributes to climate change and transformation of the environment, livelihoods and spaces at an ever-accelerating rate, and the growth of which remains largely unquestioned, if not celebrated (Higgins-Desbiolles et al., 2019).

Degrowth, the effort of finding pathways to sustain the economy without the growth imperative, has its foundation in the economic critique of the productivist societies of the 1970s (e.g. most famously through the works of Georgescu-Roegen (1971); Meadows, Meadows, Randers, and Behrens III (1972); and Gorz (1980)). Degrowth has developed into both a revolutionary worldview and an explicit political agenda for an alternative socio-economic system, relevant today more than ever (D'Alisa, Demaria, & Kallis, 2015; Gómez-Baggethun, 2020; Parrique, 2019). Degrowth has entered tourism research only recently, even though critique around inappropriate and undesirable tourism development has always been part of this field of inquiry (Andriotis, 2018; Hall, 2009). A recent special issue of the *Journal of Sustainable Tourism* focused on degrowth also indicates the growing momentum of this agenda.

Of those scarce studies that do focus directly on tourism and degrowth, the majority ponder degrowth possibilities in the tourism sector in general or tourism demand in particular (e.g., Andriotis, 2018; Canavan, 2014; Fletcher el al., 2019; Hall, 2009; Higgins-Desbiolles et al., 2019; Hollenhorst, Houge-Mackenzie, & Ostergren, 2014), whereas little attention has been paid to the role of tourism firms in degrowth. Meanwhile, small-scale tourism firms, especially lifestyle entrepreneurs, deserve more attention, as they are among the scarce businesses where some principles of degrowth business ethics can be observed empirically. In other words, lifestyle entrepreneurs are important agents of alternative business praxis and ethos, who require a closer analysis from a degrowth perspective.

This chapter lies at the confluence of two research streams, degrowth and lifestyle entrepreneurship, and is informed by the ethical perspectives of social limits to growth (Hirsch, 2005). Originating from different theoretical sources but containing similar ontological insights, both degrowth and lifestyle entrepreneurship literature provide indispensable tools for understanding socio-economic phenomena outside of the hegemonic growth paradigm. Bringing together the insights from degrowth and lifestyle entrepreneurship literature in tourism, therefore, is an important step towards integrating the theory and practice of alternative business models and ways of life.

We argue that a vast majority of nature-based tourism (NBT) entrepreneurs in Scandinavia (Norway and Sweden) can be considered lifestyle entrepreneurs, which is supported by previous literature (Andersson Cederholm & Hultman, 2010; Bredvold & Skålén, 2016; Lundberg & Fredman, 2012; Lundberg, Fredman, & Wall-Reinius, 2014) as well as the nationwide data from both countries used in this chapter. By looking at the case of small-scale (NBT) entrepreneurs we demonstrate the prerogative of lifestyle values in this business, such as the importance of working outdoors, being self-reliant, working with people with similar interests, contributing to sustainable development, utilizing local resources or educating people about nature. We discuss lifestyle entrepreneurs as viable and vigorous actors against the 'crisis of meaning', erosion of autonomy and self-limitation as well as insatiable appetite for positional goods, typical for the consumption societies of late capitalism (D'Alisa et al., 2015). Furthermore, we offer a critical discussion on the role of nature-based tourism (NBT) in the context of global drivers of change, such as environmental degradation, climate change and resource depletion, as well as a renewed interest in socio-economic relationships not based on utilitarian exchange with ever-increasing returns. The empirical data comes from a synthesis of nation-wide surveys among NBT entrepreneurs in Sweden and Norway conducted in 2013 and 2018, as well as some qualitative insights from interviews with nature-based entrepreneurs conducted in 2018.

Re-reading lifestyle entrepreneurship in tourism from a degrowth perspective

Higgins-Desbiolles et al. (2019) argue that if degrowth principles are to be achieved in tourism, resistance to corporate businesses and their perpetual growth agendas along with prioritization of local interests is crucial. Yet, there are very few examples from the tourism sector where its actors would pro-actively resist the growth-oriented modus operandi. Global tourism arrivals are growing by the annual average rate of 6%, and only a health or economic crisis (as happened in 2008 or in 2020) changes this trajectory (UNWTO, 2019). There is some evidence that the much-publicized Swedish 'flight shame' (*flygskam* in Swedish) might have had a certain role to play in the reduction of overseas travel in Sweden in 2018, which curbed a years-long growth trend, as adult Swedes took half a million fewer overseas flights than in 2017 (The Local, 2019). However, 2018 also witnessed an exceptionally weak Swedish krona and a weeks-long record heatwave, which are likely to be contributing factors to the Swedish citizens' decision to stay at home for the summer. In other words, there is still very little evidence that there are some significant *voluntary* long-term efforts to resist tourism growth and consumption. In this context, it is especially interesting to turn to lifestyle entrepreneurs.

According to Kallis, Demaria, and D'Alisa (2015, p. 6): "The foundational theses of degrowth are that growth is uneconomic and unjust, that it is ecologically unsustainable and that it will never be enough". Lifestyle entrepreneurs, it seems, have been able to define what *enough* is when it comes to the growth of their business. Lifestyle entrepreneurship emerged as a descriptive term aiming to explain business motivations and personal choices, invisible from within the traditional quantitative success frameworks of expansion and income maximization (Ateljevic & Doorne, 2000; Peters, Frehse, & Buhalis, 2009). For example, a 2019 BBC news article featured Audun, a tourism entrepreneur and musher from Svalbard, Norway, who says:

> For some reason in the world it has become natural to think that every business should double their income every single year. It is even in the deadly sins – greed! It's like a basic thought pattern, that you should *not* do this. But still, everyone thinks it is natural that every business should do it to be able to survive. And I think, as long as I can feed my dogs, as long as I can feed myself, and live in a house – I am fine.
>
> (quoted in South & Young, 2019)

This quotation captures eloquently the spirit of many micro-entrepreneurs within the tourism industry in Scandinavia and beyond. Understanding lifestyle values has emerged as an important avenue for making sense of such tourism entrepreneurs in recent decades (Williams, Shaw, & Greenwood, 1989). It has been observed that a significant portion of small-scale tourism entrepreneurs consciously choose to move away from conventional business ethics, striving towards a balance of economic interests together with personal values, as well as pursuing social and environmental goals (Ateljevic & Doorne, 2000; Lopéz, Buhalis, & Fyall, 2009; Peters et al., 2009; Shaw & Williams, 1998). Contrary to the classic understanding of entrepreneurship, where the lack of capital, skills or acceptance of suboptimal profits are considered detrimental for business, lifestyle entrepreneurs survive by carving out their distinct business niche based on the increasingly differentiated tourist experience demand. The central argument raised in tourism scholarship has therefore been that small tourism entrepreneurs cannot be understood as regular entrepreneurs only scaled down. There are not only quantitative but also crucial qualitative differences that should be taken into account. One such difference is the attitude towards business growth.

Resistance to the perpetual growth imperative, in fact, has been suggested, directly and indirectly, as key to understanding lifestyle entrepreneurs in tourism. It has been observed that the tourism sector attracts a significant share of entrepreneurs who have a wide range of personal goals and values, only weakly related to conventional business ethics. In contrast, doing business is seen primarily as a means to pursue certain personal goals, such as the ability to live in a specific area (often remote and peripheral), to work outdoors, to be physically active in nature, to meet like-minded people, to be independent and to be able to contribute to the local sustainability (Ateljevic & Doorne, 2000; Hollick & Braun, 2005; Lopéz et al., 2009; Peters et al., 2009; Shaw & Williams, 1998; Skokic & Morrison, 2011; Shepherd & Wiklund, 2005; Dobbs & Hamilton, 2007). Furthermore, it has been found that such entrepreneurs prioritize business values that are in direct conflict with growth strategies, such as keeping the business within the family, maintaining independence and full control over the operations or maintaining status as providers of authentic and unique experiences (Ateljevic & Doorne, 2000; Glancey, 1998; Hollick & Braun, 2005; Komppula, 2004).

The concept of lifestyle entrepreneurship, however, is not without criticism. First, very few studies exist that empirically demonstrate the difference between lifestyle entrepreneurs and 'normal' entrepreneurs (Masurel & Snellenberg, 2017). Second, it has been pointed out that lifestyle entrepreneurship is also characterized by less positive properties, such as conservatism, reluctance to innovate or being heavily family-oriented and inward looking (Bredvold & Skålén, 2016). Furthermore, entrepreneurship in general has been often unquestionably valorized with its disadvantages downplayed, such as inconsistent income, bad work-life balance, stress or low survival rates. The very nature of this work requires a certain set of personal characteristics and is not suitable for everyone. Nevertheless, Masurel and Snellenberg (2017) conclude that the concept of lifestyle entrepreneurship has high relevance in terms of self-image of many entrepreneurs. In our chapter we approach lifestyle entrepreneurs from the perspective of business goals and motivations, focused on achieving certain lifestyle conditions (see Table 3.1).

In order to make better sense of lifestyle entrepreneurship, it could be useful to turn to the social limits of growth. It is postulated that perpetual growth in wealthy economies is sustained by the dream of access to positional goods, i.e. goods communicating wealth status and prestige in society (Hirsch, 2005; Kallis, 2015). Economic growth can never satisfy the desire for positional goods, since their accessibility is shifting upwards along with the increasing wealth of the general population, exclusivity being in their very essence. Positional goods and conspicuous consumption, therefore, can be seen as signals of the social limit to growth, i.e. the limit of what growth can deliver for one's well-being and development in a given society (Hirsch, 2005). Chasing positional goods remains one of the key drivers for the desire of perpetual growth, even if this pursuit is ultimately a futile endeavor, being a zero-sum game, and is explicitly discouraged in the traditional religious and secular ethics across human cultures since time immemorial. Within some philosophies and religions, wealth is discussed as a means to an end, which is well-being. Take, for example, the Xenophon's Socratic dialogue *Oeconomicus* where the nature of wealth is explored, and where wealth is eventually defined as something instrumental, something that is of a clear benefit to its owner, otherwise it cannot be considered wealth (Strauss, 1970). *Ecclesiastes* (4:7–8) tells us about a man who had been working hard endlessly, never content with his wealth, but had no one to share it with, which is presented as a meaningless and miserable situation; 'For whom am I toiling, Why am I depriving myself of enjoyment?', he asks himself in despair. Similar sentiments can be found in *The Words of the High One* (*Hávamál*, the Old Norse gnomic verses of mundane wisdom) Larrington (2014), where for example stanza 40 predicates that once a man has gained enough wealth, he should not crave more (Auden & Taylor, 1981). In contrast, neoliberal laissez-faire economics proposes a new ethics of reinventing greed (by no means a purely capitalist phenomenon) as a locomotive for development. In his essay published in the *New York Times*, Milton Friedman (1970) famously declared that the social responsibility of business is to increase its profits, and business owners should be concerned with making as much money as possible without violating the basic rules of society. While this doctrine has been seemingly universally adopted by large-scale businesses (and valorized in pop culture as in the famous speech by Gordon Gekko from the 1987 movie *Wall Street*; Pressman & Stone, 1987), this is clearly not the case with lifestyle entrepreneurs.

In modern-day affluent societies, the share of income spent on positional goods increases due to progressively larger proportions of income becoming disposable (Kallis, 2015; Skidelsky & Skidelsky, 2012). Comparing oneself to one's peers might be natural, but this need for belonging is channeled towards consumerism and amplified by social norms, pop culture, media, as well as normalization of consumption as the main avenue of identity construction (Saren, 2007). Along with this, the share of resources spent on leisure, health and wellness, socialization, family, involvement with the local community and politics does not increase in proportion to wealth (Kallis, 2015). This is also true for access to high-quality public goods, such as a clean environment. For example, air pollution, a growing problem in many major global financial centres, is associated with a vast range of negative effects on human health, ranging from respiratory to neurological diseases, resulting in diminished quality of life and life expectancy in general (Fonken et al., 2011). In other words, there is no evidence for a simplistic linear relationship between the growth of wealth and well-being.

Skidelsky and Skidelsky (2012) attempt to revitalize the notion of the 'good life' as a way to move away from insatiability as the foundation of modern-day economics. The central question we should be asking is 'What is wealth for?' From this perspective, the so-called lifestyle tourism entrepreneurs can be viewed as belonging to a segment of society who consciously choose to recognize the social limits to growth, prioritizing other non-monetary personal goals over the 'rat race' of positional competition. For these entrepreneurs, business and income generation remains a *means* towards certain ends, which is well-being, rather than an end in itself. Success, therefore, is defined in terms of the ability to reach and maintain a set of goals related to the quality of life, rather than expansion and profit maximization. In other words, growth becomes 'dethroned' as the ultimate good. Based on the aforementioned discussion, it can be suggested that lifestyle entrepreneurs are important agents of alternative business praxis and ethos, who deserve a closer analysis from the degrowth perspective.

The case of NBT entrepreneurs in Scandinavia

Previous research suggests that NBT business attracts lifestyle entrepreneurs (Andersson Cederholm & Hultman, 2010; Bredvold & Skålén, 2016; Ateljevic & Doorne, 2000; Lundberg & Fredman, 2012; Lundberg et al., 2014; Tzschentke, Kirk, & Lynch, 2008). Such properties of NBT as proximity to nature, clean environment and the opportunity to educate others about nature are often brought up as reasons to become an NBT entrepreneur (Ateljevic & Doorne, 2000; Margaryan & Stensland, 2017). Although NBT and outdoor recreation have long traditions in Scandinavia, the proliferation of tourism entrepreneurs offering nature experiences commercially is a relatively new phenomenon. For example, it has been estimated that in the early 1990s the number of NBT firms in Sweden was around 150, which grew to at least 500 by the late 1990s (Fredman, Gössling, & Hultman, 2006). However, due to the lack of comprehensive databases and registers, it had been impossible to obtain reliable information about the development of the NBT sector until recently.

The research on lifestyle tourism entrepreneurship has been dominated by small-scale case studies and qualitative interviews (e.g., Andersson Cederholm & Hultman, 2010; Ateljevic & Doorne, 2000; Lundberg & Fredman, 2012; Milano et al., 2019; Lopes, 2017; Mottiar, 2007) save for rare exceptions (e.g. Claire, 2012). Due to the absence of large-scale quantitative data, it has been impossible to make any generalized claims about the size and significance of lifestyle values in the tourism sector. Aiming to fill this gap, in this chapter we offer quantitative insights from country-level surveys of NBT entrepreneurs in Sweden and Norway. In addition, we provide supplementary qualitative insights from a recent study on NBT entrepreneurs in the north of Sweden.

Based on an in-depth qualitative study in Sweden, Lundberg et al. (2014) suggested that the relationship of NBT entrepreneurs towards money and growth is complex and not fully understood. Specifically, they argue that the very identity of these entrepreneurs is not compatible with profit and growth. In addition, direct dependence on nature makes these entrepreneurs more sensitive towards environmental impacts and sustainable nature use, which is prioritized over growth (Lundberg et al., 2014). Taking previous literature as a point of departure, as well as based on the findings by Lundberg and Fredman (2012) on the success factors and constraints among NBT entrepreneurs in Sweden, we included a series of questions in our surveys that aimed to understand the motivations NBT entrepreneurs have behind running their businesses (see Table 3.1).

The quantitative data in this chapter comes from three different research projects, the details of which are published separately as technical reports (see Fredman & Margaryan, 2014; Stensland et al., 2014, 2018). These three projects aimed to capture representative national samples of NBT entrepreneurs to generate comprehensive knowledge on this sector. NBT entrepreneurs were conceptualized as entrepreneurs operating in predominately unmodified natural and/or rural areas, and commercially offering nature-related tourism experiences (e.g. hiking, biking, skiing, hunting, fishing, dog-sledding, kayaking/canoeing/rafting). In both countries, the sampling strategy was based on requesting data from regional tourist information centres, which keep track of their local tourism businesses. As a result, the following samples were collected: in Sweden – 1,841 companies in 2014, in Norway – 1,785 in 2014 and 1,614 in 2018.

Additionally, in the frames of a research project on migrant NBT entrepreneurs, qualitative face-to-face interviews were conducted in 2018 with entrepreneurs who relocated to the rural north of Sweden with an intention to start NBT operations (seven middle-aged couples running family businesses). Based on the literature on lifestyle entrepreneurship (Lundberg & Fredman, 2012; Glancey, 1998; Hollick & Braun, 2005; Komppula, 2004) and the insights from the quantitative surveys, questions about motivations to relocate and operate NBT businesses were asked during the course of two-hour in-depth interviews, such as "Why did you decide to become a tourism entrepreneur?" and "What goals do you have for your business?" All interviews were transcribed and analyzed based on the standard three-step coding procedure (Saldaña, 2015).

Results

Response rates from the surveys comprised 35.5% (n=648) in Sweden, 38% in Norway in 2014 (n=684) and 34.6% (n=585) in 2018. In both countries, the sample was dominated by micro businesses (with the overwhelming majority having less than two year-round employees). All surveys included questions regarding personal and business goals for operating an NBT business, the results of which are presented in Table 3.1.

As can be seen from the results, business goals associated with quality of life, interesting job, sustainability and access to nature are prioritized over

Table 3.1 Motivations to operate a nature-based tourism (NBT) business

Motivations to operate an NBT business in Sweden	*Mean**	*Motivations to operate an NBT business in Norway*	*Mean** 2014*	*Mean** 2018*
To offer clients good nature experiences	4.6	To offer clients good nature experiences	6.5	6.5
To have an interesting job	4.2	To convey the value of nature to the clients	6.0	6.0
To be able to use local resources	4.2	To have an interesting job	5.9	6.0
To contribute to sustainable development	4.1	To be able to use local resources	5.9	5.9
To be independent	4.0	To contribute to sustainable development	5.9	5.9
To be able to work outdoors	3.8	To have social contacts with clients	5.8	5.7
To meet people with similar interests	3.7	To be able to work outdoors	5.7	5.7
To educate people about nature	3.6	To be able to live where I live now	5.6	5.6
To have a secure and stable income	3.6	To be independent	5.5	5.5
To maximize income	3.0	To have growth in company turnover		5.2
		To maximize income	4.5	4.4
		To grow in number of full-time employees		3.8

* Measurement was done on a 5-point Likert scale (1 not important at all – 5 very important)
**Measurement was done on a 7-point Likert scale (1 not important at all – 7 very important)

growth-oriented ones, which are typically attributed to lifestyle entrepreneurship (Ateljevic & Doorne, 2000; Hollick & Braun, 2005; Lopéz et al., 2009; Peters et al., 2009; Shaw & Williams, 1998; Skokic & Morrison, 2011; Shepherd & Wiklund, 2005; Dobbs & Hamilton, 2007). When it comes to the insights from the qualitative data, the main themes emerging from the interviews with NBT entrepreneurs in the north of Sweden were the following. All of the interviewees mentioned that starting an NBT business and relocating to a remote rural area was a way to improve their quality of life, which spans beyond monetary gain. This is explained by both push and pull factors, such as avoiding unfavorable conditions inherent in their previous social context as well as pursuing a better natural environment, work-life balance and new life paths overall. Piet, for example, who used to work in an office, stated:

> I have just had enough to be honest. . . . Always being busy, never good enough, never fast enough. Always stressed out . . . I thought, why can't we just be normal, just give each other a little space and not ask impossible things from each other. Life would be a lot easier.

The interviewees presented NBT business as a way to gain access to conditions that were important for a better life, such as a clean environment, abundant natural resources, work-life balance, peace and nice conditions to raise children. Ellen, for example, says: "We came here for space, nature, silence – the quality of life". In fact, none of the interviewees mentioned getting into their NBT business to increase the family income. It can be argued that NBT emerged as the most readily available income-generating option, since business and employment opportunities in the peripheral rural areas of the north are rather limited.

Further, it became clear that the interviewed NBT entrepreneurs do not have growth and expansion as a paramount business priority. Wim, for example, explains:

> I'd like to grow in such a way to just stop worrying about my day-to-day finances. I do not need to become rich. And we won't become rich anyway. If I wanted to get rich, I would have picked a different profession.

A recurrent narrative around doing business was that it is simply instrumental for achieving personal goals, such as living in a certain area and having a certain lifestyle. Business success, in other words, is measured by these entrepreneurs against the ability to meet these goals, rather than against economic indicators of business performance. At the same time, however, it is important to emphasize that lifestyle entrepreneurship might offer a more enriching but not necessarily an easier life. In fact, our interviewees pointed out that they work longer hours than they did in their previous jobs. Operating a campsite, for example, demanded constant on-site presence beyond working hours, which contradicts the goal for work-life balance, typically ascribed to lifestyle entrepreneurs. Nevertheless, the interviewees considered that with the decision to become NBT entrepreneurs, their lives have improved in other ways.

Discussion and conclusion

In this chapter we demonstrate that the NBT sector in Norway and Sweden is overwhelmingly dominated by lifestyle entrepreneurs and help to empirically validate the definition of lifestyle entrepreneurship. We base our arguments on the countrywide statistical data from Norway and Sweden, which has not been done before. The results suggest that the goals NBT entrepreneurs set for their businesses are in line with the ethics of social limits to growth and degrowth agendas (Kallis, 2015; Skidelsky & Skidelsky, 2012). Extrinsic goals such as 'to provide good nature experiences', 'to convey nature's value', 'to contribute to sustainable development', 'to use local resources' as well as intrinsic ones such as 'to have an interesting job' or 'to be independent' have been assessed as important or very important goals in both countries, with a remarkable consistency across two measurements in Norway. What is interesting is that growing financially is somewhat important, but growing in terms of the number of employees is not, which could imply, for example, a preference to keep the business within the family. Overall, these results statistically validate previous qualitative findings on lifestyle entrepreneurship (Andersson Cederholm & Hultman, 2010; Ateljevic & Doorne, 2000; Lopes, 2017; Lundberg & Fredman, 2012; Tzschentke et al., 2008) and give additional insights into the importance of sustainability principles for these type of businesses.

Nature-based tourism entrepreneurs as agents of degrowth

In this chapter we argue that NBT entrepreneurs can largely be viewed as agents of degrowth. It has to be emphasized that degrowth does not imply zero growth, recession or a Luddite resistance to progress. While different suggestions have been made regarding promotion of degrowth principles on the macro-level (Andriotis, 2018; Canavan, 2014; D'Alisa et al., 2015; Fletcher et al., 2019; Hall, 2009; Higgins-Desbiolles et al., 2019; Hollenhorst et al., 2014), examples of micro-level degrowth applications, especially when it comes to the supply side of tourism, have been rarely discussed. Andriotis (2018), for example, focuses on individual tourists' choices, such as reducing consumption, travelling less and more slowly. Tourism industry actors, however, are usually presented as vehement defenders of growth (Fletcher et al., 2019). To operationalize degrowth, Hall (2010, p. 131) advocates for "steady-state tourism . . . that encourages qualitative development but does not aggregate quantitative growth to the detriment of natural capital". Without trying to romanticize NBT entrepreneurs, we argue that they are nevertheless rather unique examples of businesses where degrowth principles have been voluntarily practiced at least to some visible extent, which could then be statistically observed in large-scale data.

Rediscovering meaning and deprioritizing growth in tourism businesses

Due to their conscious rejection of growth-dominated business ethics, lifestyle entrepreneurs have often been labeled as 'non-economic' or 'non-entrepreneurs' and have been accused of stagnating regional development (Ateljevic & Doorne, 2000; Lopes, 2017; Williams et al., 1989). Based on our data we would like to contribute to the argument that by deprioritizing growth, lifestyle entrepreneurship brings other values to the destination and society at large. Specifically, we would like to focus on the role of meaning in tourism and business in general.

Our survey demonstrates that factors that are overwhelmingly ranked as important or very important for the NBT entrepreneurs can be understood from the perspective of meaning both for the entrepreneurs and their tourists. Erosion of meaning has been discussed as one of the unintended consequences of the modern-day capitalist system, propagating a 'money metric' as the ultimate and universal standard of value (D'Alisa et al., 2015). In the context of overtourism, where tourism becomes a global force of endless circulation of people among commodified spaces, amplified by the endless circulation of images in the digital spaces, the discussion of meaning becomes more relevant than ever. By deprioritizing growth and resisting the measurement of success in purely monetary terms, lifestyle entrepreneurs instead focus on creating and maintaining meaning both in the experiences they provide to the tourists as well as in their own lives. Perhaps finding the answer to the eternal question 'What is wealth for?' comes a bit easier for lifestyle entrepreneurs.

References

Andersson Cederholm, E., & Hultman, J. (2010). The value of intimacy – Negotiating commercial relationships in lifestyle entrepreneurship. *Scandinavian Journal of Hospitality and Tourism*, *10*(1), 16–32.

Andriotis, K. (2018). *Degrowth in tourism: Conceptual, theoretical and philosophical issues*. Wallingford: CABI.

Ateljevic, I., & Doorne, S. (2000). 'Staying within the fence': Lifestyle entrepreneurship in tourism. *Journal of Sustainable Tourism*, *8*(5), 378–392.

Auden, W. H., & Taylor, P. B. (1981). *Norse poems*. London: Burns & Oates.

Bredvold, R., & Skålén, P. (2016). Lifestyle entrepreneurs and their identity construction: A study of the tourism industry. *Tourism Management*, *56*, 96–105.

Canavan, B. (2014). Sustainable tourism: Development, decline and de-growth. Management issues from the Isle of man. *Journal of Sustainable Tourism*, *22*(1), 127–147.

Claire, L. (2012). Re-storying the entrepreneurial ideal: Lifestyle entrepreneurs as hero? *Journal for Critical Organization Inquiry*, *10*(1), 31–39.

D'Alisa, G., Demaria, F., & Kallis, G. (Eds.). (2015). *Degrowth: A vocabulary for a new era*. Abingdon: Routledge.

Dobbs, M., & Hamilton, R. (2007). Small business growth: Recent evidence and new directions. *International Journal of Entrepreneurial Behaviour and Research*, *13*(5), 296–322.

Fletcher, R., Murray Mas, I., Blanco-Romero, A., & Blázquez-Salom, M. (2019). Tourism and degrowth: An emerging agenda for research and praxis. *Journal of Sustainable Tourism*, *27*(12), 1745–1763.

Fonken, L. K., Xu, X., Weil, Z. M., Chen, G., Sun, Q., Rajagopalan, S., & Nelson, R. J. (2011). Air pollution impairs cognition, provokes depressive-like behaviors and alters hippocampal cytokine expression and morphology. *Molecular Psychiatry*, *16*(10), 987–995.

Fredman, P., Gössling, S., & Hultman, J. (2006). Sweden: Where holidays come naturally. In S. Gössling & J. Hultman (Eds.), *Ecotourism in Scandinavia: Lessons in theory and practice* (pp. 23–37). Wallingford: CABI.

Fredman, P., & Margaryan, L. (2014). *The supply of nature-based tourism in Sweden: A national inventory of service providers*. Östersund: ETOUR.

Friedman, M. (1970, September 13). The social responsibility of business is to increase its profits. *The New York Times Magazine*, 122–126.

Georgescu-Roegen, N. (1971). *The entropy law and the economic process*. Cambridge, MA: Harvard University Press.

Glancey, K. (1998). Determinants of growth and profitability in small entrepreneurial firms. *International Journal of Entrepreneurial Behaviour and Research*, *4*(1), 18–27.

Gómez-Baggethun, E. (2020). More is more: Scaling political ecology within limits to growth. *Political Geography*, *76*. doi:10.1016/j.polgeo.2019.102095.

Gorz, A. (1980). *Ecology as politics* (P. Vigderman & J. Cloud, Trans.). Montréal: Black Rose Books.

Hall, C. M. (2009). Degrowing tourism: Décroissance, sustainable consumption and steady-state tourism. *Anatolia*, *20*(1), 46–61.

Hall, C. M. (2010). Changing paradigms and global change: From sustainable to steady-state tourism. *Tourism Recreation Research*, *35*(2), 131–143.

Higgins-Desbiolles, F., Carnicelli, S., Krolikowski, C., Wijesinghe, G., & Boluk, K. (2019). Degrowing tourism: Rethinking tourism. *Journal of Sustainable Tourism*, *27*(12), 1926–1944.

Hirsch, F. (2005). *Social limits to growth* (2nd ed.). London: Routledge.

Hollenhorst, S. J., Houge-Mackenzie, S., & Ostergren, D. M. (2014). The trouble with tourism. *Tourism Recreation Research*, *39*(3), 305–319.

Hollick, M., & Braun, P. (2005). Lifestyle entrepreneurship: The unusual nature of the tourism entrepreneur. In *Proceedings of the Second Annual AGSE International Entrepreneurship Research Exchange*. Melbourne: Swinburne Press.

IPBES. (2019). *Global assessment report on biodiversity and ecosystem services: Summary for policymakers*. Retrieved April 2, 2020, from https://ipbes.net/sites/default/files/inline/files/ipbes_global_assessment_report_summary_for_policymakers.pdf

Kallis, G. (2015). Social limits to growth. In G. D'Alisa, F. Demaria, & G. Kallis (Eds.), *Degrowth: A vocabulary for a new era* (pp. 217–221). Abingdon: Routledge.

Kallis, G., Demaria, F., & D'Alisa, G. (2015). Introduction: Degrowth. In G. D'Alisa, F. Demaria, & G. Kallis (Eds.), *Degrowth: A vocabulary for a new era* (pp. 1–18). Abingdon: Routledge.

Komppula, R. (2004). Success and growth in rural tourism micro-businesses in Finland: Financial or lifestyle objectives? In R. Thomas (Ed.), *Small firms in tourism: International perspective* (pp. 115–138). Oxford: Elsevier.

Larrington, C. (Ed.). (2014). *The poetic Edda*. Oxford: Oxford University Press.

The Local. (2019, April 9). *Why people in Sweden are breaking a steady trend and travelling less*. Retrieved May 31, 2020, from www.thelocal.se/20190409/why-swedes-are-breaking-a-steady-trend-and-travelling-less

Lopes, R. (2017). *A study of the philosophic underpinnings of the motivations and behaviors of lifestyle entrepreneurs* (Unpublished Master's thesis). Maine: Honors College, University of Maine.

Lopéz, E. P., Buhalis, D., & Fyall, A. (2009). Entrepreneurship and innovation in tourism. *Pasos*, *7*(3), 355–357.

Lundberg, C., & Fredman, P. (2012). Success factors and constraints among nature-based tourism entrepreneurs. *Current Issues in Tourism*, *15*(7), 649–671.

Lundberg, C., Fredman, P., & Wall-Reinius, S. (2014). Going for the green? The role of money among nature-based tourism entrepreneurs. *Current Issues in Tourism*, *17*(4), 373–380.

Masurel, E., & Snellenberg, R. (2017). Does the lifestyle entrepreneur exist? An analysis of lifestyle entrepreneurs compared with other entrepreneurs on the basis of the development of entrepreneurial competences. *Research Memorandum*, *1*, 1–14.

Meadows, D. H., Meadows, D. H., Randers, J., & Behrens III, W. W. (1972). *The limits to growth: A report to the club of Rome*. New York, NY: Universe Books.

Milano, C., Cheer, J. M., & Novelli, M. (Eds.). (2019). *Overtourism: Excesses, discontents and measures in travel and tourism*. Wallingford: CABI.

Mottiar, Z. (2007). Lifestyle entrepreneurs and spheres of inter-firm relations: The case of Westport, Co Mayo, Ireland. *The International Journal of Entrepreneurship and Innovation*, *8*(1), 67–74.

Parrique, T. (2019). *The political economy of degrowth*. Economics and Finance. Université Clermont Auvergne; Stockholms Universitet. Unpublished doctoral dissertation. Retrieved May 31, 2020 from https://tel.archives-ouvertes.fr/tel-02499463

Peters, M., Frehse, J., & Buhalis, D. (2009). The importance of lifestyle entrepreneurship: A conceptual study of the tourism industry. *Pasos*, *7*(2), 393–405.

Pressman, E. (Producer), & Stone, O. (Director). (1987). *Wall street*. [Motion picture] twentieth century-fox film corporation. Retrieved February 17, 2020 from www.youtube.com/watch?v=VVxYOQS6ggk

Saldaña, J. (2015). *The coding manual for qualitative researchers*. London: Sage.

Saren, M. (2007). To have is to be? A critique of self-creation through consumption. *The Marketing Review*, *7*(4), 343–354.

Shaw, G., & Williams, A. M. (1998). Entrepreneurship and tourism development. In D. Ioannides & K. G. Debbage (Eds.), *The economic geography of the tourist industry* (pp. 235–255). London: Routledge.

Shepherd, J., & Wiklund, D. (2005). Entrepreneurial orientation and small business performance: A configurational approach. *Journal of Business Venturing*, *20*(1), 71–91.

Skidelsky, E., & Skidelsky, R. (2012). *How much is enough? Money and the good life*. London: Penguin.

Skokic, V., & Morrison, A. (2011). Conceptions of tourism lifestyle entrepreneurship: Transition economy context. *Tourism Planning & Development*, *8*(2), 157–169.

South, D., & Young, T. (2019, April 21). I'm the father to 110 huskies. *BBC Newshour*. [Video file] Retrieved May 30, 2020, from www.bbc.com/news/av/world-europe-47977584/i-m-the-father-to-110-huskies

Stensland, S., Fossgaard, K., Apon, J., Baardesen, S., Fredman, P., Grubben, I., . . . Røren, E. A. M. (2014). *Naturbaserte reiselivsbedrifter i Norge. Frekvens-og metoderapport: Institutt for naturforvaltning, Norges miljø-og biovitenskaplige universitet*. Ås: NMBU. Retrieved May 31, 2020, from https://nmbu.brage.unit.no/nmbu-xmlui/handle/11250/2647411

Stensland, S., Fossgard, K., Hansen, B. B., Fredman, P., Morken, I. B., Thyrrestrup, G., & Haukeland, J. V. (2018). *Naturbaserte reiselivsbedrifter i Norge: statusoversikt, resultater og metode fra en nasjonal spørreundersøkelse*. Ås: NMBU. Retrieved May 31, 2020, from https://nmbu.brage.unit.no/nmbu-xmlui/handle/11250/2648392

Strauss, L. (1970). *Xenophon's Socratic discourse: An interpretation of the Oeconomicus*. Ithaca: Cornell University Press.

Tzschentke, N., Kirk, D., & Lynch, P. (2008). Going green: Decisional factors in small hospitality operations. *International Journal of Hospitality Management*, *27*(1), 126–133.

United Nations World Tourism Organization (UNWTO). (2019). *Tourism highlights – 2019 Edition*. Madrid: UNWTO.

Williams, A. M., Shaw, G., & Greenwood, J. (1989). From tourist to tourism entrepreneur, from consumption to production: Evidence from Cornwall, England. *Environment and Planning A*, *21*(12), 1639–1653.

4 Mobility transitions and rural restructuring in Sweden

A database study of holistic simplifiers

Marco Eimermann, Urban Lindgren, Linda Lundmark and Jundan Jasmine Zhang

Introduction

As globalization and place competition continue to shape development strategies and societal planning around the world, issues of growth and decline at different geographic locations, which lead to uneven geographies of development, become increasingly pressing. Places that decline 'lose the battle' for flows of capital, people and knowledge. Meanwhile pandemic, social, economic and environmental issues remind us of the limits to growth and the urge to search for alternatives and feasible solutions for a sustainable future (e.g. Hagbert & Fauré, 2021). For instance, some scholars would say that critical to thinking beyond the current emphasis on production and growth is rethinking work and the moral value attached to work in modern societies (Singh, 2019). What is the role of the demands made on our capacities to care for one another and the world, in projects of socially and ecologically just and viable futures? (Singh, 2019). In concrete terms, this relates to questions about the values attached to different kinds of consumption enhanced by income through work and its distribution in tourism and other branches. In recent years, the discourse of transforming societies to become more sustainable has broadened its scope beyond the demands on the major political and economic players. Living simply and sustainably has therefore become a legitimate, or even trendy, way for individuals to contribute to a better society in the global North (e.g. Singh, 2019) and has become expressed in the ways in which every individual needs to change his/her way of living, e.g. through minimalism, simplicity or even Kondo-ism.

Uneven geographic development is often studied through umbrella terms like 'urban' and 'rural' areas. In discourses of transition, rural areas appear as suitable for decluttering and reconnecting with 'nature' and 'the land' (Halfacree, 2006). 'The rural' has been stereotypically represented as 'negative', e.g. in regional development studies (De Souza, 2018), as 'less favoured' (Kristensen, Dubois, & Teräs, 2019), and acting mostly as the supplier to the urban population. However, an increasing interest in doing 'green business', e.g. in Swedish rural areas, unveils hitherto understudied socio-economic potentials (Kristensen et al., 2019). In news reports and social media, it has been described as a new 'green wave', similar to the green waves of the 1970s but oriented more towards gaining profits out of environmentally friendly business operations and products. For instance, a news report from *Sveriges Radio* (the Swedish public radio channel) indicates that the environments in these regions fit the increasingly popular perceptions of 'green consumption' (Segerson, 2017). Research further suggests that many entrepreneurs return to or move to inland areas to take advantage of such public perceptions and that they develop their environmentally friendly nature-based tourism products accordingly (Zhang & Lundmark, 2021).

Regional development policies have been influenced by this trend to some extent, thus further suggesting that smart specialization, e.g. through creating and developing a 'green business profile', could be game changing for rural communities (Kristensen et al., 2019). Because of the character of rural inland areas, nature-based tourism has been debated as an option for entrepreneurs operating in rural or peripheral regions (Müller, 2015; De Souza, 2018). Rural parts of Northern Sweden, at times called *Europe's last wilderness* or *Arctic Europe*, are perhaps some of the best cases to illustrate this development (Lundmark & Müller, 2010; Müller, 2011).

There are parallels between people who quit their 'bullshit jobs' which contribute little to their and others' well-being to live more sustainable lives by working and consuming less (e.g. Tan, 2000; Bregman, 2016) and those who start working as (tourism) entrepreneurs, since this fits better with their social values (e.g. Carlsen, Morrison, & Weber., 2008; Carson, Carson, & Eimermann, 2018; Moshe Yachin & Ioannides, 2020). One Swedish study investigates whether people engaging in such lifestyle changes to decrease consumption think that this needs to go hand-in-hand with a move to the countryside in search of more natural surroundings (Eimermann, Hedberg, & Nuga, 2021). Furthermore, a review of the existing literature on tourism and degrowth tells us that geographic concepts such as travel and mobility are rarely talked about in tourism and degrowth research, even though they are central to discussing sustainability and tourism (Chapter 1, this volume). Recognizing this, the chapter at hand does focus on holistic simplifiers (as those voluntary simplicity seekers that do move to rural areas (Etzioni, 1998)) to study connections with mobility aspects.

In Sweden as elsewhere, there is a strong and positive media narrative of voluntary simplicity seekers as persons searching for the good life while reducing levels of stress and consumption (Thulin, 2020). Although far from all simplifiers become rural tourism entrepreneurs (e.g. Eimermann et al., 2021), some of them do. In tourism, this category of entrepreneur appears to be increasing, and media coverage suggests that for some remote and sparsely populated areas, the idea of selling nature and silence as a consumer good is not only good business, but it is also a way to escape from the 'rat race', the unsustainable way of urban living, or to be able to act on one's lifestyle preferences. As the lifestyle entrepreneur label indicates, growth is not a priority for those individuals. Instead, the motivation to operate a tourism business could be varying: some see tourism as a means to maintain their own and their families' way of life (Engeset & Heggem, 2015), whereas for others, it is a bridge to retirement (Ioannides & Petersen, 2003; Rogerson, 2005). For Shaw (2014), the main motivation for tourism business start-ups is the opportunity to live in an attractive tourist destination or a place that allows an alternative way of living.

It has, however, been rare to interrogate how large the phenomenon of voluntary simplicity is in terms of numbers. Therefore, this chapter uses a geo-referenced statistic database (ASTRID, comprehensive database provided by Statistics Sweden and stored at department of Geography, Umeå university) to focus on holistic simplifiers. As such, the chapter advances our understanding of macro- and micro-level changes when studying sustainability and degrowth in connection to tourism. It thus offers an example of a quantitative enquiry into voluntary simplicity, the nature of (tourism) work and transformations in value(s) associated with it (Singh, 2019).

In the next section, this chapter offers a brief specification of its main inspiration (Etzioni, 1998). It then connects mobility aspects with holistic simplifiers before presenting methods, discussing results and drawing conclusions. The chapter examines the occurrence of holistic simplifiers as a phenomenon through quantitative measures in order to establish how large the phenomenon might be in a country like Sweden. First, this will help advance the methods for researching such phenomena. Second, this can enhance discussions on the ways in which possible new flows of people can be part of rural restructuring of some (but not all) places. Finally, the incidence of holistic simplifiers as part of voluntary simplicity seeking can help us understand some of the general mobility trends in Sweden.

Sustainability and degrowth

Interconnected macro- and micro-level changes

From a sustainability perspective, the limits to growth set both physical and social boundaries on how much growth there can be without negatively affecting future possibilities (e.g. Hagbert & Fauré, 2021). Along with understanding the mechanisms behind growth and decline in various disciplines, ideas of sustainable development have also attempted to address and handle issues of limits to growth. Degrowth emerged in this context and developed from a slogan into a social movement, and then became a concept that attracted attention both within and outside academia (Demaria, Schneider, Sekulova, & Martinez-Alier, 2013). Kallis' (2013, p. 95) definition of socially sustainable economic degrowth is useful for this chapter: "a stable and equitable downscaling of society's throughput". This is not possible through mere individual reductions in energy or material consumption such as retrofitting light bulbs or driving electric cars.

Addressing directly the issues of limits to growth and aiming to enhance human well-being and ecological conditions, as well as to downscale production and consumption, degrowth is increasingly seen as a way to create sustainability at all levels (Martínez-Alier, Pascual, Vivien, & Zaccai, 2010; Schneider, Kallis, & Martinez-Alier, 2010). With the limits to growth becoming increasingly evident, it is timely to highlight and investigate macro- and micro-level processes that offer alternatives to mainstream growth development.

The development of the degrowth concept is situated in the discourses of transition and shares similarities with many transition initiatives around the world (Escobar, 2015). What is unique with the degrowth concept lies in that it criticizes the 'growth imperative', which is embedded within the Western notion of development and modernity, including sustainable development (Demaria et al., 2013). In many ways, degrowth's proposals speak to UNDP's Agenda 2030 (and the Sustainable Development Goals – SDGs). For instance, it advocates the reduction of working hours, co-housing and voluntary simplicity to promote transitions from materialistic lifestyles towards more participatory and environmentally friendly societies, while proposals on enhancing public services and redistributive taxation and including cooperative and circulative economy aim to bring more equity and socio-economic opportunities to more people (Bregman, 2016; Cosme, Santos, & O'Neill, 2017). The question remains, however, how these specific proposals may be implemented at different geographic locations to achieve the SDGs.

While degrowth is differentiated from sustainable development, it is less clear whether the concept takes considerations of the different economic status across geographic locations. Latouche (2014) argues that the word *degrowth* does not mean to stop economic growth entirely or negative growth. Instead, it means to "repoliticise the debate on the much needed socio-ecological transformation" (Demaria et al., 2013, p. 192) and signifies "a voluntary transition towards a just, participatory and ecologically sustainable society" (Research & Degrowth, 2010, p. 523). Nevertheless, the word *degrowth* has a connotation of decline and recession, and indeed what is the 'voluntary transition' can be further discussed. As Bonaiuti (2017) points out, the transitions that increase its costs or reduce its benefits to slow down the continuous growth can be called involuntary degrowth, which is already happening in many places. To this end, it is also useful and important to contextualize the degrowth concept in the theoretical discussions on the geographies of growth and decline for a more integrated plan to build an equal and sustainable society.

Taking connections between degrowth and voluntary simplicity as a starting point, we concur with Kallis (2013, p. 95) that the emphasis in degrowth should not be on voluntary simplicity alone, but rather on the 'right to simplicity' needs to be ensured through social and institutional support that makes simpler living possible (Singh, 2019). Kallis (2013, p. 95) cites Alexander (2011) when describing voluntary simplicity as "an expanded, integrated and committed package of downscaled lifestyles that goes beyond isolated consumption decisions and hence reduces . . . the scope for rebounds". Kallis (2013) sees voluntary simplicity as a necessary, but not sufficient, condition for a sustainable downscaling because the reduced demand from the resource savings of 'simplifiers' can reduce the costs of resources to others and increase overall consumption (Alcott, 2008). Thus, institutional interventions and limitations at various scales are necessary to ensure that the resource gains of simplicity are not invested in further capital accumulation and resource expansion, making simplicity materially irrelevant (Kallis, 2013).

This chapter links voluntary simplicity to individual consumption as indirectly related with the possible meanings of work for Swedish simplifiers. It does so with particular attention to issues that arise when work life is experienced as stressful due to the limits of combining high pressure at the work floor with high social demands, e.g. in family life, visiting friends and relatives and leisure time (Juniu, 2000). Some authors have suggested that 'voluntary simplicity' aims at simple living (e.g. Grigsby, 2004), 'slow living' (e.g. Andrews, 2006), 'opting out' (e.g. Belkin, 2003) or 'basic income' (e.g. van der Veen & Van Parijs, 2006; Bregman, 2016). Within this range of issues, this chapter is based on earlier accounts of voluntary simplicity (Etzioni, 1998). Discussing psychological implications and societal consequences, Etzioni (1998, pp. 621–626) described three variations of people engaging in voluntary simplicity:

- *Downshifters*: "economically well-off and secure people who voluntarily give up some consumer goods, often considered luxuries, they could easily afford, but basically maintain their rather rich and consumption-oriented lifestyle" (Etzioni, 1998, p. 622)
- *Strong simplifiers*: "people who have given up high-paying, high-stress jobs as lawyers, business people, investment bankers and so on, to live on less, often much less, income" (Etzioni, 1998, p. 623)
- *Holistic simplifiers*: who "adjust their whole life patterns according to the ethos of voluntary simplicity. They often move from affluent suburbs or gentrified parts of major cities to smaller towns, the countryside, farms and less affluent or urbanised parts of the country . . . with the explicit goal of leading a 'simpler' life" (Etzioni, 1998, pp. 625–626).

The first category (downshifters) does not lend itself to the proposed research design using register data, because this definition refers to reduced consumption of consumer goods such as fancy clothes (which we cannot measure). The second category is less difficult to operationalize, but for the purpose here to discuss mobility in relation to leading a simpler life, the third category (holistic simplifiers) is most suitable for the questions we have. Measuring this category involves factors such as occupation, trade and industry, work income, housing, neighbourhood and socio-economic composition of neighbourhoods, which can be studied using longitudinal register data covering the entire population (see method section that follows).

Tourism, rural restructuring and simplicity

As stipulated by Hall (2009), there are connections between degrowth and tourism that have yet to be fulfilled by the focus on sustainability. Instead of decreasing emissions of greenhouse gases, we have seen an increase due to increased air travel. In the context of globalization, peripheries are often identified as potential destinations for tourists. In Sweden, public stakeholders have put substantial effort into promoting tourism development because access to pristine nature is an advantage in the competition for tourists in a global economy (Almstedt, Lundmark, & Pettersson, 2016). In tourism research, a strong link between lifestyle entrepreneurship, sustainable tourism development and a sustainable way of life, sometimes referred to as Buen Vivir (Chassagne & Everingham, 2019), has been demonstrated. However, in the tourism and degrowth literature, little attention has been paid to the supply side of tourism, or the entrepreneurs. Arguably, small-scale tourism firms operated by lifestyle entrepreneurs are among those who have the potential to make those transformative changes proposed by the degrowth paradigm both as individuals and at the local community level.

As previously discussed, lifestyle entrepreneurs often do not follow corporate growth strategies, but instead their motivations lie in personal desires, such as being able to stay in a place they already live in or to be able to move to and settle somewhere rural and peripheral, and to be able to work with something in relation to the cultural and/or natural environment at that place. As shown by Zhang and Lundmark (2021), many 'green' entrepreneurs value the contribution they can make to local sustainability and how their way of doing business is perceived as more sustainable than other tourism businesses. Examples of tourism epithets that are used for sustainable tourism are ecotourism, slow tourism, nature-based tourism and alternative tourism to mention a few. These motives are mostly related to green entrepreneurship (tourism entrepreneurs selling 'slow' products). One important connection between tourism and lifestyle migration is the example of lifestyle migrants becoming tourism entrepreneurs (Carson et al., 2018).

Research on rural areas and in particular research departing from economic geography perspectives has discussed extensively the changes and transformations going on in the rural areas (Cloke, 1996; Ilbery, 2014). The interest circulates around the apparent shifts in recent decades in terms of how resources are used and distributed. Although production is still important, other values of the rural landscape have become established and are now dominating the mindset of people in general. This is the perspective of "the new rural paradigm" (OECD, 2006) or post-productivism (Wilson, 2001; Mather, Hill, & Nijnik, 2006; Lundmark, 2006). This mindset allows a view on the landscape as a landscape for consumption and recreation in situ, instead of a landscape that formerly was a production landscape (Lundmark, 2006; Mather et al., 2006). This idea is a prerequisite for other development to take place, and restructuring is connected to an economic process that in brief entails a shift from goods production to service provision (Pettersson, 2002), and also into the new knowledge economy where knowledge production and innovation are emphasized. However, one important point here is that the underlying assumption of growth or decline is still taken for granted. Post-production is thus still very much embedded in the growth paradigm. This has been challenged by results presented by proponents of the degrowth paradigm (Järvensivu, 2013) who argue, from a Finnish forestry case, that in fact some practices may disrupt habitual ways of approaching nature and of turning it into exchangeable and consumable objects.

In practice, there is a certain agreement that the reorientation of society needs to take place locally. This is the case with land-use planning practices in Finland (Lehtinen, 2018), or by individuals and communities concerning tourism (Renkert, 2019). In the degrowth literature, simplicity is one path that individuals can follow in order to contribute to transformative change (Alexander & Yacoumis, 2018; Ims, 2018; Trainer, 2019a, 2019b, 2020). However, although much of the transformation is postulated to take place at a local level, the need for transformation is global (Hall, 2009), and although there is plenty of doom and gloom in the literature surrounding the Anthropocene, the degrowth literature also demonstrates many positive examples. Krueger, Schulz, and Gibbs (2018) discuss how transformative economic practices can help places to investigate new avenues for development by using previously overlooked opportunities for creating alternative economic visions and practices in ways that are not growth oriented. The focus should instead be put on the redistribution of wealth and ecological sustainability.

The theoretical discussions on post-productivism are relevant here, showing an assumption that urban areas should grow, whereas rural areas tend to shrink and decline. The logic therefore is that rural areas need to change in order to catch up with the growth. However, recent studies suggest that people's movements between so-called rural and urban areas play a big role in the changing pattern of livelihood, employment and mobility of the rural communities, contributing to the view of a rural-urban spectrum (Carson, Carson, & Lundmark, 2014; Paniagua, 2002). What we need to understand more is how the views on 'growth' can also be challenged. The degrowth concept criticizes the norm of the growth imperative and the norm that all places should develop in the same form of growth. The degrowth concept builds on the ecologist and culturalist critiques of economics, where the former centres on Nicholas Georgescu-Roegen's bio-economics and the latter includes post-development theorists and political ecologists (Asara, Otero, Demaria, & Corbera, 2015). Activists and scholars who are in favour of the idea of degrowth identify the central problem as the current capitalist productive model and its growth imperative. Driven by theoretical debates and social movements, the degrowth paradigm therefore proposes that in order to achieve economic, ecological and social sustainability, we need to decouple the individual and society's well-being from the circle of production and consumption, so changes must be made from both top-down and bottom-up directions (Cosme et al., 2017; Hagbert & Fauré, 2021). Degrowth's critiques of the growth paradigm shed lights on the research gap of how rural areas can contribute to building sustainable societies as a whole. A series of questions are asked: what are the reasons to growth; what is regarded as growth; what are the costs of growth; and who wants to have growth? In this context, degrowth can be seen as a continuation of the discussion on sustainable development, post-productivism and the changing geographies of growth and decline.

Method

In many studies of voluntary simplicity, data was collected through interviews and surveys that provide rich information on many qualitative aspects of people's changes to obtain a less stressful life situation. The drawback with such approaches is that they are commonly based on small data sets, which makes it difficult to determine to what extent the results are representative of the population at large.

Studies based on surveys made in the US, UK and Australia report that 20% to 25% of the population aged 30 to 59 has engaged in similar stress-reducing lifestyle changes (Schor, 1998; Hamilton, 2003; Hamilton & Mail, 2003). From a Swedish point of view – where a strong Lutheran work ethic has been the dominating paradigm for ages, and where recent governments have been politically successful in national elections by pursuing a 'primacy of work' principle designed to create measures for people to find jobs and support themselves – it appears doubtful that as many as one-quarter of the population in working ages would reduce their working hours. The welfare regime of Sweden, however, although recently put under financial pressure, redistributes vast resources among individuals via a plethora of transfer systems, which enables households to obtain a livelihood not solely based on work income. From this point of view, the prerequisites for voluntary simplicity in Sweden may be more beneficial than in countries pursuing more neoliberal political agendas including less income redistribution.

In order to get a broader view of the diffusion of simplicity-seeking behaviour, register data provides a useful alternative to interviews and surveys. A positive trait of using register data for this research purpose is that this data covers the entire population over many years. This implies that there are no non-response issues frequently associated with surveys. Low response rates may negatively affect representativeness and reduce the strength of the conclusions. Having access to registers that include information on every single individual enables the study of rare attributes, i.e. the investigation of individuals whose characteristics and changes in characteristics are uncommon in the population. The identified group can also be compared with the population in general or whatever group is found relevant. The use of register data, however, is affected by some drawbacks that have to be kept in mind. Register data is not primarily produced for scientific purposes, but in accordance to needs of different bodies of government. This type of data does not carry much information about individuals' values and attitudes, but it reflects revealed preferences and choices made within different spheres of everyday life. For example, choices related to family formation/dissolution (e.g. having children, getting married), the labour market (e.g. getting a job, change jobs, earnings, occupation), geographic mobility (e.g. commuting, migration), and consumption related to housing and car ownership are covered in this data set. Since individuals can be followed longitudinally across years, it is possible to observe changes that hint at shifts towards leading a 'simpler' life.

At this point, it may be obvious that the proposed method is not useful for studying all parts of the multi-faceted concept of voluntary simplicity. People's perceptions of voluntary simplicity and their views on their life situation cannot be found in register data, but they leave traces that may be associated with simplicity in different ways. From this point of departure, we searched within the literature for definitions of the phenomenon in line with what would be possible to study by means of register data. Admittedly, there may be many possible definitions suitable for the research design proposed here, but we settled for a definition proposed by Etzioni (1998) cited earlier. Etzioni's (1998) category of 'holistic simplifiers' is operationalized by selecting individuals who in 2014 (the latest year the required register data was collected) were residing in affluent neighbourhoods in metropolitan areas or major cities (parishes with higher than average work income in each city, respectively). From this group, individuals are selected who two years later have moved to other parts of the country (i.e. outside the territory defined in the first step) and reduced their work income by at least 50%. In a subsequent step of the analysis, these individuals are compared to individuals who stay in the affluent neighbourhoods.

Results: holistic simplifiers 'keeping up with the Joneses'

Our database analysis shows that 1.2 million people changed place of residence in 2014. The number of people moving between municipalities or between counties amounted to approximately 705,000. Employing Etzioni's (1998) definition, close to 3,200 individuals (or 0.45% of the movers) were identified as holistic simplifiers.

The estimation of a logistic regression model comparing the characteristics of holistic simplifiers and non-holistic simplifiers shows there is no difference between men and women. Regarding age, the analysis indicates that in comparison to the youngest age group (people aged 30 to 34), the likelihood of becoming a holistic simplifier decreases with higher age up to the mid-fifties. Thereafter, the likelihood increases all the way to the age of 75. This means that people in their forties and early fifties are least likely to give up their prestigious jobs and places of residence in affluent neighbourhoods. Swedish-born people are more likely (25% odds ratio) than foreign-born people to take this step. In the same vein, highly educated people are less likely to become holistic simplifiers (40% odds ratio). Thus, our data suggest that mostly people with less schooling take this step. Concerning labour market-related factors, the model estimates demonstrate that lower income from work increases the likelihood of becoming a holistic simplifier, and having unemployment benefits increases the likelihood substantially (odds ratio 70%). Families with children under the age of 17 are much less likely to become holistic simplifiers.

There are differences in type of housing on the probability to make this change in lifestyle. In comparison to residing in condominiums, people residing in rented apartments or detached houses are more prone to become holistic simplifiers. It is difficult to know the reasons for these differences, but tentatively it may be due to factors related to housing satisfaction or budget constraints in cases when a move involves the purchase of a new place of residence. For people living in affluent neighbourhoods, commonly in detached houses with gardens located in tranquil and secure surroundings, there are few good reasons to leave such environments. People residing in rented apartments may be working in low-income jobs, making it difficult to afford buying new property, which usually is the main option in regions farther away from Swedish metropolitan areas and cities.

In the analysis of the holistic simplifiers, a number of variables reflect differences before and after the move. One of those variables called "from rented flat to home ownership" is a dummy variable indicating if the individual lived in rental housing before the move and owned their place of residence after the move. This variable estimate is strongly positive and significant and demonstrates that holistic simplifiers commonly end up owning their houses.

According to the literature, voluntary simplicity involves a transition to a life with less income. Since we use individual work income in the definition of holistic simplifiers (i.e. reduction of work income more than 50%), this variable cannot be used in the analysis. However, it is possible to use household disposable income, which includes work income of the partner and a number of other sources of income. The model results indicate a positive effect of this variable on becoming a holistic simplifier, which is contrary to expectations on families searching for a simpler way of living. These families have more money to spend after the move as compared to their former neighbours who chose to stay in the affluent neighbourhoods and keep their jobs as professionals or managers.

Car ownership is a big expenditure in many households. One way to reduce the cost of cars is to reduce the number of cars a household owns. Once again, we receive an unexpected result as the studied holistic simplifiers tend to have more cars after the move compared to their former neighbours who didn't move from the affluent neighbourhoods. It could be argued that an increased number of cars does not necessarily imply enhanced costs because the household can compensate the cost burden by buying older and cheaper cars. This assumption can be tested since the available data set includes new car registration date. The modelling results, however, show that there is no difference between the holistic simplifiers and their previous neighbours in the affluent neighbourhoods regarding the age of their cars, implying that these simplifiers are 'keeping up with Joneses' by having as many new cars as before. It can be concluded that individuals assumingly dedicated to voluntary simplicity do not cut down on car consumption in connection to the move.

The dwelling is another presumably big expenditure item for the households. It can be expected that simplifiers aim at reducing these costs as they make this lifestyle change. We investigate this issue by analyzing changes in living space (m^2) before and after the move and changes in the individuals' estimated housing prices (SEK/m^2). Regarding the former indicator (living space), the analysis reveals that holistic simplifiers move to bigger houses, which is not expected. From a theoretical point of view, a smaller dwelling should be less costly due to such things as lower construction costs and lower costs for heating and maintenance, among others. Still, the holistic simplifiers consume larger dwellings after the move. Although living space is related to housing costs, this measure is somewhat crude when looking at particular dwellings, since there are other important factors such as location and standard of the dwelling. Estimated housing prices (taxation values reflecting 75% of market value) in relation to living space (m^2) makes a reasonable proxy for housing costs, which is used by constructing a variable that shows the difference in those costs before and after the move. Since the typical holistic simplifier moves from an affluent neighbourhood with relatively high housing costs to less prestigious addresses in other parts of the country, we expect to get a negative parameter estimate on the probability of realizing their ambitions. However, the analysis shows insignificant parameter estimates, which means that the holistic simplifier after the move owns a new dwelling as expensive as the dwellings of the former neighbours in the affluent neighbourhood. Separate models for men and women show different results in this respect. When running the model with only men, the expected parameter estimate is received. Men who engage in holistic simplicity substantially reduce their housing costs after the move, but this is not the case for women, who consume housing to the same extent as the reference category.

Individuals who become retirees during the study period (2014–2016) are significantly more likely to be holistic simplifiers. This may not come as a surprise since retirees almost by definition quit their jobs and thus meet the criterion of more than 50% work income reduction. Moreover, according to the migration literature, migration propensity increases modestly for people in their mid-sixties (Lundholm, 2007), which is a period when most people retire.

Conclusions

To sum up, the number of people who can be identified as holistic simplifiers through our method is low in Sweden. By using register data covering the entire population, it is shown that only a small fraction of the population makes such changes in their lives in accordance with Etzioni's (1998) definition of holistic simplifiers. The results indicate that although individuals meet the overall criteria for voluntary simplicity, they do not necessarily behave accordingly by cutting down on consumption and living a simpler life. At least this is not a clear pattern in the data set used here. Maybe there are other processes at work, which should be highlighted in future studies. For example, the age distribution of the studied holistic simplifiers clearly demonstrates that middle-aged persons are least likely to quit their careers and move out of their prestigious neighbourhoods. It is more likely that the typical holistic simplifier is a person who approaches retirement age and performs this lifestyle change as a part of leaving professional life. This is why we find so many individuals who have retired during the study period and moved out from cities. A large part of the observed holistic simplifiers may actually be people who are about to retire. If this assumption is correct, then the tribe of profound holistic simplifiers is possibly even smaller than observed here.

Thus, there are far fewer than expected holistic simplifiers in this study. Although we cannot assert exactly why it is so, there are some plausible explanations. First, it might be that Sweden is not comparable to other countries such as Australia, where there are signs that this trend is much larger. This can be due to a couple reasons: a) Sweden is still a stable welfare society, which allows for generous parental leave, sick leave and medical care; and b) there is a historically strong union movement that has accomplished a lot for the working population in terms of good working conditions and salaries. Taken together, this means that people can afford to work part time, for example, during the most stressful years of child-rearing, thus offering some relief and reducing the need to 'move away from the rat race'. This is only speculation though, and the apparent difference is something that should be further investigated. Another avenue for research is following on ideas developed in Hedlund (2016) on how these changes can be seen as geographic patterns and how to describe the transformation of different rural areas in Sweden and the Nordic countries. These patterns might be useful to identify since it is a highly uneven development, and some rural and sparsely populated areas will be prosperous in attracting these mobile populations but some will not be able to. Although built on ideals of equality, justice and sustainability for all, it is highly unlikely that it will be an even transition on every geographic scale. Instead, it is a highly diverse development dependent on, for example, infrastructure, mobilities in general and tourism mobilities in particular, especially with regards to tourism demand for 'the Arctic' and the possibilities to become involved in, for example, green business (Lundmark, Müller, & Bohn, 2021; Zhang & Lundmark, 2021).

Lastly, a profound question is what this kind of quantitative endeavour can do to help overcome some of the structural geographic issues of sustainability also in places outside of urban areas? How can this kind of study contribute to not only research but also in practice, rural municipalities that are involuntarily degrowing but are willing to use degrowth as a strategy to attract more people? What attracts this small but not insignificant group of people to particular rural areas? One thing this study suggests for these municipalities is that two full-time incomes in highly qualified jobs might not be needed for attracting return migrants or in-migrants. Desirable living conditions and housing of the right sort might be good enough (Lundmark, 2021). Another suggestion is that we should perhaps not disregard the group of retirees who seem to be frequent among holistic simplifiers. Other research has shown that there is potential in terms of entrepreneurship among retirees moving to rural areas as lifestyle migrants or green wavers, for example (Widell, 2021). A third is that in Sweden, with a large amount of second homes, it is reasonable to assume that, for some, the desire to lead a simpler life without the demands and burdens of modern urban life can be met for some months every year through their second homes. What that could mean for municipalities in terms of potential contribution to the local community is an avenue for research that has not yet been linked to the ideas of voluntary simplicity or to holistic simplifiers, leaving a research gap to fill in the future.

References

Alcott, B. (2008). The sufficiency strategy: Would rich-world frugality lower environmental impact? *Ecological Economics*, *64*(4), 770–786.

Alexander, S. (2011). *Property beyond growth: Toward a politics of voluntary simplicity* (PhD dissertation). University of Melbourne, Melbourne.

Alexander, S., & Yacoumis, P. (2018). Degrowth, energy descent, and 'low-tech' living: Potential pathways for increased resilience in times of crisis. *Journal of Cleaner Production*, *197*, 1840–1848.

Almstedt, Å., Lundmark, L., & Pettersson, Ö. (2016). Public spending on rural tourism in Sweden. *Fennia, 194*(1), 18–31.

Andrews, C. (2006). *Slow is beautiful: New visions of community, leisure, and joie de vivre*. Gabriola Island: New Society Publishers.

Asara, V., Otero, I., Demaria, F., & Corbera, E. (2015). Socially sustainable degrowth as a social – Ecological transformation: Repoliticizing sustainability. *Sustainability Science, 10*(3), 375–384.

Belkin, L. (2003, October 26). The opt-out revolution. *New York Times Magazine*.

Bonaiuti, M. (2017). Are we entering the age of involuntary degrowth? Promethean technologies and declining returns of innovation. *Journal of Cleaner Production, 197*, 1800–1809.

Bregman, R. (2016). *Utopia for realists and how we can get there*. London: Bloomsbury.

Carlsen, J., Morrison, A., & Weber, P. (2008). Lifestyle oriented small tourism firms. *Tourism Recreation Research, 33*(3), 255–263.

Carson, D. A., Carson, D. B., & Eimermann, M. (2018). International winter tourism entrepreneurs in northern Sweden: Understanding migration, lifestyle, and business motivations. *Scandinavian Journal of Hospitality and Tourism, 18*(2), 183–198.

Carson, D. A., Carson, D. B., & Lundmark, L. (2014). Tourism and mobilities in sparsely populated areas: Towards a framework and research agenda. *Scandinavian Journal of Hospitality and Tourism, 14*(4), 353–366.

Chassagne, N., & Everingham, P. (2019). Buen Vivir: Degrowing extractivism and growing wellbeing through tourism. *Journal of Sustainable Tourism, 27*(12), 1909–1925.

Cloke, P. (1996). Rural life-styles: Material opportunity, cultural experience, and how theory can undermine policy. *Economic Review, 72*(4), 433–449.

Cosme, I., Santos, R., & O'Neill, D. W. (2017). Assessing the degrowth discourse: A review and analysis of academic degrowth policy proposals. *Journal of Cleaner Production, 149*, 321–334.

De Souza, P. (2018). *The rural and peripheral in regional development*. Abingdon: Routledge.

Demaria, F., Schneider, F., Sekulova, F., & Martinez-Alier, J. (2013). What is degrowth? From an activist slogan to a social movement. *Environmental Values, 22*(2), 191–215.

Eimermann, M., Hedberg, C., & Nuga, M. (2021). Is downshifting easier in the countryside? Focus group visions on individual sustainability transitions. In L. Lundmark, D. B. Carson, & M. Eimermann (Eds.), *Dipping in to the North – Living, working and traveling in sparsely populated areas*. London: Palgrave.

Engeset, A. B., & Heggem, R. (2015). Strategies in Norwegian farm tourism: Product development, challenges, and solutions. *Scandinavian Journal of Hospitality and Tourism, 15*(1–2), 122–137.

Escobar, A. (2015). Degrowth, post development, and transitions: A preliminary conversation. *Sustainability Science, 10*(3), 451–462.

Etzioni, A. (1998). Voluntary simplicity: Characterization, select psychological implications, and societal consequences. *Journal of Economic Psychology, 19*, 619–643.

Grigsby, M. (2004). *Buying time and getting by: The voluntary simplicity movement*. Albany, NY: State University of New York Press.

Hagbert, P., & Fauré, E. (2021). A future society beyond GDP growth – What would that look like? In L. Lundmark, D. B. Carson, & M. Eimermann (Eds.), *Dipping in to the North – Living, working and traveling in sparsely populated areas*. London: Palgrave.

Halfacree, K. (2006). From dropping out to leading on? British counter-cultural back-to-the-land in a changing rurality. *Progress in Human Geography, 30*(3), 309–336.

Hall, C. M. (2009). Degrowing tourism: Décroissance, sustainable consumption and steady-state tourism. *Anatolia, 20*(1), 46–61.

Hamilton, C. (2003). *Downshifting in Britain: A sea-change in the pursuit of happiness*. Retrieved from www.tai.org.au/sites/default/files/DP58_8.pdf

Hamilton, C., & Mail, E. (2003). *Downshifting in Australia: A see-change in the pursuit of happiness*. Retrieved from www.tai.org.au/sites/default/files/DP58_8.pdf

Hedlund, M. (2016). Mapping the socioeconomic landscape of rural Sweden: Towards a typology of rural areas. *Regional Studies*, *50*(3), 460–474.

Ilbery, B. (Ed.). (2014). *The geography of rural change*. London: Routledge.

Ims, K. J. (2018). Quality of life in a deep ecological perspective. The need for a transformation of the Western mindset? *Society and Economy*, *40*(4), 531–552.

Ioannides, D., & Petersen, T. (2003). Tourism 'non-entrepreneurship' in peripheral destinations: A case study of small and medium tourism enterprises on Bornholm, Denmark. *Tourism Geographies*, *5*(4), 408–435.

Järvensivu, P. (2013). Transforming market-nature relations through an investigative practice. *Ecological Economics*, *95*, 197–205.

Juniu, S. (2000). Downshifting: Regaining the essence of leisure. *Journal of Leisure Research*, *32*(1), 69–73.

Kallis, G. (2013). Societal metabolism, working hours and degrowth: A comment on Sorman and Giampietro. *Journal of Cleaner Production*, *38*, 94–98.

Kristensen, I., Dubois, A., & Teräs, J. (2019). Introduction: Unveiling the potential of less-favoured regions in regional policy. In I. Kristensen, A. Dubois, & J. Teräs (Eds.), *Strategic approaches to regional development* (pp. 1–14). Abingdon: Routledge.

Krueger, R., Schulz, C., & Gibbs, D. C. (2018). Institutionalizing alternative economic spaces? An interpretivist perspective on diverse economies. *Progress in Human Geography*, *42*(4), 569–589.

Latouche, S. (2014). Foreword. In M. Bonaiuti (Ed.), *The great transition* (pp. xi–xxiii). London: Routledge.

Lehtinen, A. A. (2018). Degrowth in city planning. *Fennia*, *196*(1), 43–57.

Lundholm, E. (2007). Are movers still the same? Characteristics of interregional migrants in Sweden 1970–2001. *Tijdschrift voor Economische en Sociale Geografie*, *98*(3), 336–348.

Lundmark, L. (2006). Mobility, migration and seasonal tourism employment. *Scandinavian Journal of Hospitality and Tourism*, *6*(3), 3–17.

Lundmark, L. (2021). Housing in SPAs – Too much of nothing or too much for 'free'? In L. Lundmark, D. B. Carson, & M. Eimermann (Eds.), *Dipping in to the North – Living, working and traveling in sparsely populated areas*. London: Palgrave.

Lundmark, L., & Müller, D. K. (2010). The supply of nature-based tourism activities in Sweden. *Tourism*, *58*(4), 379–393.

Lundmark, L., Müller, D. K., & Bohn, D. (2021). Arctification and the paradox of overtourism in sparsely populated areas. In L. Lundmark, D. B. Carson, & M. Eimermann (Eds.), *Dipping in to the North – Living, working and traveling in sparsely populated areas*. London: Palgrave.

Martínez-Alier, J., Pascual, U., Vivien, F.-D., & Zaccai, E. (2010). Sustainable de-growth: Mapping the context, criticisms and future prospects of an emergent paradigm. *Ecological Economics*, *69*(9), 1741–1747.

Mather, A. S., Hill, G., & Nijnik, M. (2006). Post-productivism and rural land use: Cul de sac or challenge for theorization? *Journal of Rural Studies*, *22*(4), 441–455.

Moshe Yachin, J., & Ioannides, D. (2020). "Making do" in rural tourism: The resourcing behaviour of tourism micro-firms. *Journal of Sustainable Tourism*, *28*(7), 1003–1021.

Müller, D. K. (2011). Tourism development in Europe´s "last wilderness": An assessment of nature-based tourism in Swedish Lapland. In A. A. Grenier & D. K. Müller (Eds.), *Polar tourism: A tool for regional development* (pp. 129–153). Montreal: Presses de l'Université du Québec.

Müller, D. K. (2015). Issues in Arctic tourism. In B. Evengård, J. Nymand Larsen, & Ø. Paasche (Eds.), *The new Arctic* (pp. 147–158). Heidelberg: Springer.

OECD. (2006). *The new rural paradigm. Policies and governance.* Paris: OECD publications.

Paniagua, A. (2002). Urban-rural migration, tourism entrepreneurs and rural restructuring in Spain. *Tourism Geographies*, *4*(4), 349–371.

Pettersson, Ö. (2002). *Socio-economic dynamics in sparse regional structures.* GERUM kulturgeografi 2002: 2. Umeå: Umeå University.

Renkert, S. R. (2019). Community-owned tourism and degrowth: A case study in the Kichwa Anangu community. *Journal of Sustainable Tourism*, *27*(12), 1893–1908.

Research & Degrowth. (2010). Degrowth declaration of the Paris 2008 conference. *Journal of Cleaner Production*, *18*(6), 523–524.

Rogerson, C. M. (2005). Unpacking tourism SMMEs in South Africa: Structure, support needs and policy response. *Development Southern Africa*, *22*(5), 623–642.

Schneider, F., Kallis, G., & Martinez-Alier, J. (2010). Crisis or opportunity? Economic degrowth for social equity and ecological sustainability. Introduction to this special issue. *Journal of Cleaner Production*, *18*(6), 511–518.

Schor, J. B. (1998). *The overspent American: Upscaling, downshifting and the new consumerism.* New York, NY: Basic Books.

Segerson, D. (2017). *Populärt att driva företag på landet.* Retrieved from https://sverigesradio.se/sida/artikel.aspx?programid=106&artikel=6723276

Shaw, G. (2014). Entrepreneurial cultures and small business enterprises in tourism. In A. A. Lew, C. M. Hall, & A. M. Williams (Eds.), *The Wiley-Blackwell companion to tourism* (pp. 120–131). Hoboken, NJ: Wiley-Blackwell.

Singh, N. M. (2019). Environmental justice, degrowth and post-capitalist futures. *Ecological Economics*, *163*, 138–142.

Tan, P. (2000). *Leaving the rat race to get a life: A study of midlife career downshifting* (PhD dissertation). Swinbourne University, Melbourne.

Thulin, L. (2020). Är tiden kommen för ett enklare liv? [Has time come for a simpler life?]. *Sveriges Radio P1*. Retrieved from https://sverigesradio.se/avsnitt/1502939

Trainer, T. (2019a). Entering the era of limits and scarcity: The radical implications for social theory. *Journal of Political Ecology*, *26*(1), 1–18.

Trainer, T. (2019b). Remaking settlements for sustainability: The simpler way. *Journal of Political Ecology*, *26*(1), 202–223.

Trainer, T. (2020). De-growth: Some suggestions from the simpler way perspective. *Ecological Economics*, *167*, 106436. https://doi.org/10.1016/j.ecolecon.2019.106436

Van der Veen, R. J., & Van Parijs, P. (2006). A capitalist road to communism. *Basic Income Studies*, *1*(1), 1–23.

Widell, B. (2021). The senior work force. In L. Lundmark, D. B. Carson, & M. Eimermann (Eds.), *Dipping in to the North – Living, working and traveling in sparsely populated areas.* London: Palgrave.

Wilson, G. A. (2001). From productivism to post-productivism . . . and back again? Exploring the (un)changed natural and mental landscapes of European agriculture. *Transactions of the Institute of British Geographers*, *26*(1), 77–102.

Zhang, J., & Lundmark, L. (2021). Selling greenness. In L. Lundmark, D. B. Carson, & M. Eimermann (Eds.), *Dipping in to the North – Living, working and traveling in sparsely populated areas.* London: Palgrave.

Part 2

Degrowth and tourism destinations

5 Diverse tourism

A poststructural view on tourism destination degrowth transition

Outi Kulusjärvi

Introduction

As I have experienced it, doing research as a critical tourism geographer means being aware of the growth logics of the tourism economy and adapting a critical stance towards such development. When reading about tourism development plans, I have a constant doubtful stance on their positive impact. Do local actors actually support this development? Are there individuals and groups whose voices have been left out from the development plans? If realized, how will tourism growth increase well-being in the destination community? Furthermore, any claim made about advancing sustainability or acting responsibly is met with suspicion; what is the sustainability they refer to? Is the aim to sustain growth or the natural environment?

The critical take is of course much needed and increasingly urgent due to the dominant, growth-focused capitalist tourism economies. Fletcher (2011) explains that over the course of history, tourism and capitalism have co-evolved, with the former often serving the needs of the latter. Such structural critiques on tourism have pointed out that rather than local communities and their resources being used for the benefit of the capitalist system, it is necessary to look in more detail at how to better work towards conditions where the transnational tourism economy enables socially just and ecologically sustainable livelihoods (Bianchi, 2009; Fletcher, 2011). Concerns over the impact that increasing international air travel has on global warming also gives another strong justification for questioning the ideal of economic growth in tourism.

Growing interdisciplinary scholarship of degrowth focuses on studying and enacting cultural and institutional changes within capitalist economies to move away from the growth rationale (D'Alisa, Demaria, & Kallis, 2015). D'Alisa et al. (2015) emphasize that to achieve changes towards societies where economy is not in conflict with social justice or environmental sustainability, it is necessary to seek new economic structures and modes of economic organization. They state "our emphasis here is on *different*, not only *less*" (Kallis, Demaria, & D'Alisa, 2015, p. 4). As the degrowth movement has its roots in critical theory and activism, the needed economic transitions refer not only to a decrease in the levels of economic activity but also to *social change* (Schmid, 2019).

In this chapter, my aim is to explore the possibilities of a degrowth perspective in the study of tourism destination development and change. The chapter is based on an ethnographically oriented case study that I conducted in the sparsely populated but tourism-intensive area of the Finnish North in 2015. Via semi-structured in-depth interviews with local tourism actors in the Ylläs tourism destination, I gained an understanding of everyday tourism realities of those individuals who produce tourism via their economic agency. Drawing on this empirical study, I build a bottom-up perspective on socio-economic change in tourism production and degrowth transitions.

In this chapter I refer to degrowth as a project of social change and not only as a project of defining a limit for tourism growth (Fletcher, Murray, Blanco-Romero, & Blázquez-Salom, 2019). It may seem unrealistic to apply such an approach to degrowth within the field of tourism economy that is capitalist and operates with a global reach. How to suddenly create a new tourism system? To seek answers for such questions, I draw on feminist economic geography that has advanced poststructural perspectives to socio-economic change (e.g. Gibson-Graham, 2006, 2008; North, 2016; Schmid, 2019). This scholarship shares the aim of moving away from growth-focused economic processes. However, to do so, they see it as necessary to rethink whether there even now actually exists *the* internally coherent capitalist economy that can be altered only by creating something that is *alternative* or *new* in relation to it. This scholarship asks to pay attention to the forms of economic practice that already deviate from the growth rationale. Within the degrowth literature, such grassroots possibilities have been identified as crucial in socio-economic change (e.g. D'Alisa et al., 2015; Fletcher et al., 2019).

My interest now is to examine destination growth aims and degrowth transitions by attending to the possibility of *economic difference*. In this chapter, I suggest that while maintaining the typical critical viewpoint that started this chapter, it is necessary to also allow for seeing the diversity in economic rationales and practices in the tourism economy. In addition, it is crucial to recognize their potential in larger-scale sustainability transformations, such as tourism destination degrowth.

Degrowth transitions in tourism economies

As Fletcher et al. (2019) note, degrowth scholarship has not engaged with tourism economies to date nor has past research on sustainability and tourism drawn on degrowth studies, with few exceptions (e.g. Hall, 2009, 2010). Hall advanced a steady-state economies perspective for tourism degrowth arguing that the externalities of tourism development (e.g. emissions and the loss of natural capital) should be paid by the producer, not by the local community. Hall (2009, p. 58) underlines that "from a degrowth perspective firms therefore have a concept of social responsibility that shifts from being 'beholden to shareholders' to one that is more stakeholder based and includes the workers and communities on which their survivability is partly based". To realize this, Hall (2010) calls for new sustainable tourism policies yet believes that the willingness of tourism planners and developers to put such policies in practice is minimal.

More recently, Fletcher et al. (2019) utilize degrowth perspectives to advance a turn in tourism economies away from a growth paradigm and towards truly sustainable tourism economy. They state that "serious degrowth likely demands pursuit of post-capitalism" (Fletcher et al., 2019, p. 1747). This shows how their project calls for radical rethinking of economic agency and structures in tourism economies as well as the current societies. They call for changes both at the micro and macro levels. Fletcher et al. (2019) emphasize that critical tourism research should also think how sustainability transformations in tourism can be achieved in practice. Indeed, in degrowth scholarship, attention has been given to envisioning how new, truly sustainable economies might look. Thus, degrowth perspectives are valuable for advancing transformative capacity of critical tourism geographies that tend to focus on criticizing the growth paradigm in tourism development, giving less attention on how to realize changes towards sustainability in practice.

Agency for degrowth

To apply and elaborate degrowth thinking in the tourism context, I draw on the degrowth perspectives advanced by D'Alisa et al. (2015). They emphasize that to achieve changes towards societies where the economy is not in conflict with social justice or environmental sustainability, it is required to seek new economic structures and modes of economic organization. While degrowth is a critique on growth, it is necessarily also a critique on capitalism as it is a system that necessitates growth. While in mainstream economics, a degrowing economy is perceived as a challenge that needs fixing, degrowth research instead shows that after economic growth reaches a certain stage, welfare and employment no longer increase (see Lehtinen, 2018). The idea has been thus to re-politicize 'the economic' and, in this way, argue for the possibility to move beyond capitalism (D'Alisa et al., 2015).

However, degrowth does not equate to simple economic decline or recession (Fletcher et al., 2019). In degrowth thinking, the point is to re-organize economies in a way that (human) life can flourish and also be stable without growth (Kallis, Kerschner, & Martinez-Alier, 2012; Kallis & March, 2015). Thus, degrowth research and practice also deals with how to move towards better societal and economic organization, i.e. 'degrowth transition'. As Schmid (2019) notes, in degrowth scholarship and practice, the idea is not only to pinpoint the causes for injustices, nor to identify how the ideal economic organization would look in a utopian future, but to think also *how* to enact social changes. Degrowth scholars share an aim to envision new ways of living and producing. This requires rethinking how to organize our economic life (one that is not to be separated from the social or the environmental) in a way that is also *socially* sustainable (D'Alisa et al., 2015).

To realize changes, degrowth scholarship has not identified one set of means as the way for change. Instead, a diversity of forms of economic practice and organizing are regarded as potential ways that could be taken. D'Alisa et al. (2015) note that new welfare institutions are needed to be in line with the increase in unemployment (as understood in capitalist societies). Political regulation towards degrowth can be implemented by tariff and tax policies, adequate environmental licensing and emission limits (Lehtinen, 2018). In addition, at the micro level, degrowth projects support grassroots economic practices such as eco-communities, cooperative firms, community currencies or solidarity economy. While such everyday practices and agency of people do not have "a built-in dynamic to accumulate and expand" (Kallis et al., 2015, p. 12), these can elude hegemonic capitalist economic processes.

When realizing such new forms of economic organization, a conflict within the capitalist system is evident (Kallis & March, 2015). Social struggle and opposition are needed to defend new degrowth economies against the presently powerful actors who benefit from the growth regimes (Koch, 2020).

Focus on the existing diversity in tourism production

While degrowth research has an interest in community-led and bottom-up grassroots organizing aside the institutional changes in realizing degrowth societies, degrowth scholarship and practice have much in common with poststructural perspectives to political economic change (see Schmid, 2019). However, as degrowth scholarship identifies the current capitalist economic system as the root cause for social and environmental unsustainabilities, it seems evident that change that has to happen is large-scale and touches the whole economic system. This may seem like a contradiction: how can a structural change be realized through place-based incremental changes?

The ongoing discussions in feminist economic geography and elsewhere study economic transitions and transformations in poststructural terms. One of their intentions has been to break down the coherent image of *the* capitalist economy and show that economies are more diverse than they are usually portrayed. One of the methods used by Gibson-Graham (2008, p. 623) is "reading for difference rather than dominance", which indicates an ethical rather than structural perspective on economy. Their approach moves the search focus from the general development paths and structural explanations to the coexisting particularities, and to the possibility of new economic futures. Drawing for instance on economic anthropology, such poststructural political economy literature asks readers to recognize that even if capitalist processes are currently hegemonic, these have not erased all economic diversity that exist and have existed across the world. Gibson-Graham (2006) bring to the fore that despite the hegemonic 'capitalocentric' representation and reproduction of the economy, there still exist diverse forms of economic organization, relations and agency that go unnoticed if attention is paid only to *the* structural forces. In other words, the poststructural perspective to political economy asks to recognize that not 'the whole system' has to change – only the reproduction of those currently hegemonic economic processes that aim at capital growth as well as the practices that reproduce their hegemonic position.

Therefore, poststructural takes help to clarify there is not a clear-cut divide between capitalism and degrowth economy (which refers to *the other* to capitalism). Social and economic transformation starts and coexists with the current forms of capitalist economic processes. By drawing attention to existing, unorthodox views that typically remain marginal or unseen, it becomes possible to increase their potential as possible objects of policy and politics (Gibson-Graham, 2008). As Gibson-Graham (1996, pp. xi–xii) argue:

> while there exists a substantial understanding of the extent and nature of economic difference, what does not exist is a way of convening this knowledge to destabilize the received wisdom of capitalist dominance and unleash the creative forces and subjects of economic experimentation.

Using feminism as an example, they argue there are as many chances for economic transformation as there are places of capitalism (e.g. occupations, workplaces, localities, or regions).

Joutsenvirta (2016), for example, draws on the practice approach to study the role of grassroots initiatives in structural degrowth transitions. She calls for recognizing that while institutions are reproduced in human agency, grassroots initiatives and institutional acts are interrelated. It is vital to pay attention to the ways in which agents can disrupt the status quo that is institutionally maintained. Joutsenvirta (2016, pp. 24–25) writes that

> non-capitalist grassroots initiatives carry the potential to rupture the unity of growth-oriented institutions and their dominance in society. But there is a need to investigate and respond to institutions that enable or constrain the functioning of alternative practices in the current economic environment.

This perspective also asks to pay attention to the currently hegemonic institutional agency, not only to the alternative agency outside it.

Although the work of Gibson-Graham focused on fostering non-capitalist forms of economic organization, their research approach of seeking economic diversity is applicable in the context of enterprises. In a commentary on Gibson-Graham (1996), Lee, Leyshon, and Gibson-Graham (2010, p. 118) highlight that “even within capitalist practice, a wide variety of notions of value are always simultaneously at work, always informing economic action”. Gibson-Graham (2014, p. 151) also argues that an economic subject can possess other motivations for economic agency and value creation than “individual self-interest, competition, efficiency, freedom, innovative entrepreneurship, exploitation, and the pursuit of private gain”. Thus, entrepreneurship, especially in small- and medium-sized enterprises (SMEs), should be regarded as a social force and not strictly as an ‘economic’ action (North, 2016).

Ethnographically oriented case study in the Ylläs destination in the Finnish North

Several studies have incorporated a poststructural political economy approach for the study of sustainability transformations in tourism research (e.g. Mosedale,

2011; Hillmer-Pegram, 2016; Cave & Dredge, 2018; Brouder, 2018). Mosedale (2011) recommends this line of research in tourism, arguing that poststructural political economy would be a way to advance the vast study of 'alternative' forms of tourism (e.g. sustainable tourism, ecotourism or green tourism) that currently are treated cautiously as environmentally sustainable and socially just forms of economic organizing. In a special issue on diverse tourism economies, Cave and Dredge (2018) draw on poststructural political economy thinking to discuss economic diversity in tourism economies. The case examples offered in the special issue illustrate that there already exist forms of economic organization that differ from the 'ideal' capitalist economic organizing. Renkert (2019) also draws on Gibson-Graham (2006) to argue that some forms of ecotourism can be used as a localized way to enact degrowth society. To value such alternative tourism practices more broadly than only in specific firms and sites, Cave and Dredge (2018, p. 473) argue that "a reworking of the traditional ways we conceive the economic organization of tourism (and the economy more broadly) is needed". This is taken to mean that research should focus on studying what the notion of economic difference entails in tourism spaces where change is currently dominated by market-driven rationales.

To explore alternative economic thinking and practice at the destination scale, the chapter draws on an ethnographically oriented case study that was conducted in the Ylläs tourism destination in the Finnish North in 2015 (Figure 5.1). During the fieldwork, the author met with 37 tourism actors who operate and primarily also live in the destinations (the group includes both native locals and immigrants). The aim was to gain an empirical understanding of the everyday tourism realities of the local agents of 'the' global tourism economy. In-depth semi-structured interviews were the main method for research material collection. The themes discussed included tourism cooperation, influence opportunities, participation in decision-making and desired destination development in the future. The discussion I had with local tourism actors shed light on how these individuals relate to the idea of destination growth and also offer information about the actual tourism practices.

The ethnographically oriented case study approach coheres with contemporary grounded theory methods. Charmaz (2006, p. 2) explains that "we [grounded theorists] try to learn what occurs in the research settings we join and what our research participants' lives are like". In this sense, the theories are 'grounded' in the empirical data. Charmaz (2006) also importantly explains that theory does not emerge out of data, nor are results discovered. Rather, that "we are part of the world we study and the data we collect. We construct our grounded theories" (Charmaz, 2006, p. 10). In line with grounded theory, I have also returned to the empirical data during the research process and re-checked whether the theoretical frame I have used captures the phenomena that I see in the data. Writing this book chapter is a continuum of such attempts; my aim is to re-present and re-conceptualize my previous analysis of the fieldwork data (Kulusjärvi, 2019) in a way that is better illustrative of the messiness of the actual tourism views and practices that escape any exact categorizations. As Clarke (2012, p. 19) states: "we need to grasp variation within data categories".

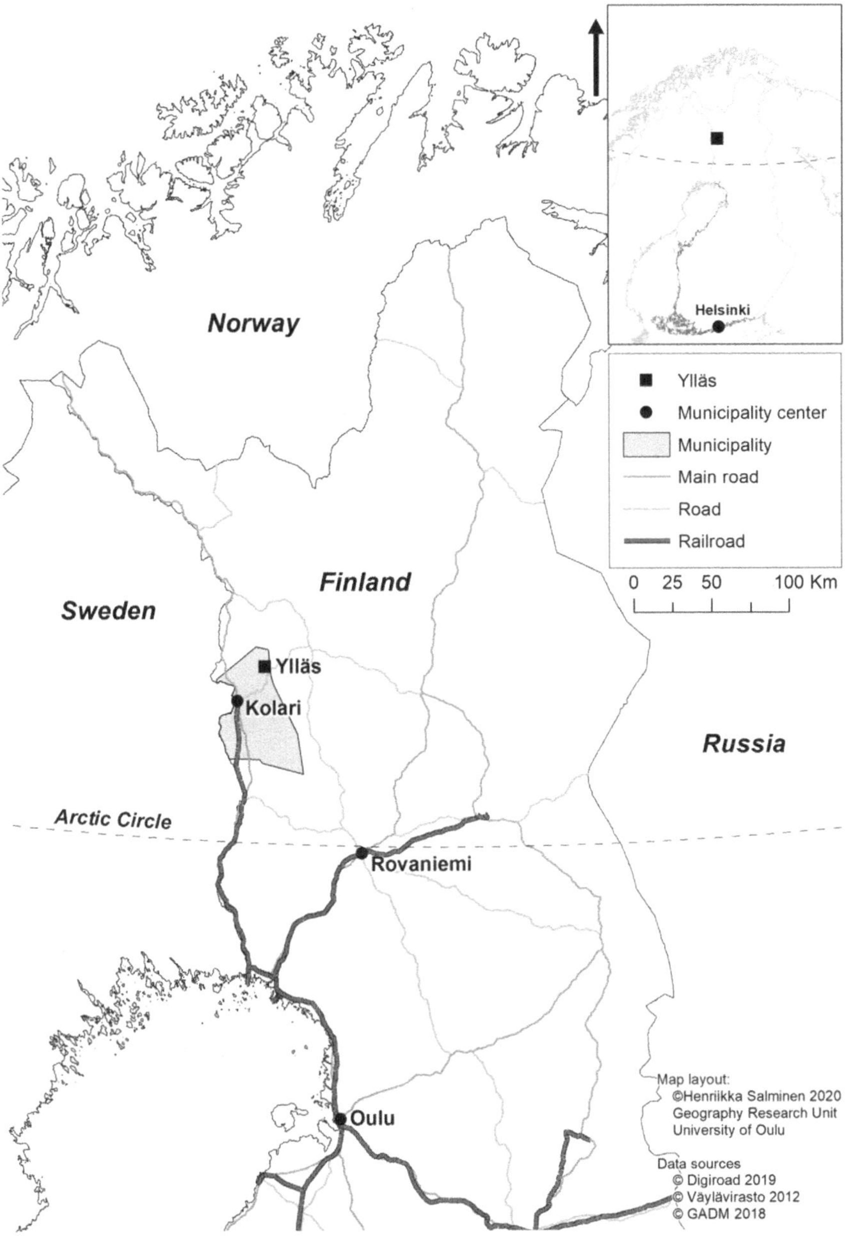

Figure 5.1 Map of Northern Finland showing case study location

Source: Map layout ©Henriikka Salminen

Diversity in local economic agency in tourism production

The Ylläs tourism destination is located at the foot of a chain of fells in the western part of the Lapland region, in the municipality of Kolari. Since the 1930s, people in Ylläs have utilized the scenic fell area to maintain their local livelihoods. In the beginning, local villagers were the main actors in small-scale tourism practices, accommodating visitors in their homes, for instance. They also owned the ski lift operations, until these were bought out by non-local actors. Since the early decades, the fell area, as well as the two local villages of Äkäslompolo and Ylläsjärvi located close by, have transformed into a tourism destination space with 332,861 overnight stays in 2018, 47% of which came from international visitors (Statistics Service Rudolf, 2019).

In critical geography approaches outside tourism research, economic actors such as entrepreneurs are usually seen as the primary actor group responsible for social injustices and environmental unsustainabilities related to capitalist tourism economy. In Ylläs also, almost each tourism actor who I met with talked about 'growth' and 'development' and *to this extent* shared the discourse of growth-focused tourism development. Some local tourism actors support the resort growth path where intensive growth in tourist numbers and income is sought through new building plans. In 2019, a master plan for building 10,000 new bed-places and increasing the ski slope capacity by 60% only into the Ylläsjärvi side of the fell was confirmed. The hope is to attract international investors such as well-known hotel chains that are able to bring in international customers. Such large-scale master plans are advanced by the ski lift companies, together with the local municipality officials through land use planning. The interviewed municipality representatives regard tourism growth as a tool for dynamism and vitality in the whole rural municipality.

However, this analysis highlights that referring to 'growth' does not mean that actors mean the same things when they use the term. The interviewed tourism actors do not agree on how much and what type of growth is desired (e.g. own customers, tourist numbers in destination, form of further construction, year-round tourism), by what means growth is to be achieved and to what degree tourism development is acceptable at the expense of the cultural and natural environment. Thus, there is diversity inside the discourse and practice of tourism development and growth. There exist several sources of economic difference among the economic actors in the destination, which are presented next. These do not represent mutually exclusive groups but rather intersecting axes of difference.

Differing position and power in the local tourism network

The large-scale projects on both sides of the Ylläs fell get support from many of the smaller tourism entrepreneurs whose businesses are dependent on alpine skiing tourism and are often located in the resort core. Yet, these actors have a different, less powerful position in their tourism supply chain, because these businesses do not have direct economic linkages with non-local tourism operators, as accommodations often do, but serve customers whose travel has been arranged by other firms. Although these actors support resort growth, they can still maintain less growth-focused intentions in their own firm-level practices. They do not in the same way contribute to growing the tourist numbers.

Many of the interviewed actors in the nearby areas also looked at tourism destination growth from a different perspective, from outside in terms of geographic location and field of economy. These actors did not engage with either the destination development work or specifically criticize the resort growth. I met with many small tourism entrepreneurs who take benefit from, as well as are dependent on, the tourism destination development in the area but may identify themselves more with other livelihoods such as reindeer herding or the horse sector. These actors are dependent either on the individual travelers who happen to visit their business or organized tourist groups who regularly visit the subcontractor. Although for some of these firms, pre-arranged international group tourists formed the clear majority of their customers, they did not share the mainstream business talk in regard to their own business practices. Rather, they seem to tap into tourism destination operations to maintain their own livelihood in the rural Finnish North.

Different tourism products and the level of tourism internationalization

Even in the destination core area in the two villages, there were differences among local tourism actors also in terms of marketing and customer segments. Many of the local tourism actors supported the tourism development path that was based on cross-country skiing tourism. This difference may look inconsequential and be indicative only of maximizing personal economic gains as well as competitive relations between actors. Yet, many of these interviewees did not relate to the internationalizing tourism core area, but their work relied on the wide skiing track network, and they also appreciated the domestic customer base. These tourism actors can be seen as maintaining more traditional Lappish tourism that is based on the local cultural and natural environments more than ski resort operations.

Different target of growth and development

For many of the locals in the two tourism villages, growth does not primarily refer to tourist numbers or income but to village growth. Tourism growth creates employment and enables living in their home village where there would otherwise be fewer employment possibilities. For many of these actors, tourism work and entrepreneurship are something they have grown into since they were children. As one interviewee explained:

> It starts when you are a kid, you are there, there is huge lack of employees, I have also worked in tourism for my dad when I was 10, my dad is also an entrepreneur, I started to clean cottages, I kind of grew up into entrepreneurship and tourism, and especially to the idea that there is work.

A few of these actors also shared the past experiences of how, alongside the resort growth, their own possibilities to influence tourism development had decreased. Some interviewees identified themselves more as village developers than tourism developers. A strong discourse was to support growth in off-seasons, although some shared that it is important that the local villages also have time off between peak tourism seasons.

Conflicts in terms of the use of local natural and cultural environments as a tourism resource

It is especially noteworthy that some economic actors openly question the means and extent of tourism growth in Ylläs and advocate for less-intensive and more nature-based tourism practices. These actors highlighted the negative effects of 'mass' or 'bulk' tourism and of 'tourism as an industry' and saw it as a contrary path to tourism that is based on local, more genuine environments in local villages. Many of the interviewees also opposed the ongoing master plans, as these require cutting down primeval forest areas. In other words, these voices oppose ongoing growth plans and maintain conservationist ideas despite being tourism entrepreneurs. Many of these actors shared that their less-growth-focused views were not appreciated in destination governance networks where alpine-style ski resort development dominated. Such voices also included some of the largest firms in the Ylläs destination offering ecotourism products (e.g. snowshoeing, winter biking) and smaller-scale accommodation or complementary services (e.g. restaurants, cafés, grocery stores) in the local villages.

Firm-level degrowth business aims

A few of these actors specifically underlined that they are not interested in growing their own business either. For instance, an originally local woman explained that "I need to make a bit of money, but I have no need to make big business. But I would like to make this *my* size". She had put up a café so that she was able to move back to her home village. To be able to preserve the traditional style in her childhood home and the land, she saw it as a positive issue that the place was located a bit farther away from the resort core area.

Voluntary joint efforts

Special attention should also be paid to the economic thinking and agency that is linked to tourism entrepreneurship but can be seen as operating outside of the monetary and individualized economy. Many of the interviewees shared that many of the tourism development efforts are based on voluntary cooperation. For instance, arranging an event in local villages is often done by local tourism actors together with village associations. Both local residents and visitors can take part. This type of event seems to be organized more based on village residency, friendship or even kinship. Not everyone is engaged in this work but still benefits from it. Another example of linkages between tourism economies and community economies is the ways in which tourism actors put effort into advertising each other's services. Some of the actors explained how it is essential that they cross-sell their services. One interviewee mentioned how she had astonished some people by advertising a 'competing' firm's services for her own customers on Facebook, without any compensation.

Conclusions: towards tourism destination degrowth

To sum up the empirical analysis, tourism actors do not similarly support, benefit from or take part in the growth of international tourist flows and new tourism construction. The economic agency of many of the local actors, both in large and small firms, deviates from the logics of growth maximization. Many of the actors do not prioritize growth in their own business, as studies on lifestyle tourism entrepreneurship have noted (see Chapters 3–5, this volume). In addition, many of the interviewees value the conservation of local natural and cultural environments as the tourism destination is for them not only a business environment but also their home. Not all local actors engage in destination growth plans, and some even actively oppose them. Diversity exists even inside 'the' growth discourse; the term does not mean the same things for every actor. For many, tourism represents a means to live in this northern rural area; tourism is viewed more as a livelihood, a means of earning one's living, than as a growth economy. There already exists economic difference inside tourism economy.

To build poststructural perspectives to destination economies, tourism destinations are better conceptualized as meeting places of diverse tourism practices and motivations rather than as coherent economic systems, as they may easily appear when viewed top-down. Local tourism economy consists of coexistent, diverse paths towards the perceived positive change in the destination community. To a certain extent, these less-growth-focused tourism practices can be seen as forms of degrowth economies. This conclusion is in line with Panzer-Krause (2019), who identified a nascent degrowth-oriented faction within a network of ecotourism operators in rural Ireland. Yet, despite the existence of economic diversity in tourism destinations, destination development as a whole currently seems to follow the growth logics. Thus, it is necessary to study how the recognition of diverse economic rationalities could work as a way to foster *destination* degrowth.

The empirical analysis indicates that it is power hierarchies in the destination community rather than active, intentional agency of the group of tourism entrepreneurs that reproduce the growth-oriented development path. The tourism views and practices that depend less on increase in international tourist arrivals or rapid increase in new tourism construction in destination do not currently become recognized as valuable for overall destination development. Thus, degrowth transitions may not require total transformation of all tourism practices in 'the whole tourism system', and some forms of tourism need to even be supported while others need to degrow.

The ethnographic fieldwork highlights that the recognition of conflict within capitalist economic processes is crucial in the field of tourism economy (Kallis &

March, 2015). To date, economic change towards more socially and environmentally sustainable tourism economies has been most often pictured as taking place through joint actions and in collaboration by tourism stakeholders (see a more detailed discussion in Kulusjärvi, 2019). In destination development contexts, the recognition of conflict means that the often-noted difficulty of engaging all stakeholders in tourism destination development is recognized as political decision-making where voices exist also for counter-hegemonic practices that deviate from the growth path (Valdivielso & Moranta, 2019).

Thus, degrowth transitions can be fostered by facilitating alternative economic subjectivities and encouraging political agency inside the economy. One such action is resigning from the destination management organization in cases where ongoing practices do not appreciate the diverse local views on local change. The case study showed that this practice has already taken place in the Ylläs tourism destination. Referring to Joutsenvirta (2016), such practice is valuable as it can disrupt the status quo that is otherwise institutionally maintained. Such a post-structural political economy take is a more hopeful, forward-looking approach than the critical stance that often concludes with the ever-existent growth focus of the powerful actors.

The present perspective to tourism destination degrowth suggests that degrowth transitions can be advanced by incremental changes from below, not only in official political structures. This is a valuable notion while governments increasingly embrace the role of economic promoter in line with neoliberal logics (e.g. Hall, 1999; Dredge, Jenkins, & Whitford, 2011). It may therefore not be wise to remain waiting for sustainability transformations in tourism economies via policy actions only. However, it seems evident that collective actions, negotiations and dialogue are needed to make a diversity of local voices turn into forms of economic organization. Fletcher et al. (2019, p. 1754) note that "transition towards tourism degrowth could comprise both top-down and bottom-up elements". More research is needed, for instance, on how to support joint efforts among actors who share alternative economic aims but may not have any organization to unite them. Research on diverse tourism could study the possibilities of alternative economic organizations such as co-operatives in the field of tourism. In addition, it is necessary to increase knowledge on diverse tourism economy among public (tourism) planners. Large-scale tourism master plans may not foster the type of tourism economy that is desired locally. By attending to such concerns, it may become possible to make the less-growth-focused practices become possible objects of policy and politics (Gibson-Graham, 2008).

By incremental changes both at the grassroots level and official destination politics, it may be possible to gradually build a network of degrowth economics in a destination space. This project would benefit from the support for degrowth economies from regional or national public bodies, a research theme that is beyond the focus of this chapter. Yet, it is noteworthy here that the diversity in economic agency in tourism economies does not need to be restricted to the local. Future research on tourism and degrowth could aim at envisioning how to contribute to building forms of tourism that are less dependent on air travel by building new scalar relations (Kallis & March, 2015). For example, how to link non-local tourism actors such as tour operators or transportation service providers with the local destination degrowth initiatives with an aim for advancing travel by land or near home?

References

Bianchi, R. (2009). The 'critical turn' in tourism studies: A radical critique. *Tourism Geographies*, *11*(4), 484–504.

Brouder, P. (2018). The end of tourism? A Gibson-Graham inspired reflection on the tourism economy. *Tourism Geographies*, *20*(5), 916–918.

Cave, J., & Dredge, D. (2018). Reworking tourism: Diverse economies in a changing world. *Tourism Planning & Development*, *15*(5), 473–477.

Charmaz, K. (2006). *Constructing grounded theory. A practical guide through qualitative analysis*. London: Sage.

Clarke, A. E. (2012). Feminism, grounded theory, and situational analysis revisited. In S. N. Hesse-Biber (Ed.), *Handbook of feminist research: Theory and praxis* (pp. 388–412). Thousand Oaks: Sage.

D'Alisa, G., Demaria, F., & Kallis, G. (Eds.). (2015). *Degrowth: A vocabulary for a new era*. Abingdon: Routledge.

Dredge, D., Jenkins, J., & Whitford, M. (2011). Tourism planning and policy: Historical development and contemporary challenges. In J. Jenkins & D. Dredge (Eds.), *Stories of practice: Tourism policy and planning* (pp. 28–43). Farnham, UK: Ashgate.

Fletcher, R. (2011). Sustaining tourism, sustaining capitalism? The tourism industry's role in global capitalist expansion. *Tourism Geographies*, *13*(3), 443–461.

Fletcher, R., Murray, I., Blanco-Romero, A., & Blázquez-Salom, M. (2019). Tourism and degrowth: An emerging agenda for research and praxis. *Journal of Sustainable Tourism*, *27*(12), 1745–1763.

Gibson-Graham, J. K. (1996). *The end of capitalism (as we knew it). A feminist critique of political economy*. Oxford: Wiley-Blackwell.

Gibson-Graham, J. K. (2006). *A post capitalist politics*. Minneapolis: University of Minnesota Press.

Gibson-Graham, J. K. (2008). Diverse economies: Performative practices for 'other worlds'. *Progress in Human Geography*, *32*(5), 613–632.

Gibson-Graham, J. K. (2014). Rethinking the economy with thick description and weak theory. *Current Anthropology*, *55*(9), 147–153.

Hall, C. M. (1999). Rethinking collaboration and partnership: A public policy perspective. *Journal of Sustainable Tourism*, *7*(3–4), 274–289.

Hall, C. M. (2009). Degrowing tourism: Décroissance, sustainable consumption and steady-state tourism. *Anatolia*, *20*(1), 46–61.

Hall, C. M. (2010). Changing paradigms and global change: From sustainable to steady-state tourism. *Tourism Recreation Research*, *35*(2), 131–143.

Hillmer-Pegram, K. (2016). Integrating indigenous values with capitalism through tourism: Alaskan experiences and outstanding issues. *Journal of Sustainable Tourism*, *24*(8–9), 1194–1210.

Joutsenvirta, M. (2016). A practice approach to the institutionalization of economic degrowth. *Ecological Economics*, *128*, 23–32.

Kallis, G., Demaria, F., & D'Alisa, G. (2015). Introduction: Degrowth. In G. D'Alisa, F. Demaria, & G. Kallis (Eds.), *Degrowth: A vocabulary for a new era* (pp. 1–18). Abingdon: Routledge.

Kallis, G., Kerschner, C., & Martinez-Alier, J. (2012). The economics of degrowth. *Ecological Economics*, *84*, 172–180.

Kallis, G., & March, H. (2015). Imaginaries of hope: The utopianism of degrowth. *Annals of the Association of American Geographers*, *105*(2), 360–368.

Koch, M. (2020). The state in the transformation to a sustainable postgrowth economy. *Environmental Politics*, *29*(1), 115–133.

Kulusjärvi, O. (2019). Towards a post structural political economy of tourism: A critical sustainability perspective on destination development in the Finnish North. *Nordia Geographical Publications*, *48*(3), Art. 107.

Lee, R., Leyshon, A., & Gibson-Graham, J. K. (2010). Gibson-Graham, J. K. 1996: *The end of capitalism (as we knew it): A feminist critique of political economy*. Oxford: Wiley-Blackwell. *Progress in Human Geography*, *34*(1), 117–127.

Lehtinen, A. A. (2018). Degrowth in city planning. *Fennia*, *196*(1), 43–57.

Mosedale, J. (2011). Thinking outside the box: Alternative political economies in tourism. In J. Mosedale (Ed.), *Political economy of tourism: A critical perspective* (pp. 93–108). London: Routledge.

North, P. (2016). The business of the Anthropocene? Substantivist and diverse economies perspectives on SME engagement in local low carbon transitions. *Progress in Human Geography*, *40*(4), 437–454.

Panzer-Krause, S. (2019). Networking towards sustainable tourism: Innovations between green growth and degrowth strategies. *Regional Studies*, *53*(7), 927–938.

Renkert, S. R. (2019). Community-owned tourism and degrowth: A case study in the Kichwa Anangu community. *Journal of Sustainable Tourism*, *27*(12), 1–16.

Schmid, B. (2019). Degrowth and postcapitalism: Transformative geographies beyond accumulation and growth. *Geography Compass*, *13*(11). https://doi.org/10.1111/gec3.12470

Statistics Service Rudolf. (2019). *Yearly nights spent and arrivals by country of residence*. Retrieved from http://visitfinland.stat.fi/PXWeb/pxweb/fi/VisitFinland

Valdivielso, J., & Moranta, J. (2019). The social construction of the tourism degrowth discourse in the Balearic Islands. *Journal of Sustainable Tourism*, *27*(12), 1876–1892.

6 Global importance, local problems

Degrowth in Italian World Heritage destinations

Alberto Amore and Bailey Ashton Adie

Introduction

Tourism is one of the most important global economic drivers (UNWTO, 2019; WTTC, 2019). In 2019, tourism garnered approximately $1.7 trillion USD worldwide (UNWTO, 2019). Contemporary tourism has moved away from the mass tourism popular in decades past and has instead shifted focus towards more niche tourism types. Heritage and cultural tourism, which is a significant focus of modern tourism demand (Park, 2013), is one of these popular niches and includes a wide range of established and emerging destinations and sites. In Europe alone it is estimated that 40% of all tourists to the continent are motivated by the destinations' cultural offerings (European Commission, 2019). For example, Venice, Italy, registered a record-high 28 million visitors, including overnight stays, day-trippers and cruise passengers in 2017 (Costa, 2018). In comparison, visits to Yellowstone National Park in the United States totaled 4.1 million in 2018 (National Park Service (NPS), 2019). The experiences of these sites reflect that of other cultural and natural heritage sites, many of which have had steady growth in visitor arrivals, site congestion and depletion of resources (Hitchcock & Putra, 2016; Kim, 2016; Lee & Rii, 2016; López-del-Pino & Grisolía, 2018). In some instances, particularly in emerging destinations, this trend can occur rapidly, as was observed in Iceland, where natural features are noted as the primary reason for visiting, contributing considerably to the boom in international arrivals (+225%) between 2014 and 2018 (Sæþórsdóttir, Hall, & Stefánsson, 2019).

Many of the previously mentioned examples are notable as UNESCO World Heritage Sites. While World Heritage status marks a site as of importance for all of humanity, inscription on the list comes with a series of managerial and planning duties that underpin the principles of sustainable development (Adie, 2017, 2019). However, the tendency among many local and national authorities is to use the World Heritage designation as part of their destination marketing strategy (Ashworth & van der Aa, 2006; Caust & Vecco, 2017; Lai & Ooi, 2015), regardless of the inconclusive findings in the literature as to whether or not the World Heritage brand functions as a tourism attractor (Adie, 2019; Adie, Hall, & Prayag, 2018). There is awareness, however, of the shortcomings of growth-driven destination policies, particularly at sites that rely heavily on the integrity and uniqueness of the natural environment (Hakim, Soemarno, & Hong, 2012). The emphasis on tourism growth can be tied to the increasing concern with overtourism at natural sites (Cruz & Legaspi, 2019), although this is not a new issue in and of itself. This in turn highlights the need for a more sophisticated approach to destination planning and site management, but as yet there has been no consensus as to how this can be achieved (Dodds & Butler, 2019). Tourism degrowth, then, provides an alternative paradigm that goes behind the rhetoric of sustainable growth and puts resource management and conservation ahead of profit and commercialization of natural heritage (Hall, 2009; Hall & Amore, 2016; Higgins-Desbiolles, Carnicelli, Krolikowski, Wijesignhe, & Boluk, 2019). To this end, this chapter critically applies the notion of tourism degrowth within natural World Heritage destinations, with evidence from two destinations in Italy, one mature and one emerging.

Tourism, World Heritage and degrowth

Regardless of geographic location, most destinations share two common features: a) the pursuit of economic, social and environmental sustainability (Edgell & Swanson, 2018) and b) an emphasis on the management of an attractive and competitive destination (Hall, 2008). For example, the New Zealand Tourism Industry Aotearoa (TIA) 2015–2025 strategy "aims to be universal so all operators are contributing to overall tourism industry sustainability . . . in a highly competitive global market" (TIA, 2015, p. 4). Similarly, Belize has implemented a 2030 Sustainable Tourism Master Plan whose goal is to be "an exclusive multicultural sustainable destination in the Central American Caribbean" (Ministry of Tourism and Civil Aviation, 2012, p. 7). The nexus between sustainable development and competitiveness has also been discussed previously in the literature, with evidence from national tourism policies (e.g. Andrades & Dimanche, 2017) and protected natural areas (e.g. Rodríguez-López, Diéguez-Castrillón, & Gueimonde-Canto, 2019). Overall, a pro-growth agenda dominates, which reinforces the vision of short-term goals, fast-track development and resource exploitation (Getz, 1987; Hall, 2008). Thus, regardless of aspirational environmental goals, the capitalist drive in global tourism development has been dominant since the 1980s and has further expanded in the last years (Amore & Hall, 2017; Britton, 1991), resulting in alarming cases of site overexploitation, environmental erosion and exceedance of carrying capacity. Some examples include Boracay Beach in the Philippines (Cruz & Legaspi, 2019), Maya Bay in Thailand (Ellis-Petersen, 2018) and the Kvennagjá geothermal pool in Iceland (Iceland Magazine, 2018). In fact, the increasing concerns in relation to 'overtourism' are a clear sign of unsustainability in the current tourism development discourse. The growing awareness of both tourists and local communities in regard to issues such as overcrowding, site congestion and resource management at tourist sites highlights the need for "a response to the unwanted effects of tourism development" (Hall, 2008, p. 10), including overdevelopment, resulting in a shift in planning and management of tourism destinations.

The inclusion of a site on the World Heritage List adds a layer of complexity to any planning system due, in part, to the intricate nature of the World Heritage governance structure (Adie, 2019; Adie & Amore, 2020). This situation is compounded in disaster-prone areas, particularly given the requirements around safeguarding and preserving the heritage assets in situ. In fact, the increasing threats posed by natural hazards and climate change resulted in the inclusion in the Operational Guidelines of a recommendation to ensure that site management plans have disaster and climate change risk management policies (UNESCO, 2019c). While these recommendations are derived from World Heritage Committee decisions, the actual governance structure of the World Heritage system places the onus of control entirely in the hands of the individual state parties (Francioni, 2008). As the actual management systems of each site are dictated by each state party, implementation of requirements and recommendations from the committee can be extremely varied, particularly when there are governance clashes between the broader World Heritage system and that of the state party (Adie & Amore, 2020). These broader governance issues have a ripple effect on management at site level with diverse interpretation of the UNESCO requirements resulting in divergent policies. This is particularly apparent when access to the site is interpreted as a license to overdevelop tourism. This can cause problems down the road when the site's periodic reporting period highlights the threat that tourism may be to the listed site.

Given these complex issues regarding site management, World Heritage governance and tourism activities, degrowth would appear to be an optimal solution. Hall (2009, p. 46) conceives tourism degrowth as an "alternative discourse to the economism paradigm that reifies economic growth in terms of GDP" and proposes a holistic destination planning framework, rooted in the steady state concept (Daly, 1991) and *décroissance* (Latouche, 2009), which restructures, redistributes, reduces, recycles and reuses limited resources. In his view, "steady state tourism is a tourism system that encourages qualitative development but not aggregate quantitative growth that unsustainably reduces natural capital" (Hall, 2009, p. 57). From a demand perspective, tourism degrowth advocates for a shift in consumer behavior that ultimately results in the reduction of service outputs and resource use (Hall, 2009). From a supply perspective, business and destination management organizations must embrace measures and solutions to regulate and reduce the anthropic pressure at the destination level and embrace technological advancements that actually contribute to resource efficiency (Hall & Amore, 2016; Hall, 2015). For example, evidence from Malaga, Spain, illustrates how local advocates embrace the principles of degrowth in contrast to the prevailing tourism development agenda in order to pursue environmentally sound policies (Navarro-Jurado et al., 2019). Overall, tourism degrowth views the reduction of environmental impact and visitor numbers as essential to the process of 'sufficiency' and 'right-sizing' of destinations.

Degrowth in the specific context of World Heritage is particularly apt given the foundational ideals of the World Heritage List. The original purpose of the Convention was to conserve and protect sites of Outstanding Universal Value, as highlighted in the emphasis on these activities in their mission statements (UNESCO, 2008). While the Operational Guidelines (UNESCO, 2019c) stress the importance of the educational elements of World Heritage Sites, this does not intrinsically require that the sites be accessible. In instances where access is very complicated or highly disruptive to the site's environment, educational elements can be transmitted via designated information centres, often museums in the case of cultural heritage sites. However, although both the Convention and the Operational Guidelines are underpinned by conservation and preservation ideals, recent shifts within the World Heritage system have resulted in a heavily politicized listing mechanism wherein the major impetus for listing appears to be driven by perceived political and economic benefits, predominantly in the form of tourism receipts (Adie, 2019; Caust & Vecco, 2017; Logan, 2012; Meskell, 2015; Plets, 2015). However, the current use of the list as a tourism brand places World Heritage at odds with the concept of degrowth.

Given that the World Heritage List is currently used as more of "a marketing device than a protection approach" (Caust & Vecco, 2017, p. 8), degrowth policies may be perceived as a hindrance of the 'benefits' which may have motivated a State Party to list a site in the first place. As Buckley (2018) has noted, tourism is an industry and as such is driven by self-interest wherein conservation and preservation are only acceptable activities if they lead to a clear benefit. In the context of World Heritage, this essentially creates an odd dichotomy wherein industry providers may, as was observed at the Great Barrier Reef, "see themselves as the key stewards" (Liburd & Becken, 2017, p. 1731) of a site while simultaneously eschewing more active environmentally driven groups. Unsurprisingly, this has also led industry members at the Great Barrier Reef to eschew climate change discussions in an attempt to avoid bad word of mouth. The avoidance of dialogue regarding climate change impacts is ironic when considering that tourism only heightens many sites' vulnerability to climate change effects (Gössling & Peeters, 2015). This is exceedingly problematic given the aforementioned conservation aims of the Convention. Moving forward, there needs to be a re-focus on these original tenets. Thus, acknowledging the important role that tourism plays for the maintenance of many sites and surrounding communities, degrowth may prove to be an optimal solution, if the managing bodies can prioritize conservation and sustainable measures.

The Italian context

Italy has a long-standing connection to the World Heritage Convention, having first ratified the document in 1978 and with its first site inscribed in 1979 at the third committee meeting. As of 2019, Italy has 55 sites inscribed on the List, sharing the title for country with the most World Heritage sites with China. While the vast majority of these sites are strictly cultural, Italy also has five natural World Heritage Sites: Ancient and Primeval Beech Forests of the Carpathians and Other Regions of Europe, Isole Eolie (Aeolian Islands), Monte San Giorgio, Mount Etna and the Dolomites. The relevance of natural heritage is a key element in the Italian Constitution (Article 9), which states that it is the State's duty to preserve the landscape and the heritage of the nation. World Heritage Sites have their own specific legislation, *Legge* 77/2006, which provides the regulations for the management of all listed sites within Italy's national borders. This legislation dictates the importance of site-specific management plans and requires the incorporation of cultural services as well as tourism within this planning process (Adie & Amore, 2020). Tourism's specific inclusion in this legislation can be viewed as essential, particularly as, based on international tourism arrivals, Italy is one of the top ten destinations in the world, with 62 million international visitors and USD$ 49 billion spent in 2018 (UNWTO, 2019). The combined total of international and domestic arrivals in natural parks and protected areas was 24.7 million in 2018, with an average tourism intensity of 7.6 tourist nights for every 100 residents (Istituto Superiore per la Protezione e la Ricerca Ambientale (ISPRA), 2019).

The issues of maturity: Cinque Terre

Cinque Terre is a 15-kilometer-wide area stretching along the Levantine coast of the Ligurian Sea, Italy. It consists of five small townships – Monterosso, Vernazza, Riomaggiore, Corniglia and Manarola – and the surrounding areas which together form the Cinque Terre National Park. This area, in combination with Porto Venere and the islands of Palmaria, Tino and Tinetto, was awarded World Heritage Site status as a cultural landscape in 1997 due to its "exceptional scenic quality that illustrates a traditional way of life" (UNESCO, 1997, p. 10) and the extensive network of dry stone terraces that were first built as part of the fortresses system designed as protection from Saracen raids (Galve et al., 2016; International Council on Monuments and Sites (ICOMOS), 1996). The landscape of Cinque Terre is a unique area that blends biodiverse ecosystems, both land and marine based, with agricultural products, particularly lemons, olive oil and the signature *Sciacchetrà* wine. Cinque Terre is considered a mature destination which is attractive to international visitors, which in turn contributes to the growth of tourism in the wider Province of La Spezia (Ministero dei Beni e delle Attività Culturali e del Turismo (MIBACT), 2016). In the early 2010s, Cinque Terre experienced a steady increase of visitor arrivals (Parco Nazionale delle Cinque Terre, 2014b), with more than 3.5 million visits in 2018 (Bompani, 2019). As Amore (2021) observes, the rise in visitor numbers is a result of the post-2011 flooding destination recovery strategy, which culminated with the opening of a visitor centre and cruise terminal in Portovenere in 2014. There are, however, growing concerns with the increasing tourist pressure in Cinque Terre, with the Mayor of Monterosso advocating for the adoption of measures that would restrict visitor access to the site and regulate the local tourism industry (Moggia, 2018).

From a destination governance perspective, the national park authority for Cinque Terre acts as the dedicated destination management and marketing organization (Amore, 2021; Lorenzini, 2011; Storti, 2005). Moreover, it is also responsible for visitor management and accessibility within the Cinque Terre park area. In 2014, the park authority and Trenitalia reached a partnership agreement wherein the park and train tickets were integrated, providing access to the several trails within the Cinque Terre site (Parco Nazionale delle Cinque Terre, 2016a). The combined ticket enables the park authority to closely monitor the number of entrants and set a cap when necessary. More recently, the park authority has announced its intention to grant access to the national park only to rail ticket holders in an attempt to drastically reduce the visitor flow down to 1.5 million (Deiana, 2018). In fact, one of the townships has identified tourism as one of the major causes of site vulnerability and subsequently has prohibited large tour groups and train passengers from entering the town during the tourist season (Comune di Riomaggiore, 2019). These decisions met with opposition from local tourism businesses and Trenitalia, with the latter appealing to the regional court in 2019 (Destefanis, 2019). Local authorities and the national park view restricted access as the only solution to the steady increase of visitors following the 2011 flooding (Camera di Commercio di Genova, 2017). In the words of the former national park authority director, Vittorio Alessandro, restricted booking "increases the appeal and indeed establishes a virtuous pact between visitor and guest that elevates the quality of hospitality" (Alessandro, 2016, author translation).

To date, resource management in the Cinque Terre area has been predominantly reactionary in nature. As Amore (2021) observes, the flooding in 2011 diverted resources to the site while also concentrating labor on the clearance of key tourist areas in Monterosso and Vernazza and the main hiking route on the coast. Additionally, the region of Liguria put in place "a ban on new building and on work to existing buildings that goes beyond mere conservation work" (UNESCO, 2012, p. 29), and the ministry of cultural heritage identified the maintenance of terraces and the promotion of agricultural production as post-flooding opportunities to repopulate abandoned areas and reduce erosion and landslides in the Cinque Terre (MIBACT, 2016). Nevertheless, the projects specifically designed to reduce or eliminate existing environmental vulnerabilities have been sporadic and mostly under the initiative of the private and voluntary sectors. This was the case at Casa Lovara, where the Fondo Ambiente Italiano (FAI) refurbished the terraces and reinstated the abandoned farm (FAI, 2016). While the area falls under the jurisdiction of the Cinque Terre National Park authority, national heritage legislation and the Cultural Heritage and Landscape Code, its governance is extremely fragmented and thus is unable to manage the highly vulnerable ecosystem. In May 2018, the Mayor of Monterosso advocated for the Cinque Terre National Park authority to become the governing body of the whole area (Moggia, 2018). He further stressed the need to establish a strategic plan which would safeguard the existing cultural landscape from the steady tourism speculation that will likely lead to the displacement of residents from the five townships (Moggia, 2018).

The current sustainable development strategy for Cinque Terre includes projects focusing on both the green economy and environmental protection. To this end, the park authority and the provincial Chamber of Commerce launched a partnership to promote green economic practices among relevant tourism businesses (Alboretti, 2015). Additionally, the park authority introduced an environmental quality certification system to persuade local businesses to proactively contribute to the promotion of sustainable development practices in the area (Parco Nazionale delle Cinque Terre, 2016b). The Cinque Terre park authority views environmental quality accreditation as vital to the achievement of economic, environmental and social sustainability within the World Heritage Site (Parco Nazionale delle Cinque Terre, 2016b). Additionally, the park authority has undertaken pilot projects like the INTERREG Programme Martittimo 2014–2020 and the LegaCoop *Alta Scuola di Turismo Ambientale* in order to promote environmental sustainability among local tourism SMEs operating in the area (ENEA, 2018; LegaCoop, 2013). There are nevertheless a series of shortcomings and incongruences in the pursuit of environmental and ecological efficiency. This is visible in the destination management strategy for Cinque Terre, which reiterates the idea of internationalization, specifically in regard to increasing international tourism arrivals in the park. For example, in both the INTERREG and the LegaCoop initiatives, the stated aim is to enhance the sustainability of tourist flows and attract lucrative foreign niche markets (ENEA, 2018; LegaCoop, 2013; Parco Nazionale delle Cinque Terre, 2014a). Furthermore, the Cinque Terre park authority and the La Spezia port authority developed a memorandum for the development and promotion of a hallmark event in the Cinque Terre site (Amore, 2021). Finally, and more importantly, the finances of the park authority rely on the number of park cards issued via Trenitalia, which in turn makes sound sustainable development practices challenging.

New is always better? the Northern Dolomites

The Dolomites are located in the northeastern part of the Italian Alps and stretch between the provinces of Belluno, Trento, Bolzano, Pordenone and Udine (UNESCO, 2019a). The site is a renowned ski destination and home to one of the largest ski areas in the world (1,256 km of groomed runs), containing 446 ski lifts (Dolomiti Superski, 2019). In 2009, the Dolomites were inscribed in the World Heritage List due to their unique geological formation and “exceptional natural beauty” (UNESCO, 2019b, p. 257) of the area. The Dolomites World Heritage area includes three national parks: Parco Nazionale delle Dolomiti Bellunesi, Parco Naturale Tre Cime di Lavaredo and Parco Naturale di Fannes-Seies-Braies. Based on estimates for the year 2015, the Dolomites had approximately 6.4 million tourist arrivals and more than 29.3 million overnight stays, with Germans and Italians representing nearly 80% of the tourist nights in the area (Istituto provinciale di statistica (ASTAT), 2016). As Elmi and Wagner (2013) report, the index of tourism density in the Dolomites World Heritage Site is concentrated in key areas (e.g. Cime di Lavaredo, Lago di Braies). This has resulted in early signs of visitor congestion across the property (ASTAT, 2016; Fondazione Dolomiti Unesco, 2014) and raised concerns among local stakeholders in regard to the anthropogenic pressure on the site (Fondazione Dolomiti Unesco, 2015b).

In terms of destination governance, the Dolomites World Heritage Site has a dedicated foundation which acts as a steering group for the relevant public and civic entities (Fondazione Dolomiti Unesco, 2019). The foundation was established following the inscription of the Dolomites on the World Heritage List in order to draft key strategic and resource management documentation, including the tourism strategy and Overall Management Strategy (OMS) (Fondazione Dolomiti Unesco, 2015b). The latter is focused on the enhancement of the site's conservation through the implementation of an integrated management strategy driven by the area's human, financial and environmental resources (Fondazione Dolomiti Unesco, 2015b). Similarly, the Parco Nazionale delle Dolomiti Bellunesi has developed and implemented dedicated strategic and operational plans for tourism. The park adheres to the European Charter for Sustainable Tourism, with relevant stakeholders agreeing on the importance of achieving economic sustainability whilst effectively managing the environmental resources of the property (Parco Nazionale delle Dolomiti Bellunesi, 2015, 2019). One aspect emphasized in the management plans concerns accessibility within the Dolomites. Stakeholders agree that there is a need to preserve the landscape of the Dolomites and use existing roads to promote slow tourism itineraries and practices (Fondazione Dolomiti Unesco, 2015a, 2015b). These slow tourism initiatives are viewed as a preferred alternative to the current visitor behavior trends and mobility patterns at and to key attractions in the area (Streifender & Omizzolo, 2017). Not surprisingly, the majority of the stakeholders believe that simple time-based restricted access for cars can actually promote sustainable mobility and, in turn, reduce the environmental footprint of tourism in the Dolomites (Fondazione Dolomiti Unesco, 2015a).

Resources within the Dolomites World Heritage Site are managed using a networked system which stresses "conservation, communication and enhancement of the Property" (Fondazione Dolomiti Unesco, 2015b, p. 5). In contrast to Cinque Terre, where management of the site is fragmented, the current management system of the Dolomites is overarching and stresses local decision-making and cooperation across governance levels in order to create a "sustainable integration between man and the natural environment" (Parco Nazionale delle Dolomiti Bellunesi, 2015, p. 24, author translation). In order to ensure that this system complies with conservation demands, particularly given the site's World Heritage status, the tourism management plan requires control of visitor levels in zones which are already heavily visited. It also aims "to prohibit intensification of infrastructure or inappropriate uses that could impact the values of the property, and to ensure effective presentation and tourism benefits compatible with the long-term conservation of the property" (Fondazione Dolomiti Unesco, 2015b, p. 7). The strategy goes further by explicitly banning the construction of any new ski resorts within the boundaries of the World Heritage Site (Fondazione Dolomiti Unesco, 2015b). Given the area's importance as a ski destination, this ban appears to illustrate a strong commitment to the conservation goals espoused in the World Heritage Convention. However, recent developments illustrate that there is a disconnect between the developed strategy and actual resource use. More specifically, as reported in *The Telegraph* (Squires, 2020), the expansion of the Dolomiti Superski area in a run-up to the 2026 Winter Olympic Games violates the conservation objectives found in the previous tourism plan. Additionally, this expansion will further complicate visitor management in the already heavily visited winter period.

Unsurprisingly, the sustainable development strategy for the Dolomites World Heritage Site espouses a vision wherein both the local community and tourists feel ownership and "responsibility for [the site's] conservation and sustainable development" (Fondazione Dolomiti Unesco, 2015b, p. 11). However, the strategy also recognizes the complexity inherent in the sustainable development plan, notably the need to "[overcome] the juxtaposition of protecting the environment, with its overtones of prohibition, and driving economic development, which is seen as exploitation of natural and landscape resources" (Fondazione Dolomiti Unesco, 2015b, p. 11). In order to balance conservation with socio-economic needs, the strategy requires that tourism be considered in tandem with the rest of the local economic environment, and any tourism planning should occur in collaboration with the local community (Fondazione Dolomiti Unesco, 2015a). Throughout the entirety of this management plan, sustainability underpins all stated aims and goals, regardless of the aspect under discussion. According to the report #Dolomiti2040, "sustainability is indicated as the way forward for a balanced relationship between growth and heritage protection" (Fondazione Dolomiti Unesco, 2015a, p. 14, author translation). Thus, even with the undercurrents of conservation and sustainability, there is still a continued focus on growth, albeit 'sustainable' growth.

Discussion

In the case of both Cinque Terre and the Dolomites, there is a significant amount of focus on sustainable concepts and, particularly in the case of the Dolomites, conservation measures. Both plans specifically espouse goals focused on balancing environmental protection and human use, which would appear to indicate a willingness to adapt in order to ensure resource efficiency, as called for by Hall and Amore (2016) and Hall (2015). It should be noted that while both sites stress sustainable principles, on paper Cinque Terre has a greater focus on economic development in comparison with the Dolomites, which may be due to its longstanding status as a tourism destination as well as its socio-economic dependency on tourism income. Regardless, both of these plans appear to subscribe, to varying degrees, to the degrowth principles of sufficiency and efficiency, prioritizing conservation ahead of profit and commercialization (Hall, 2009; Higgins-Desbiolles et al., 2019). However, in both areas, these conservation ideals are sidelined in the face of economic pressures to promote tourism growth. In the case of the Dolomites, this is particularly egregious given that the ski area's development is expressly prohibited in the tourism strategy. Therefore, the management plans can be viewed as little more than site management greenwashing.

This may in fact be a result of both areas' World Heritage status, particularly given the requirements around protection and maintenance of a site's integrity. This greenwashing can be viewed as an exercise in appeasing the site management requirements laid out by the World Heritage Committee while simultaneously engaging in otherwise disruptive and destructive on-site behavior, which is not uncommon at Italian World Heritage Sites (Adie & Amore, 2020). In the case of Cinque Terre, its status as a mature destination, both in terms of tourism flows and length of listed status, has elicited a different response than can be observed in the Dolomites. As has been noted, there have been conservationist movements within the site, predominantly driven by local non-profit and private entities. It could be supposed that the pushback to the growth mentality at this site is a direct result of long-term experience with pro-growth policies. In contrast, the Dolomites, being a significantly newer and less developed location, has little experience from which to draw the same conclusions. This has resulted in divergent responses to proposed growth policies, regardless of conservation-driven agendas. These cases emphasize Buckley's (2018) point that tourism is self-serving and will only take action when there are tangible benefits or, in the case of Cinque Terre, visible threats.

Conclusion

This chapter has presented two Italian World Heritage case studies that have policies which appear, on the surface, to adhere to specific tenets of tourism degrowth. However, as has been observed, these degrowth and sustainable policies are predominantly hollow words, most likely used to fulfill World Heritage requirements. When the sites' managing bodies do attempt to adhere to their conservation goals, actual interventions are reactionary and often undermined by a continued emphasis on economic growth. Unlike other destinations, World Heritage Sites have specific conservation and protection obligations under the World Heritage Convention. Thus, these sites, by disregarding their management plans, are also placing themselves at risk of physical damage and potential de-listing. Degrowth, as previously stated, would be an ideal solution for World Heritage areas like Cinque Terre and the Dolomites as it stresses self-sufficiency, which does not preclude economic benefits from tourism activity but instead suggests that moderation is key. In order to ensure better protection, conservation and even the continued existence of sites like Cinque Terre and the Dolomites, a degrowth mindset must be adopted at the managerial level, and there needs to be sufficient political will to enforce these policies in the face of economic pressures, both internal and external.

References

Adie, B. A. (2017). Franchising our heritage: The UNESCO World Heritage brand. *Tourism Management Perspectives*, *24*, 48–53.

Adie, B. A. (2019). *World Heritage and Tourism: Marketing and management*. Abingdon: Routledge.

Adie, B. A., & Amore, A. (2020). Transnational world Heritage, (meta)governance and implications for tourism: An Italian case. *Annals of Tourism Research*, *80*, 102844. https://doi.org/10.1016/j.annals.2019.102844

Adie, B. A., Hall, C. M., & Prayag, G. (2018). World Heritage as a placebo brand: A comparative analysis of three sites and marketing implications. *Journal of Sustainable Tourism*, *26*(3), 399–415.

Alboretti, C. (2015, July 7). Intesa tra Parco Nazionale delle Cinque Terre e Camera di Commercio della Spezia [Agreement between Cinque Terre National Park and the Chamber of Commerce of La Spezia]. *Altrimondi News*. Retrieved from http://altrimondinews.it/2015/07/17/parco-delle-cinque-terre-e-camera-di-commercio-insieme-per-una-economia-sostenibile/
Alessandro, V. (2016, February 19). Cinque Terre e turismo: Il presidente del Parco risponde alla polemica sul "numero chiuso" [Cinque Terre and tourism: The Park president responds to the controversy of the "number cap"]. *GreenNews.info*. Retrieved from www.greenews.info/rubriche/top-contributors/cinque-terre-e-turismo-il-presidente-del-parco-risponde-alla-polemica-sul-numero-chiuso-20160219/
Amore, A. (2021). Reframing sustainability and resilience in the recovery of the Cinque Terre following the October 2011 flooding. In C. M. Hall & G. Prayag (Eds.), *Tourism and flooding*. Clevedon: Channel View Publications.
Amore, A., & Hall, C. M. (2017). National and urban public policy in tourism. Towards the emergence of a hyperneoliberal script? *International Journal of Tourism Policy*, *7*(1), 4–22.
Andrades, L., & Dimanche, F. (2017). Destination competitiveness and tourism development in Russia: Issues and challenges. *Tourism Management*, *62*, 360–376.
Ashworth, G. J., & van der Aa, B. J. M. (2006). Strategy and policy for the World Heritage convention: Goals, practices and future solutions. In A. Leask & A. Fyall (Eds.), *Managing World Heritage sites* (pp. 147–158). Oxford: Butterworth-Heinemann.
Bompani, M. (2019, January 2). Assalto alle Cinque Terre, la sindaca di Riomaggiore vuole la tassa di sbarco [Assault on Cinque Terre, the mayor of Riomaggiore wants an arrival tax]. *La Repubblica (Genova Edition)*. Retrieved from https://genova.repubblica.it/cronaca/2019/01/02/news/assalto_alle_cinque_terre_la_sindaca_di_riomaggiore_vuole_la_tassa_di_sbarco-215657128/
Britton, S. (1991). Tourism, capital, and place: Towards a critical geography of tourism. *Environment and Planning D: Society and Space*, *9*(4), 451–478.
Buckley, R. (2018). Tourism and natural World Heritage: A complicated relationship. *Journal of Travel Research*, *57*(5), 563–578.
Camera di Commercio di Genova. (2017). *Il Turismo Sostenibile per lo Sviluppo della Liguria* [Sustainable tourism for the development of Liguria]. Genoa: Camera di Commercio di Genova.
Caust, J., & Vecco, M. (2017). Is UNESCO World Heritage recognition a blessing or burden? Evidence from developing Asian countries. *Journal of Cultural Heritage*, *27*, 1–9.
Comune di Riomaggiore. (2019). *Piano Speditivo di Protezione Civile* [Exploratory protection plan]. Riomaggiore: Comune di Riomaggiore.
Costa, E. (2018, June 15). Venezia, oggi sostenibili 19 milioni di turisti l'anno [Venice, today 19 million toursists are sustainable]. *CFNews*. Retrieved from www.unive.it/pag/14024/?tx_news_pi1%5Bnews%5D=5268&cHash=87cca5f2f7652d8577e41040e37b2a6a
Cruz, R. G., & Legaspi, G. F. A. (2019). Boracay beach closure: The role of the government and the private sector. In R. Dodds & R. W. Butler (Eds.), *Overtourism: Issues, realities and solutions* (pp. 95–110). Berlin: De Gruyter.
Daly, H. E. (1991). *Steady-state economics* (2nd ed.). Washington, DC: Island Press.
Deiana, S. D. (2018). *L'Ente Parco propone di limitare l'accesso di turisti alle Cinque Terre per preservare il fragile equilibrio di sentieri e borghi. Un provvedimento che farà discutere* [Park authority proposes limiting tourists' access to Cinque Terre to preserve the fragile equilibrium footpaths and villages. A measure that will be debated]. Retrieved from www.lifegate.it/persone/stile-di-vita/troppi-turisti-nelle-cinque-terre-proposto-il-numero-chiuso

Destefanis, G. (2019, July 7). Cinque Terre, troppi turisti: la richiesta di limitare gli arrivi fa litigare sindaca e Ferrovie [Cinque Terre, too many tourists: The request to limit arrivals starts a fight between the mayor and the rail line company]. *La Repubblica (Genova Edition)*. Retrieved from https://video.repubblica.it/edizione/genova/cinque-terre-troppi-turisti-la-richiesta-di-limitare-gli-arrivi-fa-litigare-sindaca-e-ferrovie/338867/339468

Dodds, R., & Butler, R. W. (2019). Introduction. In R. Dodds & R. W. Butler (Eds.), *Overtourism: Issues, realities and solutions* (pp. 1–5). Berlin: De Gruyter.

Dolomiti Superski. (2019). *Dolomiti Superski. The world's largest ski region welcomes you!* Retrieved from www.dolomitisuperski.com/en

Edgell Sr, D. L., & Swanson, J. R. (2018). *Tourism policy and planning: Yesterday, today, and tomorrow*. Abingdon: Routledge.

Ellis-Petersen, H. (2018, October 3). Thailand bay made famous by The Beach closed indefinitely. *The Guardian*. Retrieved from www.theguardian.com/world/2018/oct/03/thailand-bay-made-famous-by-the-beach-closed-indefinitely

Elmi, M., & Wagner, M. (2013). *Turismo sostenibile nelle Dolomiti. Una strategia per il Bene Patrimonio Mondiale Unesco* [Sustainable tourism in the Dolomites. A strategy for UNESCO World Heritage properties]. Trento: Eurac Research.

ENEA. (2018). *ENEA coinvolge il Parco delle Cinque Terre quale area pilota del Progetto STRATUS per l'implementazione di un Marchio di Qualità Ambientale rivolto alle imprese turistiche legate al mare* [ENEA involves Cinque Terre park as a pilot area for the STRATUS project for the implimentation of a environmental quality seal for coastal tourism organizations]. Retrieved from https://ambiente.sostenibilita.enea.it/news/enea-coinvolge-parco-cinque-terre-quale-area-pilota-progetto-stratus-limplementazione-un

European Commission. (2019). *Cultural tourism*. Retrieved from https://ec.europa.eu/growth/sectors/tourism/offer/cultural_en

Fondazione Dolomiti Unesco. (2014). *Linee guida del turismo* [Tourism guidelines]. Retrieved from www.dolomitiunesco.info/?pubblicazioni=linee-guida-del-turismo

Fondazione Dolomiti Unesco. (2015a). *#Dolomiti 2040 participatory process: Report finale* [#Dolomites 2040 participatory process: Final report]. Cortina d'Ampezzo: Fondazione Dolomiti Unesco.

Fondazione Dolomiti Unesco. (2015b). *Overall management strategy and tourism strategy*. Cortina d'Ampezzo: Fondazione Dolomiti Unesco.

Fondazione Dolomiti Unesco. (2019). *Understanding current tourism to plan for the future*. Retrieved from www.dolomitiunesco.info/understanding-current-tourism-to-plan-for-the-future/?lang=en

Fondo Ambiente Italiano (FAI). (2016). *Apre al pubblico Podere Case Lovara in Liguria* [Podere case lovara in Liguria open to the public]. Retrieved from www.fondoambiente.it/news/apre-al-pubblico-podere-case-lovara-in-liguria

Francioni, F. (2008). The preamble. In F. Francioni (Ed.), *The 1972 World Heritage Convention: A commentary* (pp. 11–21). Oxford: Oxford University Press.

Galve, J. P., Cevasco, A., Brandolini, P., Piacentini, D., Azañón, J. M., Notti, D., & Soldati, M. (2016). Cost-based analysis of mitigation measures for shallow-landslide risk reduction strategies. *Engineering Geology*, *213*, 142–157.

Getz, D. (1987). Tourism planning and research: Traditions, models and futures. In *Australian travel research workshop*. Bunbury: Western Australia.

Gössling, S., & Peeters, P. (2015). Assessing tourism's global environmental impact 1900–2050. *Journal of Sustainable Tourism*, *23*(5), 639–659.

Hakim, L., Soemarno, M., & Hong, S. K. (2012). Challenges for conserving biodiversity and developing sustainable island tourism in North Sulawesi Province, Indonesia. *Journal of Ecology and Environment*, *35*(2), 61–71.

Hall, C. M. (2008). *Tourism planning: Policies, processes and relationships* (2nd ed.). Harlow: Prentice Hall.

Hall, C. M. (2009). Degrowing tourism: Décroissance, sustainable consumption and steady-state tourism. *Anatolia*, *20*(1), 46–61.

Hall, C. M. (2015). Economic greenwash: On the absurdity of tourism and green growth. In M. V. Reddy & K. Wilkes (Eds.), *Tourism in the green economy* (pp. 361–380). Abingdon: Routledge.

Hall, C. M., & Amore, A. (2016). Turismo, sostenibilità e crescita verde: Green economy o una semplice pennellata di verde? [Tourism, sustainability and green growth: Green economy or a simple coat of green?]. In A. P. Scanio (Ed.), *Turismo sostenibile* (pp. 145–188). Rome: Aracne Editrice.

Higgins-Desbiolles, F., Carnicelli, S., Krolikowski, C., Wijesignhe, G., & Boluk, K. (2019). Degrowing tourism: Rethinking tourism. *Journal of Sustainable Tourism*, *27*(12), 1926–1944.

Hitchcock, M., & Putra, I. N. D. (2016). Prambanan and Borobudur: Managing tourism and conservation in Indonesia. In V. T. King (Ed.), *UNESCO in Southeast Asia: World Heritage Sites in comparative perspective* (pp. 258–273). Copenhagen: NIAS.

Iceland Magazine. (2018, July 11). *Hidden geothermal pool closed due to disrespectful and littering tourists*. Retrieved from https://icelandmag.is/article/hidden-geothermal-pool-closed-due-disrespectful-and-littering-tourists

International Council on Monuments and Sites (ICOMOS). (1996). *World Heritage list: Portovenere/Cinque Terre (Italy), No 826*. Paris: ICOMOS.

Istituto provinciale di statistica (ASTAT). (2016). *Andamento Turistico – Anno turistico 2014/15* [Tourist trends – Tourism year 2014/15]. Bolzano: ASTAT.

Istituto Superiore per la Protezione e la Ricerca Ambientale (ISPRA). (2019). *Annuario dei Dati Ambientali Anno 2018* [Environmental data almanac year 2018]. Rome: ISPRA.

Kim, S. (2016). World Heritage site designation impacts on a historic village: A case study on residents' perceptions of Hahoe Village (Korea). *Sustainability*, *8*(3), 258. https://doi.org/10.3390/su8030258

Lai, S., & Ooi, C.-S. (2015). Branded as a World Heritage city: The politics afterwards. *Place Branding and Public Diplomacy*, *11*(4), 276–292.

Latouche, S. (2009). *Farewell to growth*. Cambridge: Polity Press.

Lee, M. Y., & Rii, H. U. (2016). An application of the vicious circle schema to the World Heritage site of Macau. *Journal of Heritage Tourism*, *11*(2), 126–142.

LegaCoop. (2013). *Decolla in Liguria l'Alta Scuola di Turismo Ambientale con il sostegno di Legacoop* [The advanced school of environmental tourism opens in Liguria with help form Legacoop]. Retrieved from www.legaliguria.coop/decolla-in-liguria-lalta-scuola-di-turismo-ambientale-con-il-sostegno-di-legacoop/

Liburd, J. J., & Becken, S. (2017). Values in nature conservation, tourism and UNESCO World Heritage site stewardship. *Journal of Sustainable Tourism*, *25*(12), 1719–1735.

Logan, W. (2012). States, governance and the politics of culture: World Heritage in Asia. In P. Daly & T. Winter (Eds.), *Routledge handbook of heritage in Asia* (pp. 113–128). Abingdon: Routledge.

López-del-Pino, F., & Grisolía, J. M. (2018). Pricing beach congestion: An analysis of the introduction of an access fee to the protected island of Lobos (Canary Islands). *Tourism Economics*, *24*(4), 449–472.

Lorenzini, E. (2011). The extra-urban cultural district: An emerging local production system: Three Italian case studies. *European Planning Studies*, *19*(8), 1441–1457.

Meskell, L. (2015). Transacting UNESCO World Heritage: Gifts and exchanges on a global scale. *Social Anthropology*, *23*(1), 3–21.

Ministero dei Beni e delle Attività Culturali e del Turismo (MIBACT). (2016). *Porto Venere, Cinque Terre e Isole (Palmaria, Tino e Tinetto). Piano di Gestione per il sito UNESCO* [Porto venere, Cinque Terre and Islands (Palmaria, Tino and Tinetto). UNESCO site management plan]. Rome: MIBACT.

Ministry of Tourism and Civil Aviation. (2012). *National sustainable tourism masterplan for Belize 2030* [pdf]. Retrieved from http://cdn.gov.bz/tourism/National%20Sustainable%20Tourism%20Master%20Plan.pdf

Moggia, E. (2018, May 15). Un unico centro di prenotazione e più servizi per i residenti [A single reservation hub and more services for residents]. *Citta della Spezia News*. Retrieved from www.cittadellaspezia.com/cinque-terre-val-di-vara/attualita/un-unico-centro-di-prenotazione-e-piu-servizi-per-i-residenti-259758.aspx

National Park Service (NPS). (2019). *Park facts*. Retrieved from www.nps.gov/yell/planyourvisit/parkfacts.htm

Navarro-Jurado, E., Romero-Padilla, Y., Romero-Martínez, J. M., Serrano-Muñoz, E., Habegger, S., & Mora-Esteban, R. (2019). Growth machines and social movements in mature tourist destinations Costa del Sol-Málaga. *Journal of Sustainable Tourism*, *27*(12), 1786–1803.

Parco Nazionale delle Cinque Terre. (2014a). *Adesione alla Carta Europea per il Turismo Sostenibile nelle Aree Protette* [Adherence to the European charter for sustainable tourism in protected areas]. Portovenere: Parco Nazionale delle Cinque Terre.

Parco Nazionale delle Cinque Terre. (2014b). *Carta Europea del Turismo Sostenibile* [European charter of sustainable tourism] Portovenere: Parco Nazionale delle Cinque Terre.

Parco Nazionale delle Cinque Terre. (2016a). *Cinque Terre card*. Retrieved from www.parconazionale5terre.it/dettaglio.php?id=34423

Parco Nazionale delle Cinque Terre. (2016b). *Ente Parco Nazionale delle Cinque Terre: Piano della Performance* [Cinque Terre National Park authority: Performance plan]. Portovenere: Parco Nazionale delle Cinque Terre.

Parco Nazionale delle Dolomiti Bellunesi. (2015). *Carta Europea del Turismo Sostenibile: Strategia e Piano d'Azione* [European charter of sustainable tourism: Strategy and action plan]. Belluno: Parco Nazionale delle Dolomiti Bellunesi.

Parco Nazionale delle Dolomiti Bellunesi. (2019). *CEST 2020–2024*. Retrieved from www.dolomitipark.it/it/page.php?id=1352

Park, H. Y. (2013). *Heritage tourism*. Abingdon: Routledge.

Plets, G. (2015). Ethno-nationalism, asymmetric federalism and Soviet perceptions of the past: (World) Heritage activism in the Russian Federation. *Journal of Social Archaeology*, *15*(1), 67–93.

Rodríguez-López, N., Diéguez-Castrillón, M. I., & Gueimonde-Canto, A. (2019). Sustainability and tourism competitiveness in protected areas: State of art and future lines of research. *Sustainability*, *11*(22), 6296. https://doi.org/10.3390/su11226296

Sæþórsdóttir, A. D., Hall, C. M., & Stefánsson, Þ. (2019). Senses by seasons: Tourists' perceptions depending on seasonality in popular nature destinations in Iceland. *Sustainability*, *11*, 3059.

Squires, N. (2020, January 10). Italian plan to create world's largest ski network in Dolomites leaves environmentalists aghast. *The Telegraph*. Retrieved from www.telegraph.co.uk/news/2020/01/10/italian-plan-create-worlds-largest-ski-network-dolomites-leaves/

Storti, M. (2005). *Il Paesaggio Storico delle Cinque Terre* [The historic landscape of Cinque Terre]. Florence: Florence University Press.

Streifender, T., & Omizzolo, A. (2017). The dolomites UNESCO World Heritage: Visitors' mobility behaviour and acceptance of regulatory measures. *Management and Policy Issues*, *9*, 93–97.

Tourism Industry Aotearoa (TIA). (2015). *Tourism 2025 and beyond: A sustainable growth framework – Kaupapa Whakapakari Tāpoi*. Wellington, New Zealand: TIA.

United Nations Educational Scientific and Cultural Organization (UNESCO). (1997). *Convention concerning the protection of the world cultural and natural heritage. World Heritage Committee, twenty-first session, Naples, 1–6 December 1997*. Paris: UNESCO.

United Nations Educational Scientific and Cultural Organization (UNESCO). (2008). *World Heritage information kit*. Paris: UNESCO.

United Nations Educational Scientific and Cultural Organization (UNESCO). (2012). *Mission report: Portovenere, Cinque Terre, and the Islands (Palmaria, Tino and Tinetto) (Italy) (C 826) 8–12 October 2012*. Paris: UNESCO.

United Nations Educational Scientific and Cultural Organization (UNESCO). (2019a). *Dolomites – Inscribed zones*. Paris: UNESCO.

United Nations Educational Scientific and Cultural Organization (UNESCO). (2019b). *Nomination of the dolomites for the inscription on the world natural heritage list*. Paris: UNESCO.

United Nations Educational Scientific and Cultural Organization (UNESCO). (2019c). *Operational guidelines for the implementation of the World Heritage Convention*. Paris: UNESCO.

United Nations World Tourism Organization (UNWTO). (2019). *Tourism highlights – 2019 Edition*. Madrid: UNWTO.

World Travel and Tourism Council (WTTC). (2019). *How does travel & tourism compare to other sectors*. London: WTTC.

7 Opportunities and barriers for degrowth in remote tourism destinations

Overcoming regional inequalities?

Doris A. Carson and Dean B. Carson

Introduction

The purpose of this chapter is to critically examine the concept of degrowth as a viable tourism development strategy in the context of particular types of destinations exemplified here by the Top End of the Northern Territory of Australia. These are destinations in high-income countries that have variously been labelled as 'remote', 'peripheral', 'beyond periphery', and 'exotic' (Schmallegger, Carson, & Tremblay, 2010; Koster & Carson, 2019). They are typified by their isolation from source markets, their small local populations and economies, the concentration of economic and political power in one or two main cities, and their historical focus on high-volume (and international) tourism driven largely by external investors and supported by government subsidies and direct investment in both promotion and product development. In many cases, this has led to a susceptibility to 'boom and bust' cycles in visitor numbers and income from tourism and a political lock-in that continues to prioritise major investment projects within an increasingly entrenched economic growth paradigm (Carson & Carson, 2017). The consequences of this development pattern have included increasing spatial and social inequality in terms of who receives the benefits of tourism during boom times and who suffers least from the busts. This inequality is the core focus of the case study presented later, which argues that both booms and busts are increasingly managed by focusing development attention on the region's main urban growth centre (the city of Darwin) and largely neglecting the more peripheral and remote parts of the destination, even as the latter continue to provide the iconic imagery which has forged the touristic reputation of the region.

Unlike in many destinations affected by mass tourism or 'overtourism' (Fletcher, Mas, Blanco-Romero, & Blázquez-Salom, 2019; Oklevik et al., 2019), there has been no overt call for degrowth in the Top End, even as a form of involuntary re-sizing has emerged through a dramatic reduction in visitor numbers over the past decades. While there are occasional calls to move away from mass tourism to smaller-scale and alternative niche markets, these calls are rarely reflected in new government-driven investment and destination marketing strategies. This chapter considers what the benefits of a policy of degrowth could be for the sparsely populated hinterland of the Top End if tourism strategies were to follow some of the tenets in the degrowth literature (Andriotis, 2018). These include changes in priority target markets, changes to the scale and spatial concentration of infrastructure and product investment, an explicit focus on generating positive tourism spillover effects from the city to the hinterland, renewed efforts to encourage more bottom-up and local involvement in decision-making, and a re-positioning of tourism as part of broader community development agendas rather than as an export industry in itself.

The chapter further considers the conditions under which degrowth could come to be considered as a viable strategy for remote hinterland destinations. In this case it is in the context of a tourism 'crisis' where the market suddenly shrinks, threatening existing business models. Such crises are becoming more common in mass tourism destinations globally (Fletcher et al., 2019; Hall & Seyfi, Chapter 14, this volume; Prideaux & Pabel, Chapter 8, this volume) and may be linked to climate change risks, biosecurity (as witnessed in the current coronavirus scare), terrorism and political instability, sustainability concerns (particularly about long-haul air travel), and economic uncertainty. As is typical, the impacts of these threats are likely to be felt first and most severely in more remote destinations. In these locations, a turning to degrowth may not be so much about a philosophical or ideological preference, but rather a response to a market reality. The chapter consequently considers how degrowth may still achieve positive outcomes for destination communities even if it is a 'forced' strategy entered into unwillingly at the destination management level.

Finally, the chapter focuses on the central role of government in determining tourism development paths in destinations like this (Schmallegger & Carson, 2010). It considers whether and how degrowth might appeal as a strategy for a government whose 'signature' involves a certain visibility of tourism's success in terms of visitor numbers, touristic infrastructure, amenities (resorts, airports), and the potential (if not the reality) for increasing economic return on investment in the industry. Again, the intention is to demonstrate the political and institutional difficulties inherent in transitioning from one development approach to another, even when such a transition appears inevitable and market driven.

The following section provides a brief outline of the key aspects of degrowth that are relevant for understanding and improving tourism development outcomes in remote destinations. Subsequently, the case of tourism in the Top End is described, with a particular focus on illustrating and explaining recent decline, and the tensions arising between the capital city and the more remote parts of the region. The story of decline is then re-interpreted in the context of opportunities for degrowth to be embraced as a development strategy for the hinterland. The chapter concludes with a commentary on the, likely insurmountable, institutional barriers to rethinking tourism in the Top End, whether that be for degrowth or some other form of downsizing or 'right-sizing'.

Degrowth in tourism: right-sizing local industries in remote hinterland destinations?

Degrowth has emerged as a development paradigm that criticises the unconditional growth imperative of capitalist economies. It advocates for a transformation of political economies to respect social and ecological limits to growth, pursue development that emphasises higher quality of life rather than the quantity of production and consumption, and focus on reducing social and economic inequality whilst encouraging a more participatory and just society (Fletcher et al., 2019; Higgins-Desbiolles, Carnicelli, Krolikowski, Wijesinghe, & Boluk, 2019). While degrowth is often misunderstood as a concept demanding a decline or downsizing of economic production and consumption, Hall (2009) has called for degrowth to be understood as a way of 'right-sizing' the economy to aim for a 'steady state' that prioritises qualitative development instead of aggregate quantitative growth. This may include efforts to downsize and reduce economic activity in places where the limits of carrying capacity have been exceeded, while encouraging an increase in economic activity in places where new development is required to reduce poverty or socio-economic inequality. In this context, the redistribution of income and wealth between the 'haves and have nots', either between or within countries and regions, is key (Hall, 2009).

In tourism, such inequalities have been problematised as part of core-periphery discussions (Britton, 1991; Weaver, 1998), whereby tourism industries and hosting communities in the periphery are commonly dependent on not just external markets, but also powerful industry and political stakeholders located in core centres, which tend to control decision-making and reap the majority of economic and social benefits. This core-periphery dichotomy has not only been applied to tourism destinations in developing countries, but also to peripheral regions within developed nations and to peripheries within such peripheries. These tend to have highly polarised internal development patterns, with tourism, populations, and economic and political power concentrating in a few confined (and usually urban) environments (Schmallegger et al., 2010; Kauppila, 2011). The positive spillover effects from these centres to the more sparsely populated hinterland, for example, in terms of generating employment, improving socio-economic indicators, or facilitating local economic linkages, are often marginal (Kauppila, 2011). More critically, tourism in these peripheries has often been pursued by political stakeholders as one of the few promising export industries able to generate fast and large-scale investment and income. This has resulted in a form of capitalist development that is the antithesis of what development under a degrowth paradigm would look like (Andriotis, 2018) – i.e. it has relied on capital-intensive production, including large-scale infrastructure and external (and mostly long-haul) transport modes; large volumes of tourists (in relation to residents) who concentrate in time and space and engage in relatively homogenous sets of activities; an enclavic style of development revolving around a few urbanised resorts; a dominance of large industry corporations and non-local stakeholders (e.g. large hotel chains, resort operators, airlines, cruise lines); and an increasingly top-down approach to destination governance that excludes or stifles the involvement of small local industry or community stakeholders.

In contrast, a degrowth approach to tourism would aim at optimising the volume and type of tourism, rather than maximising tourist numbers, and encourage smaller-scale and more dispersed development that is within the control and capacities of local stakeholders (Oklevik et al., 2019). This may require a shift to markets that are less demanding in terms of infrastructure and amenities, appreciate simpler lifestyles, avoid mainstream or crowded resorts, are more environmentally conscious, travel more slowly, stay for longer, and value simple but in-depth encounters with local environments and cultures rather than highly commodified and capital-intensive forms of activities (Andriotis, 2018). In this context, there have been calls to redirect market attention away from an extensive export focus that revolves primarily around international or long-haul markets towards more domestic or local markets (Hall, 2009; Higgins-Desbiolles et al., 2019). These are less distant from hosting communities in a spatial and socio-economic sense and may, thus, offer 'easier' and more realistic opportunities for local tourism initiatives.

Although it is clear that previous forms of capitalist and mass-tourism development are likely to continue in destinations where these structures are already in place, a degrowth approach to tourism can shift attention to how the benefits of tourism can be redistributed in a more spatially dispersed and socially just way. For once, this would require a conscious move towards more participatory and bottom-up destination governance which respects the priorities of local stakeholders outside of the key tourism resorts and encourages their involvement in planning and decision-making. It would also require efforts to reinvest public resources into infrastructure, service, and product development to facilitate a less polarised distribution of tourists, industry players, and socio-economic benefits within the broader region, whilst making sure that new tourism development in the hinterland fits more neatly within broader community development aspirations rather than aiming at sustaining tourism as an industry in itself (Higgins-Desbiolles et al., 2019).

Tourism in Australia's Top End – ups and downs, more downs and regional inequalities

The Top End is one of currently two formal tourism regions in the Northern Territory (NT) of Australia. It comprises the tropical northern half of the NT (Figure 7.1), including the capital city of Darwin and nearby Litchfield National Park, World Heritage–listed Kakadu National Park, the Katherine region with Nitmiluk National Park, and remote Arnhem Land in the northeast (a predominantly Aboriginal-owned reserve). Darwin is the only large urban centre and had a population of around 137,000 residents in 2016, equivalent to around 60% of the NT's population (Australian Bureau of Statistics (ABS), 2019). The city is the main hub for government administration and has a relatively small Indigenous population (approximately 9%) compared to other parts of the Top End (where more than half of the population is Indigenous). (In Australia, the 'Indigenous' population comprises both Aboriginal and Torres Strait Islander peoples. In this chapter, the word 'Indigenous' is used where it is possible that Torres Strait Islander peoples or cultures are included (for example, population counts, classifications of Indigenous culture), and the word 'Aboriginal' is used where it is known that only

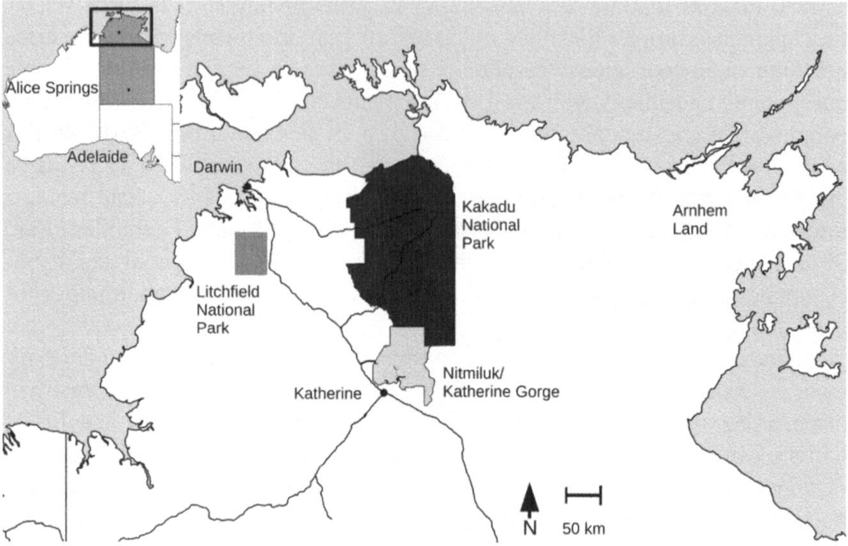

Figure 7.1 The Top End of the Northern Territory, Australia

Aboriginal peoples or cultures are referenced (for example, corporations representing specific Aboriginal nations, policies, and strategies specifically identified as 'Aboriginal'). Darwin's population grew by almost 30% between 2006 and 2016, largely on the back of major offshore gas construction projects, which also triggered a significant infrastructure and real estate investment boom in the city.

Darwin's phenomenal growth went hand in hand with a strong growth in tourism, with annual overnight visitor numbers increasing from approximately 620,000 in 2006 to 830,000 in 2016 (Tourism Research Australia (TRA), 2019). Much of this growth was driven by non-leisure tourism, including business tourism and an increasing visiting friends and relatives (VFR) market. During that period, the city saw an unprecedented level of tourism and hospitality-related infrastructure investment in the city, including the construction of a multi-billion-dollar waterfront redevelopment, a convention centre, a cruise ship terminal, and several high-rise hotel and serviced apartment complexes in the Central Business District (CBD) area (Schmallegger & Carson, 2010).

These developments joined a long list of 'major projects', mostly revolving around large-scale transport, mining and tourism infrastructure, which the NT government has historically pursued to encourage rapid economic and population growth (Carson & Carson, 2017; Pforr, 2001). Maintaining population growth has been a particular key concern for the NT government, whose budget relies to a large extent on federal government transfers providing per capita general revenue assistance (Carson, 2011). These economic and demographic constraints have meant that boosterism and an explicit economic growth paradigm have become firmly embedded in the political system (Pforr, 2001), with governments continually chasing major projects to maintain previous levels of growth.

This political imperative has resulted in spatially uneven development across the NT (Taylor, Larson, Stoeckl, & Carson, 2011). The short-term nature of investment booms has required the government to focus on development hotspots while locations and communities with more limited growth prospects often experience a relative neglect (or even abandonment) when it comes to public support for economic development. This was the case from about the mid-2000s, when Darwin's 'boomtown' growth path eclipsed any other development endeavours in the Top End region, and consequently changed tourism priorities and destination hierarchies in the NT. In the 1980s, there was substantial tourism development in regional areas around Alice Springs, Uluru-Kata Tjuta National Park, Katherine Gorge and Nitmiluk National Park, and Kakadu National Park. However, by the 2000s, Darwin had become the dominant destination and primary beneficiary of new investment, while regional and remote destinations suffered a decline in overnight stays and industry performance, reaching almost catastrophic proportions in the wake of the global financial crisis (GFC) (2009–2013). Not only were regional destinations in the Top End unable to receive positive 'spillover' effects from Darwin during this time, but Darwin's dominance of the tourism (political and economic) landscape has hindered attempts to find alternative models of development (Carson & Carson, 2019).

While Darwin has emerged as a contemporary urban resort-style destination, tourism in the Top End hinterland has remained primarily focused on passive sightseeing activities around the region's natural attractions and national parks and is characterised by tired and outdated infrastructure (Carson & Carson, 2019). Hinterland itineraries have increasingly contracted to shorter excursions and day trips undertaken from the city, instead of extended overnight stays in the region which may facilitate in-depth encounters with local people and nature- or culture-based activities.

The NT government and its tourism organisation Tourism Northern Territory have largely concentrated their marketing efforts on export markets, particularly from Europe, the US, Japan, and more recently China. The government has repeatedly spent substantial resources on lobbying national and international airlines to offer direct overseas services to and from Darwin. Similar efforts have gone into boosting international cruise tourism, with government funding the construction of a new cruise terminal along with strategies aimed at bringing large international cruise liners to the city (Schmallegger & Carson, 2010). The completion of the Ghan railway line from Adelaide to Darwin (via Alice Springs) in the mid-2000s is another example of a major tourism-related transport project that was supposed to bring more external tourists to the NT, but who mainly ended up in its capital city. While these initiatives were originally touted as strategies to improve Darwin's role as a tourist gateway to the region, they seem to have primarily benefited the city, particularly as these markets are notorious for being on tightly organised schedules that offer few opportunities for regional dispersal, slow travel, and deeper encounters with remote communities.

In contrast to Darwin, regional destinations in the Top End have lost important air and land transport services in recent years. Instead, they have become almost exclusively reliant on road-based travel, which has itself undergone dramatic changes. In the past, regional towns and attractions located along interstate highways used to be in a favourable position to capture long-haul travellers on their way to and from Darwin. However, the introduction of low-cost air services to Darwin, along with the declining popularity of long-haul driving and bus tours, have progressively reduced transit travel (Holyoak, Carson, & Schmallegger, 2009), meaning that regional destinations have become ever more dependent on tourists visiting on fly-drive itineraries or organised tours from Darwin.

At the same time, however, visitor dispersal from Darwin has been declining. This was partly driven by a 'crowding out' of leisure tourism at the peak of Darwin's construction boom, which coincided with the immediate aftermaths of the GFC. During the late 2000s, the city's accommodation sector was filled to capacity with non-leisure visitors, including construction workers, business tourists, and non-resident professionals. This drove up accommodation prices in the city and made travel to Darwin less affordable for leisure tourists on lower budgets. For example, the international backpacker market – which is often considered as a prime example of independent and slow travellers (Andriotis, 2018) – almost collapsed during that time (Terzon, 2015). Regional destinations were particularly affected during Darwin's 'boomtown' phase, with industry research (TRA, 2011) confirming that over two-thirds of visitors to Darwin at the time either stayed primarily in the city or just undertook day trips from the city to nearby areas.

Declining tourist numbers, and declining prospects for renewed tourism growth, meant that regional destinations were also losing out to Darwin in the context of regional tourism planning and decision-making. The NT's official tourism regions (supported by government funding) were consolidated from once nine to just two, with Darwin becoming the main centre for the Top End region and hosting the headquarters of both Tourism Top End (the regional destination marketing organisation, DMO) and Tourism NT (the state tourism organisation). Despite repeated promises to decentralise tourism administration to other towns in the NT, political and administrative key positions have remained firmly concentrated in Darwin where the most powerful industry stakeholders are also located. This has continued to reduce the already dwindling lobbying power of regional stakeholders within the various DMOs and government agencies (Carson & Carson, 2019).

Darwin's boom came to an abrupt (if not unexpected) end in about 2016 when the final construction projects finished and construction workers started to leave in droves. This has since triggered an unprecedented economic crisis. The city's population started to decline, the local housing market crashed, and a large number of newly constructed hotel rooms and serviced apartments were suddenly empty. Several domestic and international flight routes were cut, and many shops, restaurants/cafés, and entertainment venues in the city went out of business. The NT government, facing increasing political pressures to halt the decline and fix its budget deficit, has since started a range of initiatives to avert the crisis. The search for new 'major projects' has dominated debates around various proposals so far, and many of these have concentrated on Darwin, despite the crisis having affected the NT economy as a whole. A new multi-million-dollar 'Darwin City Deal' was announced in 2018 with plans to establish a new university campus in the CBD, along with new commercial and cultural spaces, and the construction of a water theme park and (yet another) waterfront re-development including a new luxury hotel. In addition, several 'turbocharge tourism' stimulus packages were rolled out (amounting to over AU$165 million as of 2020). This included funding to attract new airlines to Darwin (notably from China) and extensive international and interstate marketing (Abram, Murdoch, & Smith, 2018). While these packages also included support for projects in regional destinations (e.g. in Katherine and Litchfield), these were clearly dwarfed in size compared to projects proposed as part of the City Deal. More importantly, regional projects were primarily aimed at strengthening existing nature- or culture-based activities or events, meaning that the focus on Darwin as the main access and accommodation hub has remained.

Only a few locations in the Top End region are currently able to 'compete' with Darwin in terms of political attention and financial resources. One is the resort at Nitmiluk National Park, owned and managed by one of the NT's Aboriginal flagship operators, which managed to raise significant funding for a major revamp of the resort and its visitor centre. The other example concerns the mining town of Jabiru, which is located within Kakadu National Park and is about to lose its uranium mine. During election campaigns in 2018, both federal and NT governments committed a multi-million-dollar revitalisation fund towards transitioning the town into a tourist hub, including funding for infrastructure upgrades and the establishment of a visitor centre and Aboriginal cultural centre. This was also promoted as a strategy to 'save' tourism – and in particular international tourism – to Kakadu National Park, which has been in continuous decline since the late 1980s (Roberts, 2019). While these cases have brought Indigenous interests and involvement in mainstream tourism to the forefront of discussions, the scale and corporate nature of these development proposals, and the public discourses associated with them, suggest that the predominant focus on export markets and the general growth-oriented tourism pathway in the NT is likely to persist.

Indigenous cultural tourism has long been an aspired development priority for the NT government, considering its purported attractiveness to international export markets (Tremblay & Wegner, 2009). It also provides one of the few opportunities for the hinterland to differentiate its tourism offer. However, the apparent success of a few large flagship operators conceals the common experiences and realities of Indigenous tourism ventures in the Top End, with smaller operators often struggling to become 'export ready' and survive in the marketplace beyond the duration of public funding allocations (Tremblay & Wegner, 2009). Consultancy studies to support Indigenous tourism development have frequently built their business cases on assumptions of high-volume visitor flows. In other cases, there has been a focus on targeting smaller but higher-end (e.g. exclusive fly-in) markets, but these typically rely on external intermediaries, transport providers, and a certain standard of services and capital-intensive infrastructure that is difficult to provide in remote areas. The ongoing preoccupation with export tourism, which underpins government funding for Indigenous cultural tourism in the NT, has thus locked much of the NT's Indigenous tourism product into a development pathway that is closely tied to the NT's growth-dependent export focus overall.

Despite strong population growth in the city, visitors from Darwin (or any other intrastate travellers for that matter) have never been at the forefront of discussions around new tourism pathways for the NT. This is remarkable considering that regional destinations in the Top End draw considerable proportions of visitors from Darwin, including business tourists, non-resident professionals and government workers, and Indigenous travellers (Carson & Carson, 2019), many of whom also come during low season (November–March). Many of the accommodation operators around Katherine, Jabiru, and Nhulunbuy have come to rely on intrastate markets as the leisure tourism crisis has deepened. There have been some attempts over the years by various DMOs to encourage more intrastate travel and localised markets, for example, by targeting NT residents with special 'territorian rates' or as 'local ambassadors' to attract more VFRs. However, these strategies appeared more as 'gap fillers' designed to bridge the low season or a temporary crisis in export markets, and were clearly secondary to the big investment and marketing strategies rolled out as part of successive NT tourism strategies.

Benefits and risks of degrowth in remote destinations

The development of the tourism industry in the Top End over time has been driven largely by external investors who have been primarily profit-oriented and mostly had short-term ambitions and short-term associations with the destination. Large hotels and resorts have frequently changed hands, airlines have come and gone, and tour operators have changed itineraries, sometimes even omitting the destination altogether on the basis of their own financial considerations. The NT government has been complicit in this approach, investing heavily in infrastructure development particularly in the capital city of Darwin. What this approach has done, however, is to provide access for the Top End to markets and volumes of visitors that would otherwise be out of reach for a region with a small resident population and a largely under-developed economic profile. In other words, the local capacity to invest in a form of tourism that could put the Top End 'on the map' has not historically been present. The historical approach, however, corresponded with a highly industrialised era of tourism where (particularly for international visitors) large tour groups were the norm, holidays were organised for (rather than by) tourists, and few visitors had the capacity to travel independently to more remote regions. In that era, corporate 'winners' in the tourism industry were selected by intermediaries, not by the tourists themselves. This placed real restrictions on how tourism could grow spatially and how new local entrants into the marketplace could be supported.

The fundamental context for tourism in places like the Top End has since changed, and those changes present the opportunity for greater local benefit from tourism. Independent travel is now the norm rather than the exception. Many tourists engage in detailed research about the destination before they travel, often looking specifically for activities or attractions that will distinguish their trip from those of their friends or social peers. There is a cachet associated with travel that engages local businesses and products (food, craft, cultural activities), and the more exotic and 'authentic' that sense of localism (as opposed to globalised and mainstream mass tourism), the more prized is its attainment. This new environment therefore offers opportunities for smaller businesses dealing with smaller volumes of visitors in ways that encourage more in-depth encounters and longer stays, in contrast to the short-term 'tick-and-flick' sightseeing market from the past. This version of degrowth may provide some real local benefits for remote industry and community stakeholders in places like the Top End.

The principal benefit of degrowth would be to make tourism a more realistic economic development option for the small and mostly Aboriginal communities which dominate the landscape outside of Darwin. Repeated attempts to develop a thriving Indigenous tourism sector have been unsatisfactory, often because the political ambitions have been too high and demand estimates unrealistic (Tremblay & Wegner, 2009). Indigenous operators with limited experience in tourism and limited access to markets of their own choosing have been asked to accept too many visitors with too diverse cultural backgrounds. A smaller market, where Indigenous people can have more direct contact with potential visitors and manage the pace of their business development, may be more attractive and ultimately more sustainable financially and culturally, even if this requires a more patient and long-term focus on local capacity building that may not result in immediate economic returns. While the latest Aboriginal Tourism Strategy (Tourism NT, 2019) appears to reflect some of this jargon (centred around small-scale, authentic, and in-depth encounters, self-determination, local capacity building), the focus within the NT's broader mainstream marketing and development strategies has firmly remained on previous growth-oriented models, showcasing flagship attractions and operators that are able to handle large export markets.

Of course, it is not only Aboriginal communities and businesses which represent 'the local' in the Top End hinterland. Other local businesses outside of Darwin, including accommodation, hospitality, and tours, have often complained about being treated as second-class citizens when it comes to fundamentals such as government support for development, permits to access particular sites and attractions, positioning in marketing campaigns, and representation on destination marketing organisations. These businesses have seen the vast bulk of income from tourism during the good times leave the destination for the benefit of the external investors, while they have been left to bear the costs of attempting to push through more difficult times when external investors find it easy to walk away (Carson & Carson, 2017). Re-focusing Top End tourism on local businesses and local aspirations for development could help stabilise the industry and encourage more local investment. Instead of incentivising showy flagship projects in a few places, the focus needs to be more on increasing the critical mass of small operators and service providers across the region to make sure that slow and regional touring (with multiple and extended stopovers) becomes an attractive and feasible alternative to the current dominant modes of short-term transit travel or exclusive resort stays.

Encouraging more visitor dispersal and slow travel through the region would also require a more conscious focus on creating 'nice communities' in the hinterland and a realisation that many settlements in the Top End hinterland currently lack essential public and private services that are simply taken for granted elsewhere in the country, including transport and communications services, social and community infrastructure, local shops and producers, and so on. As advocated previously (Higgins-Desbiolles et al., 2019), tourism planning thus needs to be more closely integrated with broader community development agendas, rather than being treated as an industry in isolation. Instead of simply chasing large numbers of tourists or external investment that have so commonly dominated government press releases in the past (Schmallegger & Carson, 2010), tourism success in the hinterland needs to be measured and communicated in different ways, including less quantitative considerations of achieved changes to local quality of life, the social and cultural value of tourism, community empowerment, and reduced sentiments of marginalization.

Re-sizing or right-sizing Top End tourism should also be thought about in the context of stimulating and serving more local or regional (intra-destination) rather than external travel. The case study illustrates that some of this has occurred almost by default, but mostly limited to business and government travellers from Darwin. The opportunities for increased leisure travel by locals 'in their own backyards' may not only provide some more reliable income sources for a fragile industry, but also increase the social cohesion of a resident population which is widely seen as fragmented on the basis of geography (Darwin versus the rest), culture (Indigenous and not, but also various other ethnic groups), socio-economic status, and mobility (those who live in the region for short periods of time and those who are there for life). Greater interaction in a leisured environment may help break down some of the tensions among these groups, particularly if those who have had less power in tourism development politics before now can take a lead role in facilitating interactions.

These economic and social benefits of degrowth would of course be augmented by the environmental and sustainability benefits that are commonly discussed in the literature (Hall, 2009; Fletcher et al., 2019). Travel to remote places like the Top End is environmentally costly, and the infrastructure required to support large volumes of external tourism can only be provided and sustained with substantial environmental compromises. Planning for smaller numbers of more highly targeted visitors, including from more proximate source markets, can include attention to how environmental costs might be reduced. It is important to note, however, that such environmental discussions have so far largely remained under the radar within the political and public discourse, as economic survival and the retention of people and services in the NT continue to dominate local agendas. Issues around climate change threatening key natural resources and attractions have not really been present, at least not to the same extent as in other affected northern destinations (Prideaux & Pabel, this volume). Increasing the number of (international) flights is commonly portrayed and accepted as an economic necessity to maintain tourism and other industries in the NT, particularly as other modes of transport (rail or road) offer few viable alternatives and no prospects for any future expansion of public transport services.

Unlike in other sensitive nature- or culture-based destinations (Fletcher et al., 2019), there have been few outspoken concerns about the negative environmental or social issues relating to 'overtourism' (perhaps with the notable exception of Aboriginal tourism stakeholders choosing to close the famous climb of Uluru in Central Australia in 2019). This reflects in part the high acceptance of growth as being essential for the survival of the industry and the economy overall, but it may (perhaps ironically) also reflect the fact that many small and remote communities have simply had little direct engagement with tourism in the past, as tourist flows were largely channeled through external industry stakeholders. In general, though, it appears that the environmental voice and consciousness in the NT, particularly within political circles in the greater Darwin area, has traditionally been rather weak. Major construction and mining projects have repeatedly been allowed to proceed with minimal environmental concessions and at times despite outspoken opposition from traditional landowners (Carson, Govan, & Carson, 2018). Localised environmental impacts in a vast and sparsely populated territory where nature continues to be abundant, so it seems, are a sacrifice that governments in remote jurisdictions are willing to make considering the economic realities and political pressure they are under.

Conclusion

This chapter has provided some justification for applying degrowth strategies to a destination like the Top End. However, it also illustrates that there is a range of institutional barriers to even beginning a transition from the established industrial model, and there may be very sound reasons why some of the optimistic arguments in this chapter cannot be realised. The most obvious of these is the issue of political or institutional 'lock-in' and sunk costs associated with the infrastructure that has been developed over time. This includes soft as well as hard infrastructure and direct as well as indirect costs. Much work has been put into developing marketing relationships with large external companies, for example. This has come mainly from the public sector, and the political will or capacity to change is the most widely discussed barrier to rethinking tourism in this destination (Carson & Carson, 2017). Tourism, in terms of the number of visitors, the size of investment, the nature of those investors, and the visibility of touristic infrastructure, has become firmly embedded as one of the few key 'staples' industries in the north and lies at the very heart of the Northern Territory's political self-identity. Many other destinations would be envious of such a central role for tourism in the political process and the amount of public resources spent on supporting the industry. But it also brings with it a resistance to change and a protection of 'the glory days', whether it is likely they will ever be replicated or not.

Rethinking tourism in this context is not possible without rethinking economic and demographic development overall. It is clear that tourism of this particular type has played a substantial role in attracting residents and businesses and other investment outside of the tourism industry to the Top End (and primarily to Darwin). A transition to a more sustainable, smaller-scale, and more locally focused tourism model would need to be sensitive to the linkages to overall development ambitions and constraints. It is by no means clear how this could be achieved. The structural (economic, demographic, and institutional) weaknesses inherent in remote peripheries like the NT make it likely that recurring cycles of boom and bust will continue (Carson, 2011), meaning that it is a matter of when, not if, the next crisis caused by external events will occur (and coincidentally, while writing this chapter, the next tourism 'relief plan' was announced to support the industry in coping with the unfolding coronavirus outbreak; Emeck, 2020). It seems that the most recent economic crisis has reinforced, rather than dismantled, the prevalent policy stance towards boosterism and economic growth through trying to maintain or expand the same old export-oriented tourism structures. This experience is reminiscent of those in other mass tourism destinations where tourism has been part of urban growth machines and where injections into large-scale infrastructure projects have been common to revive the economy after periods of crisis (Navarro-Jurado et al., 2019). Degrowing tourism while the region's economy is already on its knees seems to be a proposition that is of little relevance to a government that is desperate to come up with a quick fix – i.e. stimulate new capital investment, cut red tape for industry, retain important transport services, prevent further outmigration, stop the real estate slump, and protect previous infrastructure investments that are simply 'too big to fail'.

Degrowing tourism will ultimately depend on degrowing other key industries in the NT, which are almost by default centred on major projects and external investors. It is unlikely that a small and more locally oriented tourism industry in the hinterland can co-exist with major investment projects occurring in other (e.g. extractive or transport) industries, and particularly within the city where such projects generate tourism growth dynamics of their own that will ultimately crowd out smaller and remote tourism players targeting alternative markets. More importantly, in the shadow of big mining or construction projects, a small and locally oriented tourism industry provides little political leverage for tourism authorities who need to justify tourism's value to the overall economy and 'compete' with other major industries for government support (Higgins-Desbiolles et al., 2019). A conscious policy move away from chasing major projects is risky, and most likely unrealistic, for a government that is under constant pressure to demonstrate economic success to not just external investors and federal government stakeholders, but also to a highly volatile NT electorate. The majority of this electorate lives in a predominantly non-Indigenous urban centre, is highly mobile (and will continue to leave if economic conditions do not improve), and may have limited personal interest in slow, small-scale, and long-term development for hinterland communities (Carson, Schmallegger, & Harwood, 2010). The fragile demography of remote northern places, which emerges from a combination of remoteness, highly temporary economies, sparse and mobile populations, and regional socio-economic imbalances (Taylor et al., 2011), thus poses a range of unique challenges for policy-makers who need to strike a balance between fast (but short-term) growth-oriented projects and more gradual and small-scale community development.

Despite the theoretical benefits of degrowth for remote communities discussed in this chapter, the risk remains that prioritising alternative markets and smaller-scale development for the hinterland may inadvertently accentuate the disparities between the city and the remainder of the region in the short to medium term. This is particularly the case if Darwin continues to act as the main growth centre and as a gatekeeper whose own interest is not in supporting a tourism economy based on smaller niche markets and the sharing of tourists and income with the region. The realities are that local actors in the hinterland may not have the know-how or resources to embrace strategies outside of the traditional export-oriented tourism model even as they see it collapse around them. In this case, who has the right to impose new models of development which might bring even greater short-term economic pain to the most vulnerable industry players? This is an issue not about the theoretical benefits of degrowth, but the moral, ethical, and political considerations in transitioning from one tourism development model to another. The conundrum then is how to enable a new approach to tourism development in parts of the destination without further marginalising the very actors such an approach seeks to benefit, and while maintaining the potential for other parts of the destination to also engage in a tourism development ideology that suits their purposes. The process of transition is therefore central to the debates about degrowth, and more research attention should be given to this issue.

References

Abram, M., Murdoch, M., & Smith, E. (2018, November 2). Northern Territory's tourism figures worst in the nation but industry hopes China will soon fill void. *ABC News*. Retrieved from www.abc.net.au/news/2018-11-02/nt-tourism-worst-in-nation-china-marketing-turbocharging/10460002

Andriotis, K. (2018). *Degrowth in tourism: Conceptual, theoretical and philosophical issues*. Wallingford: CABI.

Australian Bureau of Statistics (ABS). (2019). *Australian census data 2006–2016*. Retrieved from www.abs.gov.au/websitedbs/D3310114.nsf/Home/Census?OpenDocument&ref=topBar

Britton, S. (1991). Tourism, capital, and place: Towards a critical geography. *Environment and Planning D*, *9*(4), 451–478.

Carson, D. A., & Carson, D. B. (2017). Path dependence in remote area tourism development: Why institutional legacies matter. In P. Brouder, S. Anton Clavé, A. Gill, & D. Ioannides (Eds.), *Tourism destination evolution* (pp. 103–122). Milton Park: Routledge.

Carson, D. B. (2011). Political economy, demography and development in Australia's Northern territory. *The Canadian Geographer*, *55*(2), 226–242.

Carson, D. B., & Carson, D. A. (2019). Disasters, market changes and 'The big smoke': Understanding the decline of remote tourism in Katherine, Northern Territory Australia. In R. L. Koster & D. A. Carson (Eds.), *Perspectives on rural tourism geographies: Case studies from developed nations on the exotic, the fringe and the boring bits in between* (pp. 93–114). Cham: Springer.

Carson, D. B., Govan, J., & Carson, D. A. (2018). Indigenous experiences of the mining resource cycle in Australia's Northern Territory: Benefits, burdens and bridges? *Journal of Northern Studies*, *12*(2), 11–36.

Carson, D. B., Schmallegger, D., & Harwood, S. (2010). A city for the temporary? Political economy and urban planning in Darwin, Australia. *Urban Policy and Research*, *28*(3), 293–310.

Emeck, N. (2020, February 10). $2 million relief plan for struggling NT tourism industry. *NT News*. Retrieved from www.ntnews.com.au/business/2-million-relief-plan-for-struggling-nt-tourism-industry/news-story/55cfbd653eefc768915d131e3b6aad1b

Fletcher, R., Mas, I., Blanco-Romero, A., & Blázquez-Salom, M. (2019). Tourism and degrowth: An emerging agenda for research and praxis. *Journal of Sustainable Tourism*, *27*(12), 1745–1763.

Hall, C. M. (2009). Degrowing tourism: Décroissance, sustainable consumption and steady-state tourism. *Anatolia, 20*(1), 46–61.

Higgins-Desbiolles, F., Carnicelli, S., Krolikowski, C., Wijesinghe, G., & Boluk, K. (2019). Degrowing tourism: Rethinking tourism. *Journal of Sustainable Tourism*, *27*(12), 1926–1944.

Holyoak, N., Carson, D., & Schmallegger, D. (2009). VRUM™: A tool for modelling travel patterns of self-drive tourists. In W. Hoepken, U. Gretzel, & R. Law (Eds.), *Information and communication technologies in tourism 2009* (pp. 238–247). Vienna-New York, NY: Springer.

Kauppila, P. (2011). Cores and peripheries in a northern periphery: A case study in Finland. *Fennia*, *189*(1), 20–31.

Koster, R. L., & Carson, D. A. (2019). Considerations for differentiating among rural tourism geographies. In R. L. Koster & D. A. Carson (Eds.), *Perspectives on rural tourism geographies: Case studies from developed nations on the exotic, the fringe and the boring bits in between* (pp. 253–271). Cham: Springer.

Navarro-Jurado, E., Romero-Padilla, Y., Romero-Martínez, J. M., Serrano-Muños, E., Habegger, S., & Mora-Esteban, R. (2019). Growth machines and social movements in mature tourist destinations: Costa del Sol – Malaga. *Journal of Sustainable Tourism*, *27*(12), 1786–1803.

Oklevik, O., Gössling, S., Hall, C. M., Steen Jacobsen, J. K., Grøtte, I. P., & McCabe, S. (2019). Overtourism, optimisation, and destination performance indicators: A case study of activities in Fjord Norway. *Journal of Sustainable Tourism*, *27*(12), 1804–1824.

Pforr, C. (2001). Tourism policy in Australia's Northern Territory: A policy process analysis of its tourism development masterplan. *Current Issues in Tourism*, *4*(2–4), 275–307.

Roberts, G. (2019, August 14). Kakadu tourism lifeline for NT mining town. *The Advertiser – Cessnock*. Retrieved from www.cessnockadvertiser.com.au/story/6329339/kakadu-tourism-lifeline-for-nt-mining-town/

Schmallegger, D., & Carson, D. (2010). Whose tourism city is it? The role of government in tourism in Darwin, Northern Territory. *Tourism and Hospitality Planning & Development*, *7*(2), 111–129.

Schmallegger, D., Carson, D., & Tremblay, P. (2010). The economic geography of remote tourism: The problem of connection-seeking. *Tourism Analysis*, *15*(1), 125–137.

Taylor, A., Larson, S., Stoeckl, N., & Carson, D. (2011). The haves and have nots in Australia's Tropical North – New perspectives on a persisting problem. *Geographical Research*, *49*(1), 13–22.

Terzon, E. (2015, February 13). Illegal accommodation, tourism downturn blamed for worst Darwin hostel season in years. *ABC Radio Darwin*. Retrieved from www.abc.net.au/news/2015-02-12/darwin-budget-hostelsworst-season-in-years/6088388

Tourism NT (Tourism Northern Territory). (2019). *Northern territory Aboriginal tourism strategy 2020–2030 summary* [pdf]. Retrieved from www.tourismnt.com.au/system/files/uploads/files/2020/aboriginal-tourism-strategy-summary.pdf

Tourism Research Australia (TRA). (2011). *Strategic regional research, Northern territory – Dispersing beyond Darwin*. Retrieved from www.tourismnt.com.au/~/media/files/corporate/research/dvs-beyond-darwin-full-report_northern-territory_australia.ashx

Tourism Research Australia (TRA). (2019). National and international visitor survey data. *Tourism research Australia online student database*. Retrieved from http://traonline.australia.com

Tremblay, P., & Wegner, A. (2009). *Indigenous/Aboriginal tourism research in Australia (2000–2008): Industry lessons and future research needs*. Gold Coast: Cooperative Research Centre for Sustainable Tourism [pdf]. Retrieved from www.crctourism.com.au/wms/upload/Resources/110018%20Tremblay%20IndigenousAboriginalTRA%20WEB.pdf

Weaver, D. B. (1998). Peripheries of the periphery: Tourism in Tobago and Barbuda. *Annals of Tourism Research*, *25*(2), 292–313.

8 Degrowth as a strategy for adjusting to the adverse impacts of climate change in a nature-based destination

Bruce Prideaux and Anja Pabel

Introduction

The objective of this chapter is to examine opportunities for nature-based destinations to consider degrowth strategies as a response to changes in tourism demand in a context where key ecosystems are affected by climate change. The chapter recognizes that while destinations may have limited capacity to respond to climate change where national policy frameworks dictate overall responses, there is scope for developing and implementing local solutions at a local scale. A tourism adaptation framework that is able to assist tourism destinations with responding to changing tourism demand as key ecosystems decline is proposed later in this chapter.

Concern about the danger of excessive consumption was raised in a Club of Rome publication titled *The Limits to Growth* (Meadows, Meadows, Randers, & Behrens, 1972), which identified issues such as depletion of non-renewable resources, accelerating industrialization and a deteriorating environment. More recently, a series of conferences on degrowth have generated considerable debate leading to the 2008 Paris Declaration (Research & Degrowth, 2010) and the 2010 Barcelona Declaration (Barcelona Declaration, 2010), which outlined the central ideas of degrowth. A key concern of the degrowth movement is that rising consumption will eventually exhaust the planet's finite resources, leading to socio-economic collapse. To prevent socio-economic collapse, drastic degrowth calls for changes in consumption and production (Weiss & Cattaneo, 2017). However, political and industrial leaders and the majority of citizens, irrespective of their ideological persuasion, have mostly chosen to ignore this possibility. Arguably, the failure to gain widespread support for degrowth at political, industry and citizen levels is a reflection of the collective desire of individuals for greater levels of material comfort and their inability to see how this can lead to an eventual global ecosystem collapse, creating a wicked dilemma for policy-makers.

Until recently, the tourism literature has largely ignored the degrowth debate, although a new debate linking degrowth with overtourism (Fletcher, Mas, Blanco-Romero, & Blázquez-Salom, 2019; Milano, Novelli, & Cheer, 2019) has emerged to join earlier work on degrowth by scholars including Andriotis (2014), Hall (2009, 2013), and Scott, Gössling, and Hall (2012). The emerging debate on overtourism and concerns about the carrying capacity of some urban destinations have a strong parallel with earlier debates about tourism carrying capacity in sensitive environmental areas. It is interesting that a debate similar to overtourism concerns about urban areas has yet to re-emerge in environmental settings despite many of these areas suffering a declining ability to cope with visitor numbers that, in the past, were seen as sustainable. This may soon change, however, as glimpses of a world where growth had slowed and the environment had begun to recover emerged out of the COVID-19 pandemic (see Chapter 14, this volume).

Until there is greater citizen-level support for stronger policy measures to combat climate change, annual gross domestic product (GDP) growth based on neoliberal economic production systems will remain an important national objective and a key performance indicator (KPI) for national governments. At a destination scale, the continuing emphasis on year-on-year growth by national policy-makers and the commercial sector hinders the implementation of economic philosophies such as degrowth and the circular economy. In the absence of national policy platforms that support degrowth calls for less consumption, adoption of degrowth-aligned policies are most likely to occur where continued economic growth in its present form is no longer achievable. One example is destinations suffering declining visitor numbers through climate change–related problems. For destinations experiencing problems of this nature, degrowth strategies of the type advocated by Jackson (2009) and Andriotis (2014) offer opportunities to rebalance destination economies, including reappraising how the environment is used as a tourism experience; development of carbon-neutral industries as a substitute for tourism; introducing carbon-neutral tourism experiences and transport systems; reducing tourism numbers; large-scale recycling and reuse of resources; or adopting elements of each strategy. Irrespective of the path adopted, the measurement of success will need to change with metrics such as year-on-year growth replaced by alternatives yet to be determined.

Literature

Demaria, Schneider, Sekulova, and Martinez-Alier (2013) describe degrowth as a critique of the current neoliberal economic development hegemony. From a different perspective, Fletcher et al. (2019, p. 1745) describe degrowth as "both social theory and social movement . . . that has yet to engage systematically with the tourism industry". Fletcher et al. (2019) also state that economic growth is an obstacle to sustainability. From these perspectives, degrowth challenges the validity of sustainable growth, even of the type suggested in the United Nations Sustainable Development Goals (SDGs), as a valid policy objective. This challenge has more recently led to a reinterpretation of overtourism as a critique of the capitalist system through the lens of degrowth theory, although the debate has generally considered destination-scale problems rather than those that exist at national and international scales. Hall (2009) identified degrowth as a strategy for overcoming some of the impacts of climate change, although this connection between degrowth and climate change is not yet evident in more recent work linking overtourism with degrowth.

While the degrowth literature continues to evolve, the general characteristics were outlined in the Declaration of the 2008 Paris Conference, which describes degrowth in general terms as "a transformation of the global economic system and of its policies promoted and pursued at the national level, to allow the reduction and eradication of absolute poverty to proceed as the global economy and unsustainable economies degrow" (Research & Degrowth, 2010, p. 523). This is to be achieved by an emphasis on: quality of life rather than quantity of consumption; structural transition to service-based activities by downscaling production and consumption; investment in ecological assets; and changes in worktime as a mechanism to achieve economic stabilization (Jackson, 2009). One aim of the degrowth process is to increase human well-being through non-consumptive means, including reduced consumption and more leisure time. This could be ultimately beneficial to tourism but not in its current mass consumptivist form.

Andriotis (2014, p. 38) described degrowth as "an alternative approach to mainstream paradigms of development, (that) aims to ensure a high quality of life for people in a society where work, production and consumption are reduced". To achieve this vision of degrowth from a tourism perspective, Andriotis (2014) argued that behavioural changes are required to encourage a reduction in working hours, adoption of environmentally friendly transportation, rejection of Western-style commodified travel that results in negative environmental impacts, adoption of less organized and industrialized travel, an, adoption of labour intensification to increase the benefits for the local population. The mechanics of achieving degrowth outcomes were not discussed.

To date, the literature has failed to suggest economic or political models that demonstrate how degrowth is to be achieved (Weiss & Cattaneo, 2017). Given that alternative economic and political systems including communism, socialism, dictatorships of various kinds, monarchies and fascism have almost always adopted growth as a guiding KPI, the introduction of a citizen-supported economic/governance model able to reduce use of resources and redistribute wealth will be difficult unless external pressures such as global warming are so great that change becomes inevitable. From a destination perspective, the failure of the degrowth movement to suggest workable economic models poses the greatest challenge to its adoption.

The literature does provide some examples of where national economies have been forced to adopt degrowth-like policies to respond to extraordinary economic circumstances, as occurred in Cuba after the collapse of the USSR (Borowy, 2013). Until 1989, the USSR supplied 85% of Cuba's imports and a large percentage of its oil. The collapse of the trading agreements with the USSR resulted in Cuba losing 70% of its import capacity. As Borowy (2013, p. 18) observed:

> [Cuba] was forced to live according to degrowth rules: produce and consume locally, refrain from credits, change from energy intensive mechanized to low-energy, labour-intensive production methods, replace long-distance with face-to-face communications and live a simple, low consumption life-style.

Cuba was forced to adopt degrowth-like strategies to survive. With the possible exception of North Korea, which has struggled to cope with sanctions imposed by the United Nations (in response to the country's development of nuclear weapons), the absence of an existential threat of the type encountered by Cuba and North Korea poses problems for governments trying to implement voluntary degrowth-like strategies. Apart from the Cuban experience, there are few guides as to how degrowth strategies may be implemented.

The question yet to be addressed by the degrowth literature is how to convince the general public of the need to reduce their consumption of touristic experiences, particularly those that involve long-haul flights. Self-interest expressed through the ballot box in many countries, including the USA (2016 presidential campaign) and more recently Australia (2019 federal elections), continues to be a major barrier to implementing climate change mitigation strategies let alone reducing demand for products such as travel and other consumer products (Department of Climate Change and Energy Efficiency, 2011). Ultimately, the response of the public is likely to be determined by a mix of government policies and changes in tourism consumption patterns as individuals make the connection between their individual consumption decisions and the flow-on effect on the environment. Unfortunately, this connection may require dramatic ecosystem-related disasters such as further global-scale coral bleaching events, forest fires on the scale experienced by Australia during the summer of 2019–2020 (Evans, 2020) or pandemics such as COVID-19.

The need to change public perceptions

There continues to be a diverse range of public understanding about the dangers of carbon emissions and a continuing gap between individuals' understanding of the need for environmental action and their willingness to adopt strategies such as a low-carbon lifestyle (McKercher, Prideaux, & Pang, 2013). The concept of carbon capability may be one avenue for encouraging citizens to accept that changes are required to mitigate the impact of climate change. Whitemarsh, Seyfang, and O'Neill (2011, p. 56) described the concept of carbon capability of consumers as "individuals' abilities and motivations to reduce emissions".

Hall (2013) extended the discussion of consumers' carbon capability by outlining three approaches to achieving behavioural change in the contemporary socio-technical system: a) the neoliberal utilitarian approach based on the belief that consumers are rational utility-maximizers and government intervention is able to change behaviours (e.g. tax incentives, carbon trading); b) social/psychological approaches that utilize strategies such as nudging and social marketing to achieve desirable outcomes; and c) approaches that focus on the structural perspectives of systems of consumption and provision (e.g. alternative food networks, localism, short-supply chains and anti-consumption strategies such as voluntary simplicity). Hall (2013) further suggests that to achieve changes in policy, a three-step process of policy learning is required, where the first step is to focus on individual decision-making, followed by strategies to effect change in individual behaviours/

practices and the final outcome is a policy paradigm shift to achieve a new goal hierarchy.

Demarketing as a strategy

Demarketing is one approach that fits Hall's (2013) social/psychological approach to behavioural change (Hall, 2014). Demarketing is described as a strategy for systematically restricting public use of tourism assets (Inglis, Whitelaw, & Pearlman, 2005), especially for undesirable target markets such as high-volume mass tourism that produces overtourism. The concept was first defined by Kotler and Levy (1971, p. 75) as "marketing that deals with discouraging customers in general or a certain class of customers in particular on a temporary or permanent basis". Demarketing can address issues such as crowding in national and marine parks and reduces conflict between visitor groups and activities (Wearing, Schweinsberg, & Tower, 2016). In a nature-based setting, demarketing includes limiting access to minimize visitor flows, manipulation of prices to discourage particular segments, providing information to explain reasons for limited or restricted access (e.g. advertising that alerts customers of capacity limitations), minimizing product options to cater for a limited range of target markets, and utilizing technology to provide environmentally acceptable encounters with wildlife (e.g., virtual tours) (Wearing et al., 2016).

A number of nature-based tourism settings have successfully applied demarketing to reduce impacts and pressures on natural resources. Parks Canada has ceased to implement product-based marketing aimed at increasing visitors, focusing instead on social marketing and demarketing approaches aimed at attracting visitor markets interested in ecological integrity (Parks Canada Agency, 2000). In the Philippines, the resort island of Boracay was temporarily closed for six months in 2018 to deal with environmental issues caused by overtourism (Aquino, 2019). In Thailand, Maya Bay on Phi Phi Leh, which was made famous by the 2000 film *The Beach* starring Leonardo DiCaprio, was closed in 2018 as part of a rejuvenation program. Current plans are for the beach to reopen in 2021 with restricted visitor numbers and a bay-wide ban on boat moorings (Wipatayotin, 2019).

In each case, the decision to restrict access to the natural resource was influenced by the specific context of tourism pressures on the natural resource. Carrying capacity frameworks such as the limits of acceptable change can be used to determine the number and type of visitor segment as well as activity types that can be accommodated without incurring unsustainable impacts (Flemming & Manning, 2015). Diversion marketing (Medway, Warnaby, & Dharni, 2010) is another strategy that can be employed for demarketing at-risk places and involves redirecting visitors to alternative and less or not threatened places.

As this chapter has noted, the literature has yet to generate specific strategies to achieve the objectives of degrowth. An open letter signed by 11,258 scientists from 153 countries suggested a range of required actions (Ripple, Wolf, Newsome, Barnard, & Moomaw, 2020) that parallel many of the calls made in the degrowth literature. In their call, six specific steps were outlined to halt the problems caused by climate change. The six steps were:

1 Energy: replace fossil fuels with renewables.
2 Short-lived pollutants: swiftly cut emission of methane, hydrofluorocarbons, soot and other short-lived climate pollutants.
3 Nature: end massive land clearing and restore ecosystems such as forests, grassland and mangroves, which will assist in sequestration of atmospheric CO_2.
4 Food: reduce food waste (one-third of food ends up as garbage) and encourage a shift from animal protein to plant protein and more efficient farming methods such as minimal tillage and increase in soil carbon.
5 Economy: shift away from year-on-year growth and the pursuit of individual wealth, curtail the extraction of materials and exploitation of ecosystems to a carbon-free economy.
6 Population: stabilize global population in a way that is socially and economically fair.

While providing a clear framework of required policy actions, the six steps do not specify a mechanism for achieving these goals. This mechanism could be provided by strategies associated with the circular economy model described by Geissdoerfer, Savaget, Bocken, and Hultink (2017, p. 795) as a model of economic production that is a "regenerative system in which resource input and waste, emission, and energy leakage are minimised by slowing, closing, and narrowing material and energy loops. This can be achieved through long-lasting design, maintenance, repair, reuse, remanufacturing, refurbishing, and recycling". Adoption of the circular economy model, or similar, may provide the mechanism required to achieve the economic-related aims outlined by the degrowth movement.

The adoption of actions such as those advocated by Ripple et al. (2020) and by models such as the circular economy will introduce the disruptive system changes described by Milkoreit et al. (2018) as social-ecological tipping points, where a small change in the social-ecological system triggers a non-linear change, creating self-reinforcing positive-feedback mechanisms. One example of a social-ecological tipping point is the introduction of a carbon tax, which encourages emitters to look for non-carbon alternatives.

Cairns case study

We selected Cairns, Australia, as an example of a coastal nature-based destination facing the threat of significant decline in its key destination pull factors, the Great Barrier Reef (GBR) and the Wet Tropics Rainforests (WTR). Although both the Great Barrier Reef Marine Park Authority (GBRMPA) (2017, 2019) and the Wet Tropics Management Authority (WTMA) (2016) recognize the significant dangers posed by climate change, and both authorities have been vocal in their concerns about the threat (GBRMPA, 2017), many members of the destination's tourism industry (Cummins, Pabel, & Prideaux, 2019) have yet to acknowledge the significance of the problem. At the community level, there appears to be a relatively low level of concern, as demonstrated by two representative letters to the editor published in the *Cairns Post* (local newspaper) 28 September 2019:

> If you say you have "climate anxiety" and we are heading towards a "climate catastrophe" you need to relax a bit. Take a stroll on one of our beautiful beaches, spend a few days in the Daintree, go to the Tablelands, so you can unwind yourself from all the fearmongering that is being spread. Life isn't as bad as you think. Hop off your bandwagon of doom and gloom, enjoy the life we have around us.
>
> I see the UN's greenies have got their UN talking points to defend the UN's child climatologist. They're all going to realise how silly they were as the seas continue not rising and the air continues not to warm dangerously.

Both contributors seem unaware of the magnitude of the well-documented threats faced by the destination in future decades.

The failure of the destination to come to terms with the problems posed by climate change is not surprising given that the Australian federal government has for some time failed to provide leadership in this area. When the Liberal-National coalition parties won office in 2011, the then prime minister, Tony Abbott (a self-proclaimed climate denier), abolished the carbon trading scheme introduced by the previous government (White, 2013). The prime minister, at the time of this writing, Scott Morrison, once displayed a piece of coal in the federal parliament (in 2017) and declared that coal power stations were a logical way forward to meeting the nation's energy needs (Murphy, 2017).

The problems that climate change will cause for Cairns, and other destinations located in the tropics, are outlined in an IPCC (2018) report that assessed the changes likely to occur to global ecosystems when global temperatures rise by 1.5°C and above. Coral reefs, for example, are likely to suffer 70–90% mortality of coral species at 1.5°C. The implications for coral reef–dependent destinations are obvious. The GBR experienced three major coral bleaching events (2016, 2018 and 2020), with each event leading to a reduction in ecosystem health and insufficient time to recover before the next bleaching event. The IPCC (2018) also stated that without urgent implementation of global-scale mitigation strategies, global temperatures could rise by 1.5°C well before 2050. Many nature-based destinations including Cairns will suffer significant downgrading, or loss, of key natural resources as global temperatures rise, generating significant problems for their tourism sectors.

Cairns is one of the largest non-metropolitan tourism destinations in Australian and attracted 2 million domestic and 860,000 international visitors in 2017–2018 (Cummings Economics, 2019). The destination, which is part of a larger tourism region known as Tropical North Queensland (TNQ), has enjoyed a relatively stable 2% share of Australia's domestic visitor market between 2007–2008 and 2017–2018. In the international inbound market, the destination's market share has fallen from 15.9% in 2007–2008 to 10.3% in 2017–2018 (Cummings Economics, 2019). In commenting about the fall in market share in the inbound sector, Prideaux, Carmody, and Pabel (2018) observed that the destination appeared to have become locked into a narrative that focuses almost exclusively on the region's key environmental assets. They further noted that the region has paid little attention to alternative tourism development pathways despite the long-running decline in international market share. The following discussion highlights internal and external threats faced by the destination followed by an analysis of current destination planning strategies.

Threats

The destination has been unable, or unwilling, to acknowledge that its key pull factors based on its natural assets are not as powerful as they were in the past, particularly in the international market. This is reflected in the significant level of decline in its national market share of inbound tourists. The consequence of this decline in market share has been the loss of three international airlines in recent years and decisions by two others to downgrade services to seasonal only (Australian Aviation, 2019). The serious threats posed by climate change have been largely ignored by the destination's tourism industry. The region's destination marketing organization (DMO), Tropical Tourism North Queensland (TTNQ), continues to invest considerable resources in promoting both the GBR and the WTR while less effort has been invested in developing alternative attractions such as city tourism and food tourism.

Concerns about the long-term health of the GBR were expressed in the five-yearly *Outlook Report* produced by the GBRMPA (2019). Based on the collective findings of more than 1,000 scientific reports, the 2019 *Outlook Report* observed that many coral reefs and associated habitats were in poor condition. The report downgraded the GBR from 'poor' in the 2014 *Outlook Report* to 'very poor' in 2019. The downgrading may lead to a review of the GBR's status as a World Heritage Area (WHA) when the UNESCO World Heritage Committee meets in 2020. Given that global greenhouse gas (GHG) emissions are continuing to increase, it is doubtful that global temperatures will plateau below 1.5°C, indicating an ongoing decline in the health, and therefore attractiveness, of the GBR.

The WTMA has also been concerned about the impact of climate change for a number of years. The *2015–16 State of Wet Tropics Report* (WTMA, 2016) highlighted the decline in rainforest flora and fauna due to climate change. In a recent statement (WTMA, 2019), the authority expressed alarm about new evidence of a possible near-term extinction of the lemuroid possum, a signature Wet Tropics species, as early as 2022. Given the significance of the WTR as a destination pull factor, any adverse impacts from climate change are concerning.

Implications

In 2018, Tourism Tropical North Queensland (TTNQ) (2018) published a document titled *Tropical North Queensland Destination Tourism Plan* (TNQDTP), which reflects the views of the tourism industry in areas such as key attractions, future plans and the contemporary tourism market. The document states that the destination's hero attractions are the GBR, the WTR and Indigenous cultural experiences. The validity of these claims of the destination's so-called hero experiences (GBR and WTR) in the domestic market must be questioned given research that found that most domestic tourists rate 'to have fun' and 'rest and relax' as the most important motives for visiting the Cairns region (Prideaux et al., 2018). Conversely, in the international market, the GBR was identified as the primary travel motive. As a destination pull factor for the domestic market, the GBR declined from third position in the January to March period 2016 to twelfth position in the July to September period after the 2016 coral bleaching event and fell to ninth position after the 2017 coral bleaching event (Prideaux et al., 2018). Two years after the bleaching event, the GBR ranked as the eighth most important motive for visiting Cairns in the domestic market (Prideaux, Thompson, Pabel, & Cassidy, 2020). In the international market, the GBR remained the most important motive throughout the period 2015 to 2019, with 'to have fun' ranking second (Prideaux et al., 2020).

The TNQDTP failed to note the decade-long decline in its share of the international market. In terms of path dependence (Martin & Sunley, 2006) and locked-in theory (Hassink, 2005), the TNQDTP indicates a failure to appreciate that Cairns is continuing to follow the pathway of the past without recognizing the reality of the present. Importantly, the document failed to acknowledge the long-term threat posed to the GBR through future coral bleaching and concerns about the impending loss of keynote species in the WTR, despite the management agencies of both these WHAs raising frequent concerns about the long-term threat of climate change. The overall impression is of a DMO that has been unwilling to acknowledge the gravity of the existential threats the destination faces and unable to accept that its so-called hero experiences have become less attractive to both the domestic and international markets over time. From this perspective, the destination shows no inclination to adopt degrowth tourism policies. Ultimately, the decision may be taken away from the destination by government policy or shifts in tourists' expectations of the type of holiday experience they will prefer, or are allowed to enjoy, in a future climate-damaged world.

Discussion

The current dilemma faced by the destination, and others like it, stems from finding itself locked into a development path that has been highly successful in the past, continues to enjoy strong support in its domestic market but is losing market share in its international market. The danger of being 'locked-in' (Marèchal, 2007, 2010) to the current belief that the destination must continue to grow impedes consideration of alternative views that growth may not be a desirable or achievable objective. From this perspective, many stakeholders, including the authors of letters to the editor cited earlier, may be either unable or unwilling to accept a policy reversal that will obligate them to accept that climate change is a problem that must be confronted sooner rather than later and will require all stakeholders to accept the need for change. This situation points to policy failure (Hall, 2011), where the destination is unable or unwilling to admit that policies to grow the international market have failed (evidenced by the destination continuing to attempt to attract new international airlines despite the withdrawal of airlines cited previously), a lack of planning to develop alternative experiences to the GBR and failure to recognize that the GBR is not as important as it once was in attracting domestic visitors. In short, the destination has failed to integrate policy learning into its planning processes and has yet to consider the impact that a continued decline and/or potential loss of the GBR will have on the international market and in turn on the local economy. At some point in the future, the destination will be forced to respond to these threats and develop an alternative path for its tourism industry. Timing is important as it becomes more difficult to achieve a successful response as the magnitude of the problem increases.

Other destinations facing the loss of key ecosystems will also experience similar problems. In Australia, for example, the loss or serious degrading of many forest ecosystems as a result of the catastrophic bush fires during the spring of 2019 and the summer of 2019–2020 will force affected destinations to acknowledge policy failures in respect to ecosystem protection and to look for new tourism pathways that may include a reduction in visitation numbers.

A role for degrowth strategies

The failure to develop specific economic models to achieve degrowth is a major hurdle for destinations wishing to adopt degrowth as tourism management strategy. In some cases, destinations will be forced to accept lower visitor numbers if they are unable to offer new visitor experiences and by default enter a degrowth state. In other destinations, rising sea levels will force the abandonment of significant coastal-located infrastructure, Cairns International Airport (Coastal Risk Australia, n.d.) being only one of numerous examples. Where destinations can offer new experiences drawing on latent comparative advantages, the need for degrowth strategies may be reduced or ignored. Even when a destination's political and business elite (defined as a destination's political representatives, major investors, business associations and large destination companies) recognize the need for change, the extent to which the destination is able to respond will be contingent on the rate at which local, national and international elites decide that change is required and are able to implement required changes. Associated with a reticence to promote degrowth may be a fear of telegraphing the inevitability (Scott et al., 2012) of some form of decline and potential for loss of competitiveness.

Drawing on Whitemarsh et al.'s (2011) concept of carbon capability and Hall's (2013) strategies for achieving behavioural change in a contemporary socio-technical system, a framework capable of supporting degrowth-aligned strategies was developed and is outlined in Table 8.1. The framework recognizes that the level of tourism and community use of key natural resources will need to be adjusted downwards as key ecosystems decline and in doing so reduce the strength of the destination's pull factors. Furthermore, as a destination's current competitive advantage declines, it will need to either develop new experiences based on existing underutilized resources or accept a new equilibrium point characterized by reduced visitor numbers.

Table 8.1 outlines a suggested ten-step framework that can assist destinations with responding to ecosystem decline. Before new policies can be implemented, the destination's business and political elite must accept the need for change (step 1). Until this occurs, change is unlikely to happen unless imposed by state and/or national governments. As history has illustrated many times, at some point the political and business elite will realize that change is necessary and that the cost of inaction will be higher than the cost of action. A turning point of this type may be the result of a change in local government, a significant business collapse, ecosystem loss or degrading, a change in business leadership or new policy initiatives at state and national levels. Data collection of the type suggested in steps 2 to 5 can proceed irrespective of the views of the elite.

Step 2 refers to scientific research that investigates impacts on local ecosystems and social science research using measures such as the Tourism Climate Index (TCI; an index that measures visitor comfort based on temperature and humidity) to assess impacts on human comfort. Not all destinations have access to the depth

Table 8.1 Strategy development and data collection framework

Step	*Action required*	*Method*
1	Assess the attitude of the destination's power elite (political and business) to respond to climate change	Personal interviews, media scans
2	Assess the level of impact on the destination ecosystem at 1.5°C and 2°C levels of warming to assess impact on ecosystems and visitor to estimate impact on future carrying capacity	Local ecosystem research,Tourism Climate Index
3	Identify how markets (domestic and international) may be affected if key ecosystems undergo decline at a 1.5°C and above temperature increases	Visitor surveys to determine tourism push factors. Ideally, out-of-destination research should be undertaken to identify potential market impacts.
4	Investigate the resilience of the community and business sector to the impacts of a 1.5°C or above temperature increase	Community survey, business sector survey, focus groups, input-output analysis to determine loss of jobs, etc.
5	What level of competitive advantage will be lost at a 1.5°C temperature increase?	Industry consultation
6	Identify existing underutilized resources that may be mobilized to develop new experiences	Focus groups, business consultation, scenarios
7	Consult widely with the community and industry to ascertain views on the future of the tourism sector	Community survey, industry survey, focus groups
8	Develop strategy options that may arise from step 7	Scenarios, conversations with all stakeholders
9	Implement agreed-upon strategies	All destination stakeholders
10	Ongoing evaluation of impacts and responses	Begin at step 2 and proceed to step 9

of research available on their key environmental assets that is enjoyed by Cairns and may find it difficult to determine the extent of the change that may occur. Without a detailed understanding of ecosystem changes, it may be difficult to assess how visitors will respond to ecosystem change (step 3). The findings outlined in step 2 will also assist in understanding issues related to community resilience to climate change impacts such as job losses (step 4) and impact on business (step 5). Employing climate scenarios such as those produced by the IPCC (IPCC, 2018) and indices such as the TCI can be used to construct destination-specific future scenarios. Based on the findings of steps 3 to 5, alternative experiences including reduced visitor numbers and reappraisal of ecosystem use can be evaluated (step 6). Step 7 provides the opportunity for the community, including the political and business elite, citizens and external stakeholders (investors, government and non-government agencies and tourism partners), to determine the future pathway options for tourism development. Futures scenarios can assist in the search for suitable strategies. At this point, serious consideration can be given to the adoption of degrowth-inspired strategies such as a reduction in tourism numbers. Once this has been determined, strategies can be developed to transition to new destination experiences (step 8). Agreed-upon strategies can then be implemented (step 9). Uncertainty about the extent of ongoing climate change and how tourists, the industry and the community may respond in the long-term require an ongoing process of re-evaluation of each step in the framework (step 10).

Destinations concerned about overtourism, or interested in adopting degrowth strategies, can use the framework to develop an understanding of the extent of problems they may face as global temperatures climb to 1.5°C and beyond, both in terms of impacts on their local ecosystems as well as problems likely to be encountered by future changes in weather patterns (changed propensity for floods, wind storms, fire and drought). Developing scenarios of possible futures (van Balen, Dooms, & Haezendonck, 2014) is one approach that can be used to gain an understanding of how visitors will respond to ecosystem change over time. Used in this manner, scenarios can provide destinations with insights that may be used to explore various alternative tourism pathways in both the short-term and the long-term. In relation to Cairns, future increases in sea levels predicted by the IPCC (2018) to be between 61 and 110 centimeters by 2100, or the more alarming estimation by Bamber, Oppenheimer, Kopp, and Cooke (2019, p. 11196) of 2 meters by 2100 are likely to have a catastrophic impact on the tourism economy. A sea level rise of 0.74 meters (Coastal Risk Australia, n.d.) will result in regular high tide inundation of the city's airport and downtown area.

In Cairns, options for responding to a future coral bleaching–generated decline in international tourism should commence with a more detailed understanding of the current visitor cohort (step 3), particularly repeat domestic visitors. Ultimately, the resources identified in step 6 and the results of community consultations in step 7 will determine how the destination fashions a response to climate change. However, future unknowns such as government policy and shifts in consumer demand in a climate-damaged world may override destinations' views and impose a solution that may look very different from that developed at the destination level.

Conclusion

The aim of this chapter was to explore opportunities for applying degrowth strategies in circumstances where nature-based destinations are facing long-term decline of their key ecosystem attractions due to global warming. The suggested ten-step framework provides at-risk destinations with a pathway for collecting data and insights, and a plan for responding to climate change threats using degrowth strategies. The framework may also be used as an adaptation tool.

While degrowth principles have been adopted by groups concerned about overtourism, consideration of how to employ degrowth strategies of the type suggested by Jackson (2009) and Andriotis (2014) on a broad scale has yet to occur. This may be the result of the failure of degrowth advocates to adopt specific economic models, the hesitancy of elites to accept that change is necessary and the breaching of ill-defined and poorly understood social-ecological tipping points.

Although the framework outlined in Table 8.1 will assist some destinations to implement degrowth-type strategies, the question of how the global tourism industry will be affected by a future transition to a carbon-neutral economic production system remains unanswered. Global tourism flows continue to increase year-on-year (unless interrupted by crisis events such as COVID-19), adding to global GHG emissions and making a mockery of the belief that tourism can grow in a sustainable manner. Given the need to transition to a carbon-neutral production system in the near future, it can be expected that there will be negative impacts on global tourism flows as governments adopt policies to reorientate national economies away from the neoliberal linear model, and the cost of mitigation and adaptation is reflected in increased government spending and associated taxation. The growing alignment of concerns about overtourism and the adoption of degrowth as a solution may provide useful strategies for city destinations experiencing overtourism, but the larger question of where these tourism flows will be redirected to remains.

References

Andriotis, K. (2014). Tourism development and the degrowth paradigm. *Turističko poslovanje*, *13*, 37–45.

Aquino, M. (2019). *Why Boracay is closed for tourism: Popular beach island's closure was a long time coming*. Retrieved from www.tripsavvy.com/boracay-philippines-island-closed-for-tourism-4165709

Australian Aviation. (2019, April 29). Cathay pacific ending flights to Cairns. *Australian Aviation*. Retrieved from https://australianaviation.com.au/2019/04/cathay-pacific-ending-flights-to-cairns/

Bamber, J., Oppenheimer, M., Kopp, W., & Cooke, R. (2019). Ice sheet contributions to future sea-level rise from structured expert judgement. *PNAS*, *116*(23), 11195–11200.

Barcelona Declaration. (2010). [pdf]. Retrieved from www.degrowth.info/wp-content/uploads/2015/05/Degrowth_Declaration_Barcelona_2010.pdf

Borowy, I. (2013). Degrowth and public health in Cuba: Lessons from the past. *Journal of Cleaner Production*, *38*, 17–26.

Coastal Risk Australia. (n.d.). *Cairns inundation map 2010, Corporative research Centre spatial information*. Retrieved from http://coastalrisk.com.au/#

Cummings Economics. (2019). *Trends in tourism in tropical North Queensland*. Cairns: Cummings Economics.

Cummins, T. L., Pabel, A., & Prideaux, B. (2019). Stakeholders' perspectives of factors impacting long – Term sustainability of the recreational scuba diving tourism sector in Cairns Australia. In A. Pabel, E. Konovalov, L. Cassidy, & P. Jose (Eds.), *CAUTHE2019: Sustainability of tourism, hospitality and events in a disruptive digital age*. Proceedings of the 29th CAUTHE Conference, Council for Australasian Tourism and Hospitality Education (CAUTHE), Central Queensland University, 11–14 February, Cairns.

Demaria, F., Schneider, F., Sekulova, F., & Martinez-Alier, J. (2013). What is degrowth? From an activist slogan to a social movement. *Environmental Values*, *22*(2), 191–215.

Department of Climate Change and Energy Efficiency. (2011, December 23). *Barriers to effective climate change adaptation a submission to the productivity commission submission department of climate change and energy efficiency*. Canberra: Department of Climate Change and Energy Efficiency. Retrieved from www.environment.gov.au/system/files/resources/c8e1fa00-3d14-4e3b-98a4-b5b338700042/files/barrierstoadaptation.pdf

Evans, D. (2020). Bushfires: Can ecosystems recover from such dramatic losses of biodiversity. *The Conversation*. Retrieved from www.resilience.org/stories/2020-01-21/bushfires-can-ecosystems-recover-from-such-dramatic-losses-of-biodiversity/

Flemming, C., & Manning, M. (2015). Rationing access to protected natural areas: An Australian case study. *Tourism Economics*, *21*(5), 995–1014.

Fletcher, R., Mas, I., Blanco-Romero, A., & Blázquez-Salom, M. (2019). Tourism and degrowth: An emerging agenda for research and praxis. *Journal of Sustainable Tourism*, *27*(12), 1745–1763.

Geissdoerfer, M., Savaget, P., Bocken, N., & Hultink, E. (2017). The circular economy – A new sustainability paradigm. *Journal of Cleaner Production*, *143*, 757–768.

Great Barrier Reef Marine Park Authority (GBRMPA). (2017). *Final report 2016 coral bleaching event on the Great Barrier Reef*. Townsville: GBRMPA.

Great Barrier Reef Marine Park Authority (GBRMPA). (2019). *Great Barrier Reef outlook report 2019*. Townsville: GBRMPA.

Hall, C. M. (2009). Degrowing tourism: Décroissance, sustainable consumption and steady-state tourism. *Anatolia*, *20*(1): 46–61.

Hall, C. M. (2011). Policy learning and policy failure in sustainable tourism governance: From first-and second-order to third-order change? *Journal of Sustainable Tourism*, *19*(4–5), 649–671.

Hall, C. M. (2013). Framing behavioural approaches to understanding and governing sustainable tourism consumption: Beyond neoliberalism, "nudging" and "green growth"? *Journal of Sustainable Tourism*, *21*(7), 1091–1109.

Hall, C. M. (2014). *Tourism and social marketing*. Abingdon: Routledge.

Hassink, R. (2005). How to unlock regional economies from path dependency? From learning region to learning cluster. *European Planning Studies*, *13*(4), 521–535.

Inglis, J., Whitelaw, P., & Pearlman, M. (2005). *Best practice in strategic park management: Towards an integrated park management model*. Technical Reports. Gold Coast: Cooperative Research Centre for Sustainable Tourism (CRC).

Intergovernmental Panel on Climate Change (IPCC). (2018). Special report: Global warming of 1.5ºC. Summary for policymakers. In V. Masson-Delmotte, P. Zhai, H.-O. Pörtner, D. Roberts, J. Skea, P. R. Shukla, . . . T. Waterfield (Eds.), *Global warming of 1.5°C. An IPCC special report on the impacts of global warming of 1.5°C above pre-industrial levels and related global greenhouse gas emission pathways, in the context of strengthening*

the global response to the threat of climate change, sustainable development, and efforts to eradicate poverty. Geneva: World Meteorological Organization.
Jackson, T. (2009). *Prosperity without growth. Economics for a finite planet*. London: Earthscan.
Kotler, P., & Levy, S. J. (1971). Demarketing, yes, demarketing. *Harvard Business Review*, *49*(6), 74–80.
Marèchal, K. (2007). The economics of climate change and the change of climate economics. *Energy Policy*, *35*(10), 5181–5194.
Maréchal, K. (2010). Not irrational but habitual: The importance of "behavioural lock-in" in energy consumption. *Ecological Economics*, *69*(5), 1104–1114.
Martin, R., & Sunley, P. (2006). Path dependence and regional economic evolution. *Journal of Economic Geography*, *6*(4), 395–437.
McKercher, B., Prideaux, B., & Pang, S. (2013). Attitudes of tourism students to the environment and climate change. *Asia Pacific Journal of Tourism Research*, *18*(1–2), 108–143.
Meadows, D. H., Meadows, D. L., Randers, J., & Behrens, W. (1972). *The limits to growth*. New York, NY: Universe Books.
Medway, D., Warnaby, G., & Dharni, S. (2010). Demarketing places: Rationales and strategies. *Journal of Marketing Management*, *27*(1–2), 124–142.
Milano, C., Novelli, M., & Cheer, J. (2019). Overtourism and degrowth: A social movements perspective. *Journal of Sustainable Tourism*, *27*(12), 1857–1875.
Milkoreit, M., Hodbod, J., Baggio, J., Benessaiah, K., Calderón-Contreras, R., Donges, J. F., . . . Werners, S. E. (2018). Defining tipping points for social-ecological systems scholarship – An interdisciplinary literature review. *Environmental Research Letters*, *13*(3), 033055.
Murphy, K. (2017, February 9). Scott Morrison brings coal to question time: What fresh idiocy is this? *The Guardian*. Retrieved from www.theguardian.com/australia-news/2017/feb/09/scott-morrison-brings-coal-to-question-time-what-fresh-idiocy-is-this
Parks Canada Agency. (2000). "Unimpaired for future generations"? Protecting ecological integrity with Canada's National Parks. In *Setting a new direction for Canada's National Parks. Report of the panel on the ecological integrity of Canada's National Parks, Vol. II*. Ottawa: Parks Canada [pdf]. Retrieved from http://parkscanadahistory.com/publications/R62-323-2000-2E.pdf
Prideaux, B., Carmody, J., & Pabel, A. (2018). *Impacts of the 2016 and 2017 mass coral bleaching events on the Great Barrier Reef tourism industry and tourism-dependent coastal communities of Queensland. Report to the reef and rainforest research Centre limited*. Cairns [pdf]. Retrieved from http://rrrc.org.au/wp-content/uploads/2018/06/RRRC-Impacts-2016-17-Coral-Bleaching-on-GBR-Digital.pdf
Prideaux, B., Thompson, M., Pabel, A., & Cassidy, L. (2020). *Visitor concerns about the state of the great barrier reef*. Paper presented at the 30th annual CAUTHE Conference, 10–13 February, Auckland.
Research & Degrowth. (2010). Degrowth declaration of the Paris 2008 conference. *Journal of Cleaner Production*, *18*(6), 523–524.
Ripple, W., Wolf, C., Newsome, T., Barnard, P., & Moomaw, W. (2020). World scientists' warning of a climate emergency. *BioScience*, *70*(1), 8–12.
Scott, D., Gössling, S., & Hall, C. M. (2012). International tourism and climate change. *WIRES Climate Change*, *3*(3), 213–232.
Tourism Tropical North Queensland (TTNQ). (2018). *Tropical North Queensland destination tourism plan, October 2018*. Retrieved from https://cdn2-teq.queensland.com/~/media/0a155e2bd3124b52b11e930c27193765.ashx?vs=1&d=20181102T094647

van Balen, M., Dooms, M., & Haezendonck, E. (2014). River tourism development: The case of the port of Brussels. *Research in Transportation Business & Management*, *13*, 71–79.

Wearing, S. L., Schweinsberg, S., & Tower, J. (2016). *Marketing national parks for sustainable tourism*. Bristol: Channel View Publications.

Weiss, M., & Cattaneo, C. (2017). Degrowth – Taking stock and reviewing an emerging academic paradigm. *Ecological Economics*, *137*, 220–230.

Wet Tropics Management Authority (WTMA). (2016). *State of wet tropics report 2015–2016: Ancient, endemic, rare and threatened vertebrates of the wet tropics* [pdf]. Retrieved from www.wettropics.gov.au/site/user-assets/docs/sowt2015-16b5-lres.pdf

Wet Tropics Management Authority (WTMA). (2019). *A statement from the board of the wet tropics management authority regarding serious climate change impacts on the wet tropics of Queensland World Heritage Area* [pdf]. Retrieved from www.wettropics.gov.au/site/user-assets/docs/2019.04.29%20WTMA%20board%20climate%20change%20statement.pdf

White, A. (2013, September 18). Why Tony Abbott wants to abolish the carbon tax. *The Guardian*. Retrieved from www.theguardian.com/environment/southern-crossroads/2013/sep/18/tony-abbott-abolish-carbon-price

Whitemarsh, L., Seyfang, G., & O'Neill, S. (2011). Public engagement with carbon and climate change: To what extent is the public 'carbon capable'? *Global Environmental Change*, *21*(1), 56–65.

Wipatayotin, A. (2019, May 9). Maya Bay to remain closed until mid-2021. *Bangkok Post*. Retrieved from www.bangkokpost.com/thailand/general/1674364/maya-bay-to-remain-closed-till-mid-2021

Part 3

Degrowth and tourism policy

9 Sustainable growth in tourism?

Rethinking and resetting sustainable tourism for development

Jarkko Saarinen

Introduction: system reset?

> A factory reset, also known as master reset, is a software restore of an electronic device to its original system state by erasing all of the information stored on the device in an attempt to restore the device to its original manufacturer settings. Doing so will effectively erase all of the data, settings, and applications that were previously on the device. This is often done to fix an issue with a device, but it could also be done to restore the device to its original settings.
>
> (Wikipedia, 2020)

Recently, Martin Wolf (2019), the chief economics commentator of the *Financial Times*, discussed why the current mode of capitalism is damaging liberal democracy. As a way out, he called for a reset to the global economic system. According to him, the reason why the current economy is not delivering the expected growth and benefits lies with the rise of so-called rentier capitalism, in which the current market economy allows a few privileged individuals and businesses to extract a great deal of 'rent' (i.e. surplus above what is required to induce the desired supply of goods and services) from everybody else. This has led to weakened competition and growth and, eventually, an increase in inequality in the world. What he suggests is "a dynamic capitalist economy that gives everybody a justified belief that they can share in the benefits" (Wolf, 2019). Similarly, Dambisa Moyo (2018), a former economist at the World Bank and Goldman Sachs investment firm, has connected the challenges of liberal democracy to the current problems of global economic structures and development. She calls for a retooling of the economy by emphasising growth as imperative for improving lives and reducing poverty. In addition, Moyo (2018, p. 8) strongly argues that addressing "environmental and climate concerns" will be much harder "without growth".

Both Moyo (2018) and Wolf (2019), among many other economic commentators, emphasise the key role of growth for our future. According to them, this future growth should be more responsible and better governed than it is in the current neoliberal or rentier capitalism system. However, other voices are also calling for a more radical reset or retooling of our economic system, which refer to the limits to the growth discourse (Meadows, Meadows, Randers, & Behrens, 1972), and degrowth or post-growth in development thinking (Fournier, 2008; Hall, 2009; Koch, 2020). From these perspectives the question is not to call for less or a different kind of growth and related production and consumption: the argument, although highly debated, is that "economic degrowth is the only viable alternative goal to the growing economy" (Kerschner, 2010, p. 554). What constitutes common ground between the emphasis on a new kind of more responsible growth and degrowth/post-growth is the understanding of the political nature of the economy, calling for a more detailed and nuanced analysis on the politics and rhetoric of economic development. This includes tourism and encounters between global and local economies, core and periphery, hosts and guests, and the industry's global supply chains in general (Bianchi, 2018; Gössling, Scott, & Hall, 2020).

There is no full consensus about what growth and development precisely means in general or in a tourism planning and policy context (Gössling, Ring, Dwyer, Andersson, & Hall, 2016; Hall, 2000), but there is no disagreement about their importance and societal relevance (see Saarinen, Rogerson, & Hall, 2017). Growth has a complex multiscalar nature. For example, we can interpret that the major reason for the success of the United Nation's Millennium Development Goals project (UN MDGs) (United Nations, 2000) for reducing global poverty was based on economic growth in 1990–2015, largely in South-East Asia, and especially in China (Kiš, 2018). Indeed, in this period the number of people living in extreme poverty declined by more than half. However, while the economic growth in South-East Asia has had a global positive impact on poverty reduction, the rural areas outside of the flourishing (but increasingly polluted) cities in the region "continue to live in poverty to a larger degree" (Greve, 2020, p. 45). In addition, there are currently more people than ever before living in poverty in sub-Saharan Africa, and the number is rising (see Carmody, 2019; Fosu, 2015). Therefore, paradoxically, the global economic growth resulting in success in the UN MDGs in halving the number of people living in extreme poverty not only decreased but also increased inequality in the world. Indeed, as recently noted by Sharpley (2020), an integral irony of economic growth is that it often enhances inequality between those who have and those who have not (Wilkinson & Pickett, 2010).

All of this analysis contrasts with the simplistic growth imperative and related optimism thinking and, thus, questions whether growth should be so centralised in our equity/inequality, well-being and development thinking. After all, growth is only one aspect of development (Daly, 1996; Sen, 1992; see also Akanbi, 2014), which involves economic but also social and environmental elements. Traditionally, this wider context beyond economic issues alone has been approached by the idea and policies of sustainable development (Spangenberg, 2014). In tourism, the need for sustainability has been based on the growing impacts of global tourism from the 1960s onwards and has intensified calls for environmental protection and environmentally sound forms of production and consumption (Hall & Lew, 1998; Saarinen, 2006). Since the early 1990s, the issue of sustainability has been the policy discourse guiding the economic, social and political structures and processes that constitute the contemporary operative contexts of the tourism industry (Bramwell & Lane, 2011; Hall, 2011; Saarinen, 2014; Sharpley, 2000).

As a result, sustainability thinking is nowadays integrated into management and governance models in various planning scales that emphasise the positive role that tourism could play for destination environments, communities and development (Bramwell, 2011; Butler, 1999; Lu & Nepal, 2009). This prospective role of the industry is highlighted in many international tourism policy documents and declarations. In this respect, Hall (2011, p. 649) has stated that, from the perspective of use of the term by researchers, industry and decision-makers, *sustainability* is "one of the great success stories of tourism research". While it is easy to agree with this statement, there has been criticism of the idea of sustainable tourism or how it has been practiced (Hall, Gössling, & Scott, 2015; Liu, 2003; Sharpley, 2000). In particular, the connections and misconnections between tourism as a growth industry and the idea of sustainable development have been critically debated and challenged (Gössling et al., 2020; Saarinen & Gill, 2019). While this has created critical calls for moving beyond sustainability in tourism (Sharpley, 2009), it has also resulted in a search for alternative approaches, such as responsible tourism (see Blackstock, White, McCrum, Scott, & Hunter, 2008; Smith & Eadington, 1992), or modifying sustainability thinking to better suit the growth emphasis of the industry. The latter is connected to a timely and highly influential sustainable growth discourse at various policy-making levels (see United Nations Conference on Trade and Development (UNCTAD), 2012; United Nations Development Programme (UNDP), 2020; UNWTO, 2012), which is increasingly influencing global, supra-national and national tourism policy-making (Hall, 2000, 2019; Puhakka & Saarinen, 2013).

This chapter discusses the idea and transformation of sustainable development thinking in tourism and the need to rethink the role of growth in sustainable tourism development. First, the chapter starts by introducing the background to and key aspects of sustainable development. After that, sustainability thinking in tourism is discussed, followed by critical perspectives to the emerging idea of sustainable growth and its problematic nature in tourism development. Here, growth is seen as a marked driven approach supporting the industry, while sustainability represents (or should represent) an institutional system-level approach in setting the limits to growth for the industry. Finally, the need for sustainable development in tourism is discussed in relation to the increasing calls for degrowth and post-growth in the economy.

Sustainable development

Sustainable development refers to a policy process initiated by the UN in the 1980s (see Redclift, 2005; World Commission on Environment and Development (WCED), 1987), but as an overall idea it has a much longer history. In general, sustainable development thinking has a normative foundation that aims to define and guide our actions towards good and desired outcomes. At the same time, it tries to exclude practices and processes that are undesirable in relation to the use of natural and cultural resources and the limits to growth. In this respect, there is a balanced approach built into sustainability. According to Higgins (2015, p. 186), sustainability requires multiple types of "balance among economy, society, and environment; balance between present and future; balance in the ways we achieve happiness and well-being; and balance between our small selves and the welfare of society". As a balanced approach it has connections to a degrowth movement and a steady-state economy (Hall, 2009). The former refers to a situation in which the economic wealth produced does not increase, or it may even decrease. The latter indicates the same, i.e. that a local, regional or national economic system does not grow (Kerschner, 2010).

Both sustainability and development are complex terms. On a general level, development has been linked with various meanings. According to Peet and Hartwick (2015, p. 1), development can be defined as an evolutionary process in which human capacity increases towards "a better life". For Sharpley (2020), development has become associated with the notion of well-being, which can refer to a different scale, including societal, community or individual levels (see Sen, 1999). This process takes place in terms of initiating new structures, coping with and adapting to problems and change, and striving towards new goals. Reyes (2001) links development to sustainability by emphasising that development is a social condition in which the needs of people are satisfied by the rational and sustainable use of natural resources. This closer connection to sustainability also differentiates development from growth. According to the US Local Government Commission's practical definition, "growth means to get bigger, development means to get better – an increase in quality and diversity" (cited by Pike, Rodríguez-Pose, & Tomaney, 2007, p. 1253). Thus, compared to growth, the idea of development is more focused on human well-being and qualitative dimensions in social and economic processes.

Various theories aim to explain the idea and role of development (Carmody, 2019; Holden, 2013; Reyes, 2001). These include modernisation, dependency, world systems, globalisation and post-development theories, for example, which are beyond the scope of this chapter. In the relatively long list of development theories, however, one of the key alternative theories is sustainable development. There is no consensus on the historical background and evolution of the general idea of sustainable development (Mensah & Casadevall, 2019). With regard to a 'sustainable' use of various natural or cultural resources, one can probably find supporting verses from different religious sources or national epics, such as the *Kalevala* (Finland) or *Manas* (Kyrgyzstan), that resonate with the modern idea of sustainable development. More scientific or management-oriented antecedents of sustainable development, however, have been linked to the German Hans Carl von Carlowitz, who published *Sylvicultura Oeconomica* . . . in the early eighteenth century, which led him to be known as the father of sustainable forestry (see Werlen, 2015). Similarly, Hall (1998) and Crober (2015) have indicated that the early principles of sustainability thinking have their origins in the eighteenth-century idea of 'sustained yield' in forestry, which were further cultivated in the nineteenth-century conservation movement (Enders & Remig, 2015a). Also, Gifford Pinchot's 'wise use' thinking for forestry and natural resource management in the United States has been linked to the historical evolution of sustainable development (see Butler, 1998). In addition, some scholars have associated Malthusian population theory from the turn of the nineteenth century with the early phases of sustainability thinking (Eblen & Eblen, 1994; Kates et al., 2001). According to this theory, a rapid population growth rate needed to be managed, or otherwise a "depletion of natural resources would occur, resulting in misery for humans" (Mensah & Casadevall, 2019, p. 7).

However, in these early connections the wider socio-cultural and environmental elements (beyond economics) of the sustainable use of resources were not fully considered. For example, von Carlowitz's and Pinchot's 'sustainable use' of forest (or wood material) resources were largely economically driven and mainly designed to serve extractive industries and their needs (McCarthy, 2002). Still, in addition to the aforementioned, other individual authors (see Hall, 1998), such as George Perkins Marsh (1865) and Aldo Leopold (1949), also promoted what we could now call 'sustainability ethics'. These ethical concerns became more urgent and visibly presented in the 1960s and 1970s when Rachel Carson's (1962) *Silent Spring* and, especially, the Club of Rome's report *The Limits of Growth* (Meadows et al., 1972) were published. In relation to the latter seminal work, the authors stated:

> We are searching for a model output that represents a world system that is: 1. sustainable without sudden and uncontrolled collapse; and 2. capable of satisfying the basic material requirements of all of its people.
>
> (Meadows et al., 1972, p. 152)

According to Crober (2015), this was the first time that the term 'sustainable' was used in the meaning of 'sustaining' or 'maintaining' a socio-ecological system. In this sense, the idea of the limits to growth is embedded into our sustainability thinking and the sustainable use of resources (Enders & Remig, 2015a). For Meadows et al. (1972), this connection refers to the fact that the earth's resources are not infinite. Because of this, a growing use of resources would eventually reach the limits of system utilisation.

The idea of sustainable development was also implicitly recognised in 1972 at the UN Conference on the Human Environment, held in Stockholm (Daly, 1996), with its emphasis on the balanced approach between economic development and the environment. The term was more explicitly raised in the World Conservation Strategy in 1980, in which the third specific objective of the strategy stated that the aim was "to ensure the sustainable utilisation of species and ecosystems . . . , which support millions of rural communities as well as major industries" (International Union for Conservation of Nature (IUCN), 1980, p. 7). Again, the balanced relations and interdependency among ecological, social and economic environments were clearly acknowledged.

Despite a long and rich historical background, the conceptual formulation of sustainable development is a relatively recent process (Saarinen & Gill, 2019). In 1982 the IUCN and the United Nations Environment Programme (UNEP) established a World Commission on Environment and Development (WCED), which formed a targeted commission chaired by Gro Harlem Bruntdland in 1983. This commission published the highly influential report *Our Common Future* in 1987, which outlined and defined the notion of sustainable development in the policy and academic uses as we know it: "Sustainable development is development that meets the needs of the present generations without compromising the ability of future generations to meet their own needs" (WCED, 1987, p. 43). This definition has been widely used but also criticised.

Obviously, the formulation was a result of various political compromises (see Enders & Remig, 2015b). The core idea of sustainable development is also a deeply ideological one, with an anthropocentric 'human needs' perspective to development (Saarinen & Gill, 2019). According to the Bruntdland Commission (WCED, 1987), there are three core principles to sustainable development: holism and intergenerational and intragenerational equity, which are derived from the basic definition of sustainable development that refers to the needs of the present generations without compromising the needs of future generations. In addition to these principles, there are widely cited basic elements of ecological, social and economic sustainability in development. Again, these principles and elements should be regarded as balanced, i.e. equal, in the development process. This means that sustainable development that prioritises economic growth, for example, and neglects ecological and/or social limits is seen as both unrealistic and unsustainable in the long term.

Sustainable tourism

Tourism is one of the major economic sectors that has aimed to integrate sustainable development into its management processes and practices. Indeed, since the early 1990s, sustainable development has been a major focus for tourism policy-makers, planners and tourism researchers (Clarke, 1997; Hall, 2011; Saarinen et al., 2017). The World Tourism Organization (UNWTO), which is the key international tourism policy-maker, considers itself as being "responsible for the promotion of responsible, sustainable and universally accessible tourism" (UNWTO, 2016). The UNWTO (1993, p. 7) also made an early conceptual formulation, stating that sustainable tourism aims to meet "the needs of current tourists and host regions while protecting and enhancing opportunities for the future", which imitates the Brundtland Commission's definition of sustainable development with a tourism-specific accent. Since then, however, the concept of sustainable tourism has been defined in various ways (see Clarke, 1997; McCool & Bosak, 2016; Saarinen, 2006; Sharpley, 2000). While early definitions were less economy- (i.e. tourism-)oriented, the current hegemonic thinking in sustainable tourism emphasises the key role of the industry (Saarinen, 2014; see also United Nations Environment Programme (UNEP), 2005; UNWTO, 2017). Swarbrooke (1999, p. 13), for example, has defined sustainable tourism as "tourism which is economically viable but does not destroy the resources on which the future of tourism will depend, notably the physical environment and the social fabric of the host community". While it is a solid definition with references to ecological and social aspects of sustainability, definitions like this centralise and emphasise the role of the industry and its (future) resource needs (Burns, 1999; Hardy, Beeton, & Pearson, 2002), not the needs of the hosts or environment.

Industry-driven sustainability can be highly problematic if the needs of the industry, hosts and/or environmental protection are in conflict with each other (Hunter, 1995). This kind of situation can very likely be the case in tourism development, which is based on current market-driven neoliberal and growth-oriented approaches (Saarinen, 2019). Due to evolved tourism-centric approaches in sustainable tourism development, many tourism scholars have raised critical notions towards the whole idea of sustainability in tourism (Liu, 2003; Sharpley, 2000, 2009). As a result, some researchers have increasingly focused on other alternative tourism management frameworks such as responsibility in tourism (Caruana, Glozer, Crane, & McCabe, 2014). As early as the late 1980s, Krippendorf (1987) connected tourists' consumption patterns with environmental responsibility. For him, responsibility was characteristically individualistic (see also Sin, 2014). However, it was also tourist-centric, as the key argument for increasing responsibility was based on the emancipation of new holiday-makers (Saarinen & Gill, 2019).

Although the responsibility discourse has recently evolved strongly in tourism development and management policies and studies, e.g. in the form of corporate social responsibility (Coles, Fenclova, & Dinan, 2013; Kimaro & Saarinen, 2020), there is no clear or definite transformation from sustainability to responsibility in tourism. Both ideas are also increasingly characterised by an emphasis on market-driven development processes and individualisation supported by a growth ideology. Instead of referring to the limits to growth thinking, current tourism management policies have increasingly turned towards a relatively new idea in development discourse: a sustainable growth in tourism.

Sustainable growth: sustaining tourism growth

Several scholars have critically evaluated that sustainable development has become an oxymoron, i.e. a term that is made up of self-contradictory elements (Place, 1995; Spaiser, Ranganathan, Swain, & Sumpter, 2017). The problem is not that sustainability and development are conflicting goals per se. Obviously, this depends on how we understand the terms, but if we define development as a qualitative measure and process in which human capacity increases towards a better life (Peet & Hartwick, 2015; see also Pike, Rodríguez-Pose, & Tomaney, 2007), there is no major contradiction between the basic elements of sustainability and development. However, the problem has evolved based on how we understand and have redefined the combination and the overall aim of these two terms.

The combination of the terms, as was defined by the Brundtland Commission (WCED, 1987), is inherently anthropocentric, with a focus on human needs. From that perspective, nature conservation and biodiversity protection, for example, are ultimately selfish acts for us and societies aiming to save humanity and serve the needs of the current and future generations. Therefore, the use of the natural and social environment and the economy is to contribute to human development and well-being, both now and in the future (although we do not know the future, including future needs). This is what Pietarinen (1987) calls a humanism perspective to nature, in which the use of an environment should promote human development attaining ethical, aesthetic and mental equilibrium, and not only a source of raw materials for economic growth (Saarinen, 1998; see also Holden, 2003). In contrast to strict biocentric views involving an intrinsic value of the environment, humanism allows for the use of resources but in a sustainable way.

However, the original emphasis on needs in sustainable development has changed. In this respect, Redclift (2005, p. 212) has pointed out that the first Earth Summit in 1992 represented a fundamental transformation of sustainable development "to focus on rights, rather than needs, as the principal line of enquiry" (see also Blühdorn, 2017). For him, this new attention to rights was related to the neoliberal economic agenda of the late 1980s and especially the 1990s, emphasising the aspect of growth in different sectors and economic development. In tourism, this changing orientation resulted in the aforementioned tourism-centric views in sustainable tourism development (Burns, 1999; Saarinen, 2006; Schilcher, 2007), highlighting the role and needs of the businesses and markets in general (Saarinen & Gill, 2019). At the same time, requirements to reduce institutional regulations in natural (and cultural) resource use, management and governance have supported the emphasis on rights in sustainable development (Jessop, 2004; McCarthy, 2002; Redclift, 2005). Thus, the focus in sustainable tourism management has gradually moved onto market-oriented and corporate social responsibility directions (Visser, 2010; Luke, 2013; Saarinen, 2014; Scheyvens, 2011; Hall et al., 2015). Along this process, the balanced approach among environmental, social and economic elements of sustainability has turned towards a different kind of triple bottom line, advocating for a balanced focus on the planet, people and making a profit (for shareholders!) (Savitz, 2006).

However, this change of focus has not been the only outcome of the evolution of sustainability discourse. During the past decade, the term *sustainable development* has been increasingly replaced by or used synonymously with the notion of sustainable growth. While it would be appropriate to sideline such a term as an impossible statement and theorem, as the ultimate oxymoron 2.0, current sustainability thinking and related policies do not allow us to do so. Despite clear and theoretically well-grounded arguments stating that "sustainable growth is impossible" (Daly, 1990, p. 45; see also Hall, 2009), the term is increasingly used by various international development policy agencies, including the UN and the European Union (EU). The United Nations Development Programme (UNDP, 2020), for example, aims to support the achievement of sustainable growth that helps people to benefit from economic growth. In order to do so, their activities include establishing evidence-based analysis for national plans, promoting economic diversification and sustainable growth, and effective natural resource management.

The sustainable growth rhetoric is also strongly built into the EU's development strategies. There are numerous policy documents emphasising the central role and need for sustainable growth in the EU, including the European Commission's action plan on financing sustainable growth (European Commission, 2018) and the Europe 2020 Strategy, which is currently the EU's growth strategy for smart, sustainable and inclusive growth (European Commission, 2010). The European Commission's (2019) annual sustainable growth strategy 2020 states that:

> first, our efforts should focus on leading the transition to a nature-friendly and climate-neutral continent by 2050, while ensuring that everyone can take advantage of the opportunities that this will bring along. Second, by developing new technologies and sustainable solutions, Europe can be at the forefront of future economic growth.

Thus, while there are references to carbon-neutral policies and sustainable solutions, the bottom line is the need for future growth. These aims are mutually very challenging to achieve, and the situation reflects the oxymoronic nature of a sustainable growth paradigm, in which both the UN and the EU have devoted their major development policies and strategies. UNCTAD (2012), for example, provides an illustrative example of what this sustainable growth discourse means for tourism and how the industry positions its role in practising sustainable growth. Based on the UNCTAD meeting in Doha, Qatar, in 2012, the World Tourism Organization (UNWTO, 2012) outlined how the tourism sector could contribute to inclusive and sustainable growth:

> For many decades, international tourism has experienced dynamic growth and continued expansion. Cross-border travel for recreational, leisure or business purposes has become one of the fastest growing economic activities worldwide. The number of international tourist arrivals rose by almost forty times from 25 million in 1950 to 980 million in 2011. Even between 2000 and 2010, which was a decade of boom but also grave economic and financial crises, severe pandemics and large-scale natural disasters; international tourist arrivals continued to grow at an average annual rate of 3.4 per cent. As growth has been particularly dynamic in developing and emerging regions, their share in international tourist arrivals rose from 31 per cent in 1990 to 47 per cent in 2010.

Indeed, despite political and economic crises and natural disasters, the global tourism industry has been resilient in its growth orientation. However, it is puzzling that the UNWTO, being responsible for the promotion of sustainable tourism (UNWTO, 2017), aims to contribute to sustainability by strongly emphasising the increasing volumes and capacity of international tourist arrivals. These arrivals are mainly based on air travel, and related emissions of carbon dioxide, a major greenhouse gas, have been very clearly noted as highly problematic for climate change and the overall sustainability of the tourism industry (Scott, Hall, & Gössling, 2012).

What is important to note in respect of these sustainable growth policies in the tourism context is the very element of growth. Currently, air travel may account for 2–3% of global carbon dioxide emissions, which is less than the emissions from passenger cars. However, due to the estimated growth of international tourism, as also promoted by the UNWTO, aviation emissions may triple by 2050, and emissions based on that growth might account for 25% of the global carbon budget (Owen, Lee, & Lim, 2010; Graver, Zhang, & Rutherford, 2019). Furthermore, a recent study states that aviation may have contributed almost 5% to global radiative forcing (i.e. the difference between sunlight absorbed by the earth and energy radiated back to space) by 2005 and that carbon dioxide emissions from air travel could grow by up to 360% between 2000 and 2050 (Warnecke, Schneider, Day, Theuer, & Fearnehough, 2019).

Based on this data, it is mind-boggling that global policy-makers, who fluently use sustainability rhetoric with calls for inclusive development and equity, do not acknowledge the inconsistency between growth and sustainability. Tourism growth has a capacity to deliver many local-scale benefits for destination communities and economies, but as noted by Sinclair (1998), the positive economic aspects of tourism should be placed in an equation consisting of both the advantages and disadvantages of tourism-related growth. In the sustainability context, this should be preferably processed at a global tourism-system level. It seems, however, that growth represents a political imperative (Daly, 1996) in (sustainable) tourism, and both policy-makers and businesses are increasingly interested in the "sustainability of profitable corporate growth rather than sustainable development" (Luke, 2013, p. 89). Neoliberal capitalism is increasingly reliant on growth, and this element is firmly integrated into today's sustainable tourism growth mantra, which is sustaining the unsustainable, indicating an urgent need for resetting sustainability thinking in tourism.

Conclusions: towards sustainable degrowth or post-growth

Sustainable development has turned out to be both a successful and a problematic idea (Hall, 2011). It is an idealistic approach to development with challenging practicability. Long-term and holistic sustainable development needs have been directly linked to various economic sectors, creating terms like *sustainable forestry*, *sustainable mining* and *sustainable tourism*, which all have their own typically independent sustainability charters and indicators. This alone creates critical questions in the sustainable development context. First, different economic sectors are not 'islands', i.e. they are embedded in wider regional economies and ecosystems (Meadows et al., 1972). This applies to the tourism industry and destinations that are often the scale for evaluation in tourism studies and management (Saarinen, 2006). However, tourism destination systems do not have clear or static boundaries: instead of being fixed territorial units, destinations are relational spaces. This relationality is indicated in Leiper's (1979) tourism system approach, which involves tourist-generating and tourist-destination regions. In that system, destinations are connected to each other and the tourist-generating regions in multiple and complex ways, highlighting the need to understand sustainable development beyond a destination scale in tourism (Saarinen, 2014). In addition, as there are no real boundaries between a certain regional economy and its operational context and larger system scale, the isolated management of an individual economic sector may not have the capacity to deliver sustainable development in a holistic sense. Instead, we may end up having resource-use conflicts between different sectors that claim to operate based on their own sustainability indicators. At the same time, they justify their existing use and competition for resources by being sustainable for us, but conflicts are rarely sustainable.

Second, economic sectors often plan their activities and operations on a very short-term basis. This contrasts with the inter-generational perspective of sustainable development, making sustainability governance a challenging task in practice (Saarinen & Gill, 2019). In addition, compared to forestry or mining, for example, tourism operates on a much shorter-term planning cycle when aiming towards sustainability. Still, tourism is generally seen as having great prospective potential to meet the call for sustainability in development, especially on regional and destination scales. Furthermore, during the past two decades, the positive developmental role of the tourism industry has been increasingly linked to solving global-scale challenges (UNWTO, 2006; see also Holden, 2013; Saarinen, Rogerson, & Manwa, 2013). Recently, the prospective role of tourism is highlighted with regard to the UN's Sustainable Development Goals (SDGs) of the 2030 Agenda for Sustainable Development (Hall, 2019; Saarinen, 2020; Scheyvens, 2018). SDGs are broadly based on the idea that we should 'leave no one behind' in development. Both implicitly and explicitly, this aim includes the need for economic growth in development, specifically in the Global South but not necessarily in the Global North. Indeed, SDGs and their numerous targets indicate the need to rethink the globally unequal economic growth ideology and balance. While this need for rethinking or resetting the current economic system is recognised by scholars and influential commentators, such as Moyo (2018) and Wolf (2019), the pathway towards more sustainable development with a new kind of growth involved is not yet clear or specified.

Obviously, the basic idea in SDGs and in these calls for a new kind of growth is that economic growth should not be an end in itself. In tourism, this would require stronger governance and politics guiding the industry. For this, the COVID-19 crisis for the tourism sector may offer an opportunity to rethink growth in tourism and "question the volume growth tourism model advocated by" tourism policy-making organisations (Gössling et al., 2020). The crisis has resulted in massive public sector subventions for the industry and its survival, including the aviation sector. Unless the COVID-19 crisis is not seen as a force of 'creative destruction' that removes businesses that are neither fit nor resilient out of the free markets, it is rather challenging to justify why the tourism industry should return to the same low regulative landscape it was in before the crisis. The aviation industry in particular and its estimated growing emissions should be able to be better governed, so that it could meet previous (such as the Paris Climate Change Agreement 2015 (COP21)) and future international agreements.

Stronger public sector involvement and control in tourism may also support existing calls for a steady-state economy or degrowth in the sector (Milano,

Novelli, & Cheer, 2019). In recent years, degrowth in particular has been increasingly debated in tourism studies, but as much as a decade ago, Hall (2009, p. 46) pointed to the need for tourism to seek alternative pathways to economic development by integrating sustainable development with "degrowth processes that offer an alternative discourse to the economism paradigm that reifies economic growth." He further stated that sustainable tourism is tourism development without growth in throughput of matter and energy beyond regenerative and absorptive capacities. Therefore, degrowth thinking offers a fundamental change to tourism development processes by questioning "the assumptions which have been behind the continual expansion of the industry" (Higgins-Desbiolles, Carnicelli, Krolikowski, Wijesinghe, & Boluk, 2019, p. 1928).

Alternatively, the resetting of sustainability may shift its direction towards a post-growth process. It shares many similarities with degrowth (or a steady-state economy), and they are often seen as the same. Still, an overall post-growth process may be a more realistic step to start with, as in general terms it aims to solve problems by utilising existing ideas, practices and technologies in a better (i.e. more sustainable) way in specific contexts, in contrast to what is often a declarational or 'absolutely' limiting degrowth approach based on strict limits to growth thinking. Furthermore, post-growth has connections with post-development, which promotes alternative and localised (i.e. contextualised) ways to create and have human well-being (Escobar, 2017; Adityanandana & Gerber, 2019). This could offer more possibilities for sustainable future in the Global South, especially.

Ultimately, however, both post-growth and degrowth aim for the same goal of not having growth at a system level. In tourism, this calls for a radical reset in the ways the industry and the guiding policies frame, understand and practise sustainable tourism on a global scale. Obviously, this cannot be done independently of wider calls to change the global economic system. In the tourism context, however, the first and probably easiest step is to get away from the current and misleading sustainable growth rhetoric, and then focus on the original needs-based idea of sustainable development in tourism governance and management. By doing so, the tourism industry could be able to contribute positively to global challenges, such as SDGs, instead of mainly justifying its own existence and growth needs, as it seems to be doing at the moment.

References

Adityanandana, M., & Gerber, J.-F. (2019). Post-growth in the tropics? Contestations over Tri Hita Karana and a tourism megaproject in Bali. *Journal of Sustainable Tourism*, *27*(12), 1839–1856.

Akanbi, O. A. (2014). Structural and institutional determinants of poverty in Sub-Saharan African countries. *Journal of Human Development and Capabilities*, *16*(1), 122–141.

Bianchi, R. (2018). The political economy of tourism development: A critical review. *Annals of Tourism Research*, *70*, 88–102.

Blackstock, K. L., White, V., McCrum, G., Scott, A., & Hunter, C. (2008). Measuring responsibility: An appraisal of a Scottish national park's sustainable tourism indicators. *Journal of Sustainable Tourism*, *16*(3), 276–297.

Blühdorn, I. (2017). Post-capitalism, post-growth, post-consumerism? Eco-political hopes beyond sustainability. *Global Discourse*, *7*(1), 42–61.

Bramwell, B. (2011). Governance, the state and sustainable tourism: A political economy approach. *Journal of Sustainable Tourism*, *19*(4–5), 459–477.

Bramwell, B., & Lane, B. (2011). Critical research on the governance of tourism and sustainability. *Journal of Sustainable Tourism*, *19*(4–5), 411–421.

Burns, P. (1999). Paradoxes in planning: Tourism elitism or brutalism? *Annals of Tourism Research*, *26*, 329–348.

Butler, R. W. (1998). Sustainable tourism: Looking backwards in order to progress? In C. M. Hall & A. A. Lew (Eds.), *Sustainable tourism: A geographical perspective* (pp. 25–34). New York, NY: Longman.

Butler, R. W. (1999). Sustainable tourism: A state-of-the-art review. *Tourism Geographies*, *1*(1), 7–25.

Carmody, P. (2019). *Development theory and practice in a changing world*. Abington: Routledge.

Carson, R. (1962). *Silent spring*. Boston, MA: Houghton Mifflin Harcourt.

Caruana, R., Glozer, S., Crane, A., & McCabe, S. (2014). Tourists' accounts of responsible tourism. *Annals of Tourism Research*, *46*, 115–129.

Clarke, J. (1997). A framework of approaches to sustainable tourism. *Journal of Sustainable Tourism*, *5*(3), 224–233.

Coles, T., Fenclova, E., & Dinan, C. (2013). Tourism and corporate social responsibility: A critical review and research agenda. *Tourism Management Perspectives*, *6*, 122–141.

Crober, U. (2015). The discovery of sustainability: A genealogy of a term. In J. C. Enders & M. Remig (Eds.), *Theories of sustainable development* (pp. 6–15). London: Routledge.

Daly, H. E. (1990). Sustainable growth: An impossibility theorem. *Development, 3/4*, 45–47.

Daly, H. E. (1996). *Beyond growth*. Boston, MA: Beacon Press.

Enders, J. C., & Remig, M. (Eds.). (2015a). *Theories of sustainable development*. London: Routledge.

Enders, J. C., & Remig, M. (2015b). Theories of sustainable development: An introduction. In J. C. Enders & M. Remig (Eds.), *Theories of sustainable development* (pp. 1–5). London: Routledge.

Eblen, R. A., & Eblen, W. R. (1994). *Encyclopedia of the environment*. Boston, MA: Houghton Mifflin Harcourt.

Escobar, P. (2017). *Designs for the pluriverse: Radical interdependence, autonomy, and the making of worlds*. Durham, NC: Duke University Press.

European Commission. (2010). *Europe 2020 strategy* [pdf]. Retrieved from https://ec.europa.eu/eu2020/pdf/COMPLET%20EN%20BARROSO%20%20%20007%20-%20Europe%202020%20-%20EN%20version.pdf

European Commission. (2018). *Action plan on financing sustainable growth*. Retrieved from https://ec.europa.eu/info/publications/180308-action-plan-sustainable-growth_en

European Commission. (2019). *Annual sustainable growth strategy 2020*. Retrieved from https://ec.europa.eu/info/publications/2020-european-semester-annual-sustainable-growth-strategy_en

Fournier, V. (2008). Escaping from the economy: The politics of degrowth. *International Journal of Sociology and Social Policy*, *28*(11/12), 528–545.

Fosu, A. K. (2015). Growth, inequality and poverty in Sub-Saharan Africa: Recent progress in a global context. *Oxford Development Studies*, *43*(1), 44–59.

Graver, B., Zhang, K., & Rutherford, D. (2019). *CO_2 emissions from commercial aviation, 2018*. ICCT Working Paper 2019–16.

Greve, B. (2020). *Poverty*. Abingdon: Routledge.

Gössling, S., Ring, A., Dwyer, L., Andersson, A. C., & Hall, C. M. (2016). Optimizing or maximizing growth? A challenge for sustainable tourism. *Journal of Sustainable Tourism, 24*(4), 527–548.

Gössling, S., Scott, D., & Hall, C. M. (2020). Pandemics, tourism and global change: A rapid assessment of COVID-19. *Journal of Sustainable Tourism*. doi:10.1080/09669582.2020.1758708

Hall, C. M. (1998). Historical antecedents of sustainable development and ecotourism: New labels on old bottles? In C. M. Hall & A. A. Lew (Eds.), *Sustainable tourism: A geographical perspective* (pp. 13–24). New York, NY: Longman.

Hall, C. M. (2000). *Tourism planning: Policies, processes and relationships*. Harlow: Pearson.

Hall, C. M. (2009). Degrowing tourism: Décroissance, sustainable consumption and steady-state tourism. *Anatolia, 20*(1), 46–61.

Hall, C. M. (2011). Policy learning and policy failure in sustainable tourism governance: From first- and second-order to third-order change? *Journal of Sustainable Tourism, 19*(4–5), 649–671.

Hall, C. M. (2019). Constructing sustainable tourism development: The 2030 agenda and the managerial ecology of sustainable tourism. *Journal of Sustainable Tourism, 27*(7), 1044–1060.

Hall, C. M., Gössling, S., & Scott, D. (Eds.). (2015). *The Routledge handbook of tourism and sustainability*. Abingdon: Routledge.

Hall, C. M., & Lew, A. A. (Eds.). (1998). *Sustainable tourism: A geographical perspective*. New York, NY: Longman.

Hardy, A., Beeton, R., & Pearson, L. (2002). Sustainable tourism: An overview of the concept and its position in relation to conceptualisations of tourism. *Journal of Sustainable Tourism, 10*(6), 475–496.

Higgins, K. L. (2015). *Economic growth and sustainability: Systems thinking for a complex world*. Amsterdam: Elsevier.

Higgins-Desbiolles, F., Carnicelli, S., Krolikowski, C., Wijesinghe, G., & Boluk, K. (2019). Degrowing tourism: Rethinking tourism. *Journal of Sustainable Tourism, 27*(12), 1926–1944.

Holden, A. (2003). In need of new environmental ethics for tourism. *Annals of Tourism Research, 30*(1), 94–108.

Holden, A. (2013). *Tourism, poverty and development*. Abingdon: Routledge.

Hunter, C. (1995). On the need to re-conceptualize sustainable tourism development. *Journal of Sustainable Tourism, 3*(3), 155–165.

International Union for Conserving Nature (IUCN). (1980). *World conservation strategy*. Morges: IUCN.

Jessop, B. (2004). Hollowing out the 'nation-state' and multilevel governance. In P. Kenneth (Ed.), *A handbook of comparative social policy* (pp. 11–25). Cheltenham: Edward Elgar.

Kates, R. W., Clark, W. C., Corell, R., Hall, J. M., Jaeger, C. C., Lowe, I., . . . Dickson, N. M. (2001). Sustainability science. *Science, 292*(5517), 641–642.

Kerschner, C. (2010). Economic de-growth vs. Steady-state economy. *Journal of Cleaner Production, 18*(6), 544–555.

Kimaro, E., & Saarinen, J. (2020). Tourism and poverty alleviation in the global South: Emerging corporate social responsibility in the Namibian nature-based tourism industry. In M. Stone, M. Lenao, & N. Moswete (Eds.), *Natural resources, tourism and community livelihoods in southern Africa* (pp. 123–142). Abingdon: Routledge.

Kiš, A. D. (2018). *The development trap: How thinking big fails the poor*. Abingdon: Routledge.

Koch, M. (2020). The state in the transformation to a sustainable postgrowth economy. *Environmental Politics*, *29*(1), 115–133.

Krippendorf, J. (1987). *The holiday makers: Understanding the impacts of leisure and travel*. London: Butterworth-Heinemann.

Leiper, N. (1979). The framework of tourism. *Annals of Tourism Research*, *6*(4), 390–407.

Leopold, A. (1949). *A sand county almanac: And sketches here and there*. Oxford: Oxford University Press.

Liu, Z. (2003). Sustainable tourism development: A critique. *Journal of Sustainable Tourism*, *11*(6), 459–475.

Lu, J. Y., & Nepal, S. K. (2009). Sustainable tourism research: An analysis of papers published in the journal of sustainable tourism. *Journal of Sustainable Tourism*, *17*(1), 5–16.

Luke, T. W. (2013). Corporate social responsibility: An uneasy merger of sustainability and development. *Sustainable Development*, *21*(2), 83–91.

Marsh, G. P. (1865). *Man and nature: Or, physical geography as modified by human action*. New York, NY: Charles Scribner. Retrieved from https://archive.org/details/manandnatureorp02marsgoog

McCarthy, J. (2002). First World political ecology: Lessons from the wise use movement. *Environment and Planning A*, *34*(7), 1281–1302.

McCool, S., & Bosak, K. (2016). *Reframing sustainable tourism*. Berlin: Springer.

Meadows, D. H., Meadows, D. L., Randers, J., & Behrens, W. W. (1972). *The limits to growth: A report for the Club Rome's project on predicament of mankind*. London: Earth Island Limited.

Mensah, J., & Casadevall, S. R. (2019). Sustainable development: Meaning, history, principles, pillars, and implications for human action: Literature review. *Cogent Social Sciences*, *5*(1), https://doi.org/10.1080/23311886.2019.1653531

Milano, C., Novelli, M., & Cheer, J. M. (2019). Overtourism and degrowth: A social movements perspective. *Journal of Sustainable Tourism*, *27*(12), 1857–1875.

Moyo, D. (2018). *Edge of chaos: Why democracy is failing to deliver economic growth and how to fix it*. New York, NY: Basic Civitas Books.

Owen, B., Lee, D. S., & Lim, L. (2010). Flying into the future: Aviation emissions scenarios to 2050. *Environmental Science & Technology*, *44*(7), 2255–2260.

Peet, R., & Hartwick, E. (2015). *Theories of development: Contentions, arguments, alternatives* (3rd ed.). New York, NY: Guilford Press.

Pietarinen, J. (1987). Human and forest: Four basic attitudes (in Finnish). *Silva Fennica*, *21*(4), 323–331.

Pike, A., Rodríguez-Pose, A., & Tomaney, J. (2007). What kind of local and regional development and for whom? *Regional Studies*, *41*(9), 1253–1269.

Place, S. (1995). Ecotourism for sustainable development: Oxymoron or plausible strategy? *GeoJournal*, *35*, 161–173.

Puhakka, R., & Saarinen, J. (2013). New role of tourism in national park planning in Finland. *Journal of Environment and Development*, *22*(4), 412–435.

Redclift, M. (2005). Sustainable development (1987–2005): An oxymoron comes of age. *Sustainable Development*, *13*(4), 212–227.

Reyes, G. E. (2001). Four main theories of development: Modernization, dependency, word-system, and globalization. *Nómadas: Revista Crítica de Ciencias Sociales y Jurídicas*, *4*(2), 109–124.

Saarinen, J. (1998). Wilderness, tourism development and sustainability: Wilderness attitudes and place ethics. In A. E. Watson & G. Aplet (Comps.), *Personal, societal, and ecological values of wilderness: Sixth world wilderness congress proceedings on research, management, and allocation* (Vol. 1, pp. 29–34). Ogden, UT: USDA Forest Service, Rocky Mountain Research Station.

Saarinen, J. (2006). Traditions of sustainability in tourism studies. *Annals of Tourism Research*, *33*(4), 1121–1140.

Saarinen, J. (2014). Critical sustainability: Setting the limits to growth and responsibility in tourism. *Sustainability*, *6*(1), 1–17.

Saarinen, J. (2019). Communities and sustainable tourism development: Community impacts and local benefit creation tourism. In S. F. McCool & K. Bosak (Eds.), *A research agenda for sustainable tourism* (pp. 206–222). Cheltenham: Edward Elgar Publishing.

Saarinen, J. (Ed.). (2020). *Tourism and sustainable development goals: Research on sustainable tourism geographies*. London: Routledge.

Saarinen, J., & Gill, A. M. (2019). Tourism, resilience and governance strategies in the transition towards sustainability. In J. Saarinen & A. M. Gill (Eds.), *Resilient destinations: Governance strategies in the transition towards sustainability in tourism* (pp. 15–33). London: Routledge.

Saarinen, J., Rogerson, C. M., & Hall, C. M. (2017). Geographies of tourism development and planning. *Tourism Geographies*, *19*(3), 307–317.

Saarinen, J., Rogerson, C. M., & Manwa, H. (Eds.). (2013). *Tourism and millennium development goals: Tourism, local communities and development*. London: Routledge.

Savitz, A. W. (2006). *The triple bottom line: How today's best-run companies are achieving economic, social and environmental success – and how you can too*. Chichester: John Wiley.

Schilcher, D. (2007). Growth versus equity: The continuum of pro-poor tourism and neoliberal governance. *Current Issues in Tourism*, *10*(2–3), 166–193.

Scheyvens, R. (2011). *Tourism and poverty*. London: Routledge.

Scheyvens, R. (2018). Linking tourism to the sustainable development goals: A geographical perspective. *Tourism Geographies*, *20*(2), 341–342.

Scott, D., Hall, C. M., & Gössling, S. (2012). *Climate change and tourism: Impacts, adaptation and mitigation*. London: Routledge.

Sen, A. (1992). *Inequality re-examined*. Cambridge, MA: Harvard University Press.

Sen, A. (1999). *Development as freedom*. New York, NY: Anchor Books.

Sharpley, R. (2000). Tourism and sustainable development: Exploring the theoretical divide. *Journal of Sustainable Tourism*, *8*(1), 1–19.

Sharpley, R. (2009). *Tourism development and the environment: Beyond sustainability?* London: Earthscan.

Sharpley, R. (2020). Tourism, sustainable development and the theoretical divide: 20 years on. *Journal of Sustainable Tourism*, *28*(11), 1932–1946.

Sin, H. L. (2014). Realities of doing responsibilities: Performances and practices in tourism. *Geografiska Annaler: Series B*, *96*, 141–157.

Sinclair, T. (1998). Tourism and economic development: A survey. *Journal of Development Studies*, *34*(5), 1–51.

Smith, V. L., & Eadington, W. R. (Eds.). (1992). *Tourism alternatives: Potentials and problems in the development of tourism*. Philadelphia, PA: University of Pennsylvania Press.

Spaiser, V., Ranganathan, S., Swain, R. B., & Sumpter, D. (2017). The sustainable development oxymoron: Quantifying and modelling the incompatibility of sustainable development goals. *International Journal of Sustainable Development & World Ecology*, *24*(6), 457–470.

Spangenberg, J. H. (2014). Institutional change for strong sustainable consumption: Sustainable consumption and the degrowth economy. *Sustainability: Science, Practice and Policy*, *10*(1), 62–77.

Swarbrooke, J. (1999). *Sustainable tourism management*. Oxon: CABI.

United Nations (UN). (2000). *Millennium summit goals*. New York, NY: UN.

United Nations Conference on Trade and Development (UNCTAD). (2012). *UNCTAD annual report* [pdf]. Retrieved from https://unctad.org/en/PublicationsLibrary/dom2013d1_en.pdf

United Nations Development Programme (UNDP). (2020). *Inclusive sustainable growth*. Retrieved from www.undp.org/content/undp/en/home/2030-agenda-for-sustainable-development/prosperity/development-planning-and-inclusive-sustainable-growth.html

United Nations Environment Programme (UNEP). (2005). *Making tourism more sustainable – A guide for policy makers*. Paris: UNEP.

UNWTO. (1993). *Sustainable tourism development: Guide for local planners*. Madrid: Word Tourism Organization.

UNWTO. (2006). *UNWTO's declaration on tourism and the millennium goals: Harnessing tourism for the millennium development goals*. Madrid: UNWTO.

UNWTO. (2012). *Towards inclusive & sustainable growth & development: How can the tourism sector contribute?* Retrieved from www.unwto.org/archive/global/event/towards-inclusive-sustainable-growth-development-how-can-tourism-sector-contribute

UNWTO. (2016). *About UNWTO*. Retrieved from www.unwto.org/who-we-are

UNWTO. (2017). *Tourism and the sustainable development goals – Journey to 2030, highlights*. Madrid: UNWTO.

Visser, W. (2010). CSR 2.0: From the age of greed to the age of responsibility. In W. Sun, J. Stewart, & D. Pollard (Eds.), *Reframing corporate social responsibility: Lessons from the global financial crisis* (pp. 231–251). Bingley: Emerald.

Warnecke, C., Schneider, L., Day, T., Theuer, S. L. H., & Fearnehough, H. (2019). Robust eligibility criteria essential for new global scheme to offset aviation emissions. *Nature Climate Change*, *9*, 218–221.

Werlen, B. (2015). From local to global sustainability: Transdisciplinary integrated research in the digital age. In B. Werlen (Ed.), *Global sustainability* (pp. 3–16). Cham: Springer.

World Commission on Environment and Development (WCED). (1987). *Our common future*. Oxford: Oxford University Press.

Wikipedia. (2020). *Factory reset*. Retrieved January 22, 2020, from https://en.wikipedia.org/wiki/Factory_reset

Wilkinson, R., & Pickett, K. (2010). *The spirit level: Why equality is better for everyone*. London: Penguin.

Wolf, M. (2019, September 18). Why rigged capitalism is damaging liberal democracy. *Financial Times*. Retrieved from www.ft.com/content/5a8ab27e-d470-11e9-8367-807ebd53ab77

10 Rethinking tourism

Degrowth and equity rights in developing community-centric tourism

Karla A. Boluk, Chris Krolikowski, Freya Higgins-Desbiolles, Sandro Carnicelli and Gayathri Wijesinghe

Introduction

The current way tourism is packaged and practiced is unsustainable. Glossy magazines, brochures and advertisements on television and social media, prioritising images of unflawed, pristine beaches and often sparsely populated and glistening aqua-coloured seascapes, portray only one side of the tourism narrative. Images reflecting marginalised local populations at tourism destinations, their stories including how their lives have been and continue to be affected by tourists, and more broadly the tourism industry, are absent from tourism promotional materials. Furthermore, if and/or how local people would like to invite tourists into their communities in which they live are absent from tourism narratives (Higgins-Desbiolles, Carnicelli, Krolikowski, Wijesinghe, & Boluk, 2019). In November 2019, 11,258 scientists unequivocally argued that the earth faces a climate emergency (Ripple, Wolf, Newsome, Barnard, & Moomaw, 2020) and, unfortunately, the tourism industry is a culprit. The authors capture the breadth of human activities that have had a detrimental effect on the earth since 1979. Air transportation was identified as one of 15 indicators significantly contributing to climate change (Ripple et al., 2020). Importantly, marginalised populations often suffer the impacts of the predatory practices of tourism. The neoliberal emphasis on production and consumption has created an industry that fails to prioritise the rights of communities where tourism takes place and ignores its planetary impacts.

Contemporary mainstream magazine and newspaper articles have drawn attention to the concerns caused by tourism and scholarly research by scientists and social scientists alike. McKibben (2018), in *The Guardian* for example, stated: "we have to realize that global warming stems from the fact that we are a world without atmospheric borders, where the people who have done the least to cause the problem feel its horrors first and hardest". Furthermore, headlines signalling Ripple et al.'s (2020) evidence of a climate emergency swept the globe on 6 November 2019 one day after the *BioScience* journal article publication. Clearly, the scientists involved in the study highlight that the "climate crisis is closely linked to excessive consumption of the wealthy lifestyle [and] the most affluent countries are mainly responsible for the historical GHG [Greenhouse Gas] emissions and generally have the greatest per capita emissions" (Ripple et al., 2020,

p. 8). Klinsky et al. (2017) argue that a consideration of equity and justice are essential to our ability to understand the dynamics of political claims, actions and trade-offs, and more academic research is needed in this area. Hence, the mobility of the privileged and (in)equity rights of host communities are significant issues affecting social sustainability. We agree with Ripple et al. (2020), who put forth that all stakeholders must realign their priorities in order to alleviate climate change if we are really concerned with promoting justice and well-being for all.

The tourism industry supports the mechanics of neoliberalism (Wearing & Wearing, 2006; Mosedale, 2016; Higgins-Desbiolles, 2018a), which is not conducive to supporting long-term sustainability (Boluk, Cavaliere, & Higgins-Desbiolles, 2019). As such, an emphasis on capitalism concentrating on production and consumption is prioritised above the needs, wants or desires of local communities. Discussions about defining the limits to growth and responsibilities in tourism and how to achieve 'sustainable tourism' within a 'sustainable development' framework has a long history, starting with *The Limits to Growth* report (Meadows, Meadows, Randers, & Behrens, 1972) and the *Brundtland Report* (World Commission on Environment and Development (WCED), 1987). Sharpley (2000, p. 14) investigated the tension between the concept of 'sustainable tourism' with 'sustainable development', stating that "sustainable tourism does not appear to be consistent with the developmental aspects of sustainable development".

To the frustration of many scholars, the fact that tourism is employed as an economic activity and embedded as a component of capitalism has obstructed progress to achieve sustainable tourism outcomes. In fact, tourists are sought in the competitive tourism marketplace to drive the endless growth that is the key to contemporary politics in many countries. Obrador (2017, p. 208) states:

> the advancement of aggressive forms of tourism development has met an increasingly radicalised response. New forms of activism have emerged which advocate the need for 'degrowing' tourism economies and restricting tourist arrivals, prompting a shift in public discourse back to issues of saturation.

The saturation point of tourism in certain destinations conflicts directly with the necessity of growth, as well as with the capitalist and consumerist modus operandi of Westernised societies. Specifically, Boluk et al. (2019, p. 857) highlight a growing number of destinations that offer 'extreme examples' (the Galapagos Islands, Machu Picchu, Mount Everest, Majorca, Barcelona and Venice) feeling the impact of overtourism, signalling an imperative to more closely regulate tourism impacts given the failure of the laissez-faire approach to tourism management.

The growth trajectory sought by developing countries with the aspiration of attaining Western levels of consumption mutually challenges notions of equity and growth in tourism (Higgins-Desbiolles et al., 2019). Hall (2009a) reminds us that it is imperative to consider the environmental damages that may occur to tourism destinations if equity for all is to be implemented in tourism and travel. However, Wheeller's (1993) critique of sustainability in tourism suggests that assuming real responsibility is currently absent. Specifically, because of a commitment to capitalism, growth has superseded discussions of necessary sacrifice, behavioural changes and different actions in order to respond to sustainability imperatives (Wheeller, 1993). Wheeller (1993) dismissed some of the attempts to 'curb' unsustainable practices as superficial greenwashing. Building on Wheeller's (1993) concern, Mihalic (2016) argued that the sustainability discourse is not enough anymore unless it is integrated with responsible practice, but the discourse of her concepts still places a central focus on the tourist. It is this discourse and focus we aim to challenge.

Hall (2009a, p. 53) was one of the first to link tourism sustainability to the larger degrowth movement, suggesting that "sustainable tourism development is tourism development without growth in throughput of matter and energy beyond regenerative and absorptive capacities". Thus, degrowth thinking offers a fundamental challenge to tourism processes, as it questions the assumptions which have been behind the continual expansion of the industry since the post-war period. "To seriously pursue degrowth at both global and most national levels would, therefore, likely require drastic transformation of the tourism industry and its metabolism" (Fletcher, Murray Mas, Blanco-Romero, & Blázquez-Salom, 2019, p. 1746). Saarinen's (2014, p. 1) work examining the complexity surrounding conceptual dimensions of sustainable tourism concluded that "there is a need to re-frame i.e., rescale and decentralize tourism in policy frameworks and practices aiming towards sustainability", advocating for more host-friendly practices (see also Chapter 9, this volume). As such, we believe that fairness and justice are key facets to achieving degrowth that is socially sustainable (Muraca, 2012). In this chapter we argue that it is necessary to re-frame tourism through a degrowth strategy, and building on our previous work redefining tourism (Higgins-Desbiolles et al., 2019), we continue to argue the imperative to place the rights of local communities above the rights of tourists for holidays and the rights of tourism corporations to make profits. Specifically, we will highlight a number of examples illustrating the lack of responsibility presented by tourism corporates in responding to sustainability concerns. Drawing on the overtourism and 'last chance' tourism literature, we consider the importance of pursuing a degrowth and community-centric tourism pathway.

Current issues in sustainable tourism

Overtourism and last chance tourism

Overtourism refers to "the impact of tourism on a destination, or parts thereof, that excessively influences perceived quality of life of citizens and/or visitors in a negative way" (UNWTO, 2018, p. 4). Such concerns are not new and parallel a phrase used in earlier writings related to carrying capacity. Higgins-Desbiolles (2018b) makes this connection when stating that overtourism happens when destinations exceed carrying capacities, leading to diminishing experiences for locals and/or visitors, and may result in serious consequences for some of the world's most popular destinations. The effects of unimpeded growth of tourism may also impact the endangering of tourism destinations through phenomena such as 'last chance tourism' (LCT). The fact that those in privileged countries are continuing to travel for pleasure and leisure, while those in these endangered host communities must move in order to survive, is creating an environment of hostility, highlighting the inequity that underpins the operations of the tourism industry. One way to examine overtourism is through the wider context of tourism development, fostered by the capitalist economic system, prioritising profit accumulation of multinational corporations and the global elite (Fletcher, 2011; Higgins-Desbiolles, 2008, 2018a). As a global industry centred on maximising profits, the tourism industry has capitalised on existing travel demand, as well as growing new market niches, despite increasing environmental and social concerns. The case of overtourism demonstrates the destructive impact of such an approach on host communities, while LCT highlights the industry's role in fear mongering, exacerbating tourism impacts contributing to the vulnerability of destinations.

A significant impact caused by travellers relates to the LCT market. LCT is defined as attracting tourists who "explicitly seek vanishing landscapes or seascapes, and/or disappearing natural and/or social heritage" (Lemelin, Dawson, Stewart, Maher, & Lueck, 2010, p. 478). Specifically, travellers are often drawn to visit iconic features, species or landscapes at risk of disappearing (Dawson, Stewart, & Lemelin, 2012). LCT destinations tend to be remote areas, as part of the appeal is to witness the decline of pristine nature (Groulx, Boluk, Lemieux, & Dawson, 2019). Initially, LCT was used to describe the insurgent interest generated by tourists to visit polar regions (e.g., Eijgelaar, Thaper, & Peeters, 2010) due to the perception that impending vulnerability, as a consequence of climate change, may encumber future opportunities to experience a place in its pristine form (Lemelin, Dawson, & Stewart, 2012). Such travellers, although often aware of climate change, seem compelled to visit such places due to a strong sense of place attachment. As such, they are often able to self-justify the damage caused by their travel to the LCT destination (Groulx et al., 2019). This has led some scholars to identify a disconnect and, in fact, an ethical paradox of LCT, as travellers seemingly value the place of interest, but simultaneously contribute to the climate impacts which threaten its existence (Dawson, Stewart, Lemelin, & Scott, 2010; Groulx et al., 2019).

Dawson et al. (2012) signal early examples of travellers who had the impetus to push themselves physically in order to conquer destinations, such as Sir Edmund Hillary's successful summit of Everest. Contemporaneously, they identified a shifting lens to pursue 'lasts', specifically, pushing elite travellers to experience attractions and landscapes (Dawson et al., 2012) or vulnerable Indigenous populations (Johnston, Viken, & Dawson, 2011) before they disappear. Clearly, recognising emancipatory processes via tourism (e.g., overcoming physical challenges such as what may be required to summit Everest) and/or challenges one may face to reach particular 'far-reaching' destinations (e.g., the Arctic circle) has put such places at risk. In the case of LCT, the physical (personal or geographic) has evidently created intrigue and thus a desire to go there and benefit from being able to tell others that they have been there and done that. Dawson et al. (2011, p. 257) ask the question: "is it morally appropriate for the tourism industry or local communities to market vulnerable attractions as a tactic to achieve increased tourist visitation and revenues?"

Implied in the LCT literature is that this form of tourism undermines the places in which it occurs, rather than contributing to the destinations in a socially responsible way. Specifically, some of the literature draws attention to the inability of travellers to draw on their moral compass when choosing an LCT holiday (Groulx et al., 2019). As such, LCT prioritises the interests of privileged travellers over the well-being of the peoples and environments they are consciously choosing to visit. LCT as a niche contributes to the very same problems as overtourism, and LCT reinforces the preoccupation with relentless expansion of tourism demand, regardless of the impacts which it engenders. Exacerbating the endemic problems of broader tourism, LCT and overtourism emphasise the urgent need to re-frame tourism, as an industry which can no longer sustain further growth.

Revisiting the degrowth paradigm

From the onset of the modern tourism industry in the nineteenth century, and its rapid expansion in the post-war era, tourism became an integral part of the industrial economic structure. In recent decades, paralleling the move to neoliberal governance, the deregulation of markets and the continuous growth agenda, tourism has consolidated its role as one of the key engines of the world's economic growth. To suggest degrowth of the tourism industry, thus means to question the neoliberal growth paradigm in which it is embedded (Higgins-Desbiolles et al., 2019). Yet, as pointed out by Hall (2013, p. 1101), "in order to encourage sustainable tourism consumption the system itself, and the 'rules of the game', need to change". Explaining such 'degrowth' positioning to private and public organisations that derive benefits from tourism consumption is likely to be more than problematic. Yet, if no action is taken, the predicted growth in tourism, due to increasing democratisation of travel, and the resulting social and environmental impacts will undermine host communities and the ecosystem to an extent not previously experienced. At this stage, enforcing any viable change will be a reactive crisis management, with possible action restricted by the extent of the damage already caused.

The fundamental question remains: how may we go about achieving a degrowth agenda mutually considering the urgent call for action and the equity issues that are integral to ensuring sustainable and just tourism? In our earlier writings, we proposed a framework that would tackle some of these issues and support local communities in playing a central role in the tourism process (Higgins-Desbiolles et al., 2019). We argue here that tourism based in equity rights would allow local communities to decide the number of tourists, as well as when and how to welcome them. Focusing foremost on the benefits for them and for their land with a shift in the power structures will be essential for the sustainable future of tourism. The following sections will explore some of the steps required in this power shift, as well as examples of tourism based on equity and social justice and the role of self-reflecting academic processes which could contribute to the development of degrowth tourism.

Tourism industry, local communities and the power shift

The response from the tourism industry regarding the growing sustainability imperative has been mixed and inconsistent, not only across the different sectors of the industry, such as airlines and accommodation, but also within these sectors. As argued by Williams and Ponsford (2009), to some degree this could be attributed to the fragmentation of the tourism industry, as well as its geographic dispersion under various political and regulatory systems. Some of the inaction has also been attributed to the diverging interests with governments, industry and stakeholders looking to each other for leadership regarding sustainability (Williams & Ponsford, 2009).

The tourism sector continues to be primarily focused on sustainable tourism, failing to engage with the overall impact the travel sector is having on the sustainability of development at a much larger scale. Consequently, much of the measures that have so far dominated the industry's response to growing impacts of tourism have involved more micro-level initiatives, including self-regulation, corporate social responsibility (CSR) and accreditation schemes (De Grosbois, 2016; Inoue & Lee, 2011). With the increasing recognition of the negative impacts of tourism among consumers and broader stakeholders, CSR measures were employed to demonstrate a proactive approach in dealing with such issues. It can be argued that there is much greater awareness of the negative impacts of the tourism industry, by the industry, but the policy-action gap remains one of the main obstacles to enforcing real and substantial change (Hall, 2009b). It has been argued that CSR, just as ecolabels and the promotion of ecotourism, have become a "reputational green-wash" (Williams & Ponsford, 2009, p. 398), or a marketing ploy aimed at building the brand and creating a point of differentiation for the businesses. This model has not questioned the fundamental rationale underlying the continuous growth agenda, but rather has been used as a means of developing new 'green' market niches and generating positive publicity, and hence contributing to further expansion of tourism.

Much of the efforts to address the negative impacts caused by tourism have been based on raising consumers' awareness of sustainability issues, but there are questions whether such an approach can generate urgently needed action and rethinking the way tourism functions (Higgins-Desbiolles, 2018b), specifically, considering the needs and interests of local populations. Hall (2013) highlights the problem wherein the industry's response to crises, such as climate change, is framed by the very paradigm that has led to the issue in the first place. Integral to this debate is the perception of consumers as rational decision-makers who seek to extend the benefits of their consumption. The role of government and industry in the context of the sustainability debate is to affect the attitudes, behaviours and choices of existing and potential tourists (Hall, 2013). However, this framework hardly engages with the problematic nature of consumption itself and the role consumption plays in a neoliberal society that has been pulverised by increasing inequality, weakened communities and social support – and where travel has increasingly been used to escape this reality. The fundamental issue here is that the focus on the consumer agency and the industry self-regulation has not delivered the necessary change:

> What, after all, is the point of encouraging governance mechanisms such as partnerships, network development, self-regulation and individual responsibility if they continue to have no practical effect on the sustainability of tourism and consumption? If the ethical value of 'individual choice; leads to increased emissions from lifestyle and travel actions and worsening environmental change then how ethical is it?
>
> (Hall, 2013, p. 1104)

Recently, a few examples have surfaced from the transportation and travel segment, which demonstrate the focus on individual responsibility as the panacea to the adverse impacts generated by the tourism industry. The first example is a channel on Air New Zealand's inflight entertainment system dedicated to 'Tiaki – Care for New Zealand', educating visitors about the importance of local peoples, travelling responsibly and safely (Air New Zealand, 2019). This channel emerged from the launch of the Tiaki Promise in November 2018 by the airline, along with a few other partners including the Department of Conservation and Tourism New Zealand (Tourism New Zealand, 2019). Specifically, the Maori value of *tiaki* means to care for people and place. The video is narrated by an intergenerational duo including a young girl and millennial man. The message in the video empowers visitors to demonstrate a responsibility for the way in which they conduct themselves by protecting nature, driving carefully, being prepared and showing respect. Unfortunately, given the growth of tourism, producers' emphasis on the bottom line and entitled behaviours demonstrated by elite travellers, such reminders are needed. Interestingly, Air New Zealand's campaign also sets up a paradox. A mode of transportation recognised as being high in CO_2 emissions calls on travellers to make socially conscious choices when travelling.

A group of climate change activists challenged some of the corporate sponsors, such as Air France (refer to www.thisiscolossal.com/2015/11/brandalism-fake-ads-paris/), leading up to the UN COP21 Climate Conference. Posters were secured behind glass on bus stop billboards. One photoshopped image revealed a female flight attendant with her left index finger over her mouth as if she was "shhhing" those watching. The dissenting text read:

> TACKLING CLIMATE CHANGE? OF COURSE NOT. WE'RE AN AIRLINE. We're sponsoring the UN Climate Conference so we look like we're part of the solution and to make sure our profits aren't affected. Economic growth is far more important than saving the planet. So we'll keep on bribing

> politicians and emitting green house gases. Just keep it to yourself. AIR-FRANCE PART OF THE PROBLEM.
>
> (Sierzputowski, 2015)

The billboard clearly encouraged commuters to pay closer attention to the content revealing the unpleasant truths, and highlighting the hypocrisy of many companies publicly acknowledging the concerns of climate change but not necessarily addressing such concerns within their own business (Sierzputowski, 2015). Air France's sponsorship of the climate talks may have signalled their intent to be part of the solution; however, they fail to take action perpetuating the problem. This is typical in sustainability discussions in tourism that have been ongoing for decades, and yet there has been limited progress because there has been a lack of understanding regarding what actions have been taken (see Boluk et al., 2019).

Another airline, KLM, projects a similar responsibility message in a video (KLM Royal Dutch Airlines, 2019). However, in this context potential travellers are encouraged to think responsibly before they leave home. In the KLM video the narrator draws attention to the fact that consumers have become increasingly comfortable with flying to destinations for their holidays when indicating that "people have really gotten the hang of it" (KLM Royal Dutch Airlines, 2019). The video has an image of a glass of ice water emptying and filling again, potentially signalling the excessive consumption of water at tourism destinations (and resorts via pools, golf courses and long showers). This imagery may also draw one's attention to the melting of glaciers as a consequence of global warming aggravated by tourism; this is reinforced with the message: flying has "changed our world forever" (KLM Royal Dutch Airlines, 2019). Western entitlement and overconsumption are both explicit and implicit messages delivered in KLM's video. The next point is that if we want the next generation to experience our 'beautiful world', then 'we' (consumers) need to consider flying more responsibly. While the narrator quickly states that KLM is exploring ways to improve flying, the narrator mainly tasks consumers to make "more responsible decisions" (KLM Royal Dutch Airlines, 2019) when it comes to flying. Within a neoliberal context, KLM's choice to pass on the responsibility to consumers is obvious, easy for them to do and demonstrates a lack of critical reflection regarding how they may do their part. All three of the examples described here could be argued as "reputational green-wash" (Williams & Ponsford, 2009, p. 398) as we described earlier, as the (minimal) efforts made may be an attempt to build their brands.

Neoliberal rhetoric has convinced individuals to bear the burden of the social and environmental issues caused by tourism and to take action when really the fossil-fuel-burning and socially exploitative corporations must actively reduce their output in order for change to be realised. Indeed, corporate ads, as evidenced here, commonly appeal to individual action. However, as *Guardian* journalist Lukacs (2017) suggests, taxing individuals to change their behaviour is similar to "flapp[ing] towels in a burning house". The Carbon Majors Report (Griffin, 2017) highlights that 100 companies are responsible for 71% of global greenhouse emissions. This report suggests that a small set of fossil fuel producers could contribute to systemic change. Moreover, Cheer, Goldsworthy, Mathews, and Kanodia (2017) point out the multiple ways in which tourism organisations have been exploiting local communities in areas including sex tourism and orphanage tourism but also in the service supply chain. Clearly, large tourism businesses must make significant changes if tourism is to have a future.

KLM's video comes at a time when the world-renowned Swedish teenager Greta Thunberg, an environmental activist, has gained media attention for boycotting air travel. Thunberg's activism has seen her instigate more aggressive climate action from governments and the international community (Watts, 2019). She quickly became an international role model in her pursuit of a zero-carbon journey (from London, UK to New York, US) to attend the Climate Action Summit (Milman, 2019), thus walking (or in this case, sailing) the talk. Thunberg's activism and advocacy of the zero-carbon economy have led to the 'flight shame' (*flygskam*) movement in Sweden and beyond, encouraging rail travel rather than air travel (Henley, 2019). Clearly, individual actions supported by environmental activists such as Thunberg should not be undermined. However, what Thunberg has role modelled and advocated for is collectively calling on corporations to take responsibility and make changes to their operations in light of climate change. Specifically, Thunberg called on governments to recognise the collapse of ecosystems and called leaders out regarding their obsession with economic growth. This call could and should link the climate crisis with the social crisis taking place in exploiting local communities and Indigenous groups around the world. The social injustices (which are also based on environmental injustices) must be tackled by the tourism industry, such as the climate change strike organised by Thunberg on 28 September 2019, which finally witnessed the support of businesses. For example, in Canada, Mountain Equipment Co-op (selling outdoor equipment) closed for the day (Pittis, 2019). Such participation could be critiqued as minimal and perhaps expected given that such businesses are reliant on stable environmental conditions in order for consumers to continue purchasing their products. However, emphasising action on behalf of producers is where we feel more attention is required. Such action supports Jackson's (2017) notion of sustainable prosperity, whereby prosperity is possible without causing destructive growth.

Proposals for action

Understanding the magnitude of the crisis we are facing – a crisis in which tourism plays a very central role – requires a fundamental and paradigmatic shift that includes not only individual action but also broader reconfiguration of the governance structure and the role consumption plays in the overall economic system. Business as usual is no longer an option, and as such, corporations need to challenge the status quo, recognising that well-being should take precedence over profit. While our optimism is diminishing, we present a number of ideas to support the essential change that is necessary.

In line with the Paris Agreement signed in 2016 regarding greenhouse gas emissions mitigation and adaptation and Ripple et al.'s (2020) identification of air transport being one of the 15 indicators contributing greatly to climate change, the most obvious suggestion would be to reduce the number of flights on offer. Recognising the impacts of LCT and overtourism specifically, we feel it is incumbent upon airlines to immediately take action on reducing and ideally eliminating flights to destinations where tourism is causing irreparable harm. Airlines could consider investing more in programmes supporting the local communities that have been negatively affected by tourism. Going one step further, those airlines that are serious about reducing their impacts may consider a *reverse* reward programme rewarding those who fly less in some capacity. Building on this idea, airlines may work more closely with businesses to determine ways to reduce the amount of (unnecessary) travel required for their employees; this form of consultancy could surface as an alternative way to generate revenue. Given that Ripple et al. (2020) also identify concerns regarding the global consumption of animal products, perhaps airlines may consider providing mostly, if not all, locally (from departure point) produced vegetarian/vegan meal/snack options during inflight service. Furthermore, perhaps airlines could give a second thought to the mass-produced items sold in their inflight shopping and instead recognise and prioritise the livelihoods of those at the destinations they are bringing tourists to. Additionally, it may be in the realm of possibility for airlines to prioritise hiring local Indigenous peoples and/or develop partnerships with local guides. These examples may create a movement to support local food, experiences and guides that could have a resounding impact, transcending the transportation segment alone.

Building on some of the ideas presented here and borrowing the idea of Air New Zealand's inflight channel devoted to educating travellers about *tiaki*, perhaps a tourist education tethered with a reflexive approach on behalf of airlines regarding how they may play better is needed. Such reflexivity may move beyond solely encouraging the traveller to make responsible choices, but also allow airlines to recognise their negative impacts, and consider ways they may actively pursue better practices in light of their responsibilities to the planet and peoples. Such inflight channels may be one step in the direction of encouraging airlines to recognise their impact and shift the power toward local smaller-scale entrepreneurs and businesses that could benefit from a spotlight being shone on their efforts. In this way, such suggestions offered here are not anti-travel, but rather provide room for businesses to consider how they may demonstrate responsible leadership and sustainable prosperity.

Local community-centred tourism

Elsewhere we have argued for a locally community-centred tourism to replace the current pro-growth corporatized form of tourism (Higgins-Desbiolles et al., 2019). Placing local communities at the centre of tourism is essential because their communities receive, host and are impacted by tourism. Examples from Venice, Dubrovnik and Barcelona (see Chapter 1, this volume) suggest the possibilities, and it is now the task of tourism scholars to seize the moment to think through the frameworks which would support such a transformation. The framework starts at the centre with empowered and interested local communities (Figure 10.1). This then connects outward to the transformed roles of tourists, the tourism industry, destination marketing organisations and governments (Higgins-Desbiolles et al., 2019).

Starting at the centre of our reformed vision for tourism is an agenda to empower and involve citizens in important tourism development policy and planning. Admittedly, this will not be easy, as most societies have moved away from a vision of tourism as a matter for exercising civic-mindedness because tourism has been corporatized and recast as a mere business sector (see Higgins-Desbiolles, 2006). But we see the emerging possibilities for this situation to be challenged and overturned. Its possibilities are evident in places such as Bhutan with its policy of gross national happiness and in New Zealand with its well-being agenda and the associated *tiaki* promise programme in tourism (see Higgins-Desbiolles et al., 2019). There is also evidence in communities that are angered by the damages

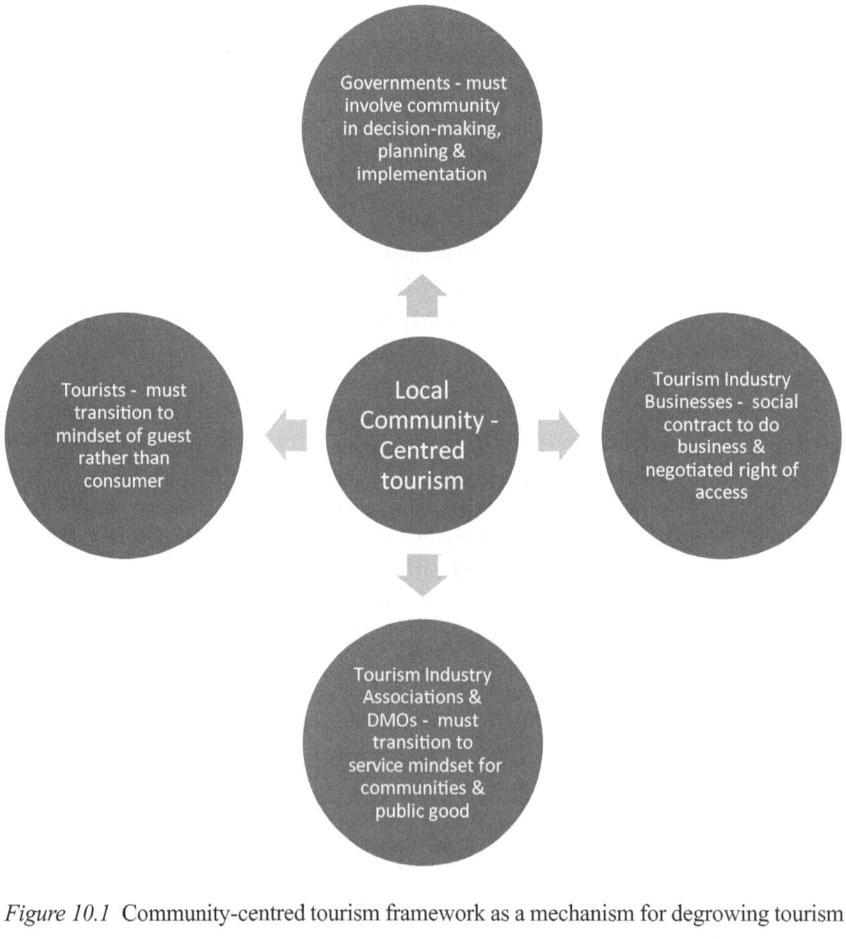

Figure 10.1 Community-centred tourism framework as a mechanism for degrowing tourism

Source: (Higgins-Desbiolles et al., 2019)

of overtourism that there is an untapped appetite for citizen action to take hold of tourism and wrest it from the control of corporate power and indifferent governments (see Burgen, 2018). The actions of the Kangaroo Island community of South Australia in creating and engaging with the Tourism Optimisation Management Model is an invaluable precedent for this (Higgins-Desbiolles, 2011; Jack, 2000). Those communities, from Barcelona to Venice to Amsterdam, are taking actions that reveal the possibilities for re-engaging citizenry in the shaping of tourism to their needs. We would propose here a role for local communities, citizenry of all categories, to take a leading role in tourism related decision-making that ensures that the only tourism allowed in their communities are those forms that they benefit from and invite. Building on the experience of Kangaroo Island, we would advocate a development of a process of citizen assemblies to participate at strategic tourism decision-making points, such as for the development of tourism strategic plans. Learning from recent thinking on empowering citizen assemblies in political decision-making can inform this process undertaken for locally governed tourism decision-making (e.g. Pratchett et al., 2009).

As part of a reformed tourism, it is essential that tourists are compelled to transform their attitudes from one of privileged consumers to one of responsible guests. As we abandon the old mindset that the places where tourism happens are 'tourism destinations', we can remember that these places are in fact the homes of local communities. We need to reinvigorate responsible tourism education, codes and guidelines and incentivise the tourists to engage with them. This can be accomplished positively through efforts to demonstrate how warm welcomes and meaningful experiences result from such an approach to touring. A positive recent example is Amsterdam's 'marry a local' tourism campaign, which invited visitors to experience off-the-beaten-path parts of the community with the guidance of willing locals (Adams, 2019). There are also more negative tools to be employed to reorient the tourists, including penalties, such as closure of certain places, fines for infringements, banning and even deportation. The seeds for this transition are already evident with New Zealand's *tiaki* promise programme working to alert visitors to the need to be prepared to drive safely on New Zealand's roads (Air New Zealand, 2019). It has been a serious source of anger to New Zealanders that international tourists have caused death and injury due to a lack of road readiness; the *tiaki* promise pushes visitors who may not be experienced in driving on the left-hand side of the road to make responsible choices such as not undertaking self-drive holidays. This serves as one illustrative example of how the concerns of the local community can shape the conduct of tourism so that visitors conform to local conditions, expectations and needs.

The next pillar of the process is re-embedding governance of tourism so that it is answerable to local community concerns and requirements. Many have noted how the advance of neoliberal agendas have reduced the role of government authorities in controlling, regulating, legislating and limiting tourism (Fletcher, 2011; Higgins-Desbiolles, 2008, 2011). In fact, governments currently act as facilitators of development against community wishes. A recent case in point comes from Australia, where the departments of state governments in states such as Tasmania,

Queensland and South Australia are facilitating luxury 'eco-lodge' developments within the national parks they administer despite considerable local community opposition (Gartry, 2019). The key problem to overcome is that in many instances the government agencies concerned with conservation and environmental protection are being forced to be made hospitable to tourism developments (see Higgins-Desbiolles, 2011). Simultaneously, the government agencies concerned with tourism have been restructured from public service to simple marketing bodies (Higgins-Desbiolles, 2018b). As a result, these tourism agencies prioritise ceaseless growth in tourism and facilitate the agendas of powerful tourism industry interests, even when this arouses considerable public anger and opposition. If the damages of tourism overdevelopment are going to be overturned, this aspect of hijacking tourism governance is going to have to be halted. More positively, we would encourage governance of tourism for community well-being through well-tested social tourism initiatives that have at their heart equity and inclusivity (see Higgins-Desbiolles et al., 2019).

Tourism businesses, particularly large multinationals and powerful business players, will have to abandon modes of operating that see destinations as opportunities to be mined for extraction of profits. However, as pointed out by Milano, Novelli, and Cheer (2019, p. 1861), "the aversion to degrowth stokes fears of economic recession and threats to political longevity of governments and business viability of corporations, making ensuing contentions prone to politicisation and fear mongering". It is increasingly apparent that one major catalyst to the crises of overtourism has been the way corporations have been operating; in analysing overtourism, accusations have been levelled at the low-cost airlines, the major cruise companies and the disrupters such as Airbnb (Oklevik et al., 2019). In the realm of corporate governance, increasing attention is being paid to reining in the powers of corporates and to disciplining corporates to the demands of social licenses to operate. We advocate an active agenda to make this meaningful in tourism destinations around the world. The tourism corporations that send tourists to local communities will have to actively engage with local communities to access their resources and deliver benefits that result in improvement and well-being rather than profit extraction. We see examples of this approach at a small scale, and there is no reason that this cannot be made realistic and mandatory at a larger scale. Additionally, active support and facilitation of locally owned and locally responsible tourism businesses should be prioritised. We can identify examples of this in campaigns such as buy local and in segments such as slow tourism (Caffyn, 2012).

Final thoughts

In tourism research, some but not abundant attention has been given to the concept of equity. Areas such as volunteer tourism, for example, have referred to the concept. Burrai, Font, and Cochrane (2015, p. 452) pointed out that "if individuals perceive that there is equity and justice within the exchange relationships they are involved in, they experience feelings of contentment that results in positive reactions and perceptions". Previously we drew attention to some strategic changes in the narrative taken by airlines, and we then discussed in detail the opportunities for local communities if the power dynamic shifted and their interests were prioritised. The examples used throughout acknowledge that a local voice is important; however, in the case of the travel and transportation sector, it seems they fail to share and ultimately shift power based on an equity rights concept. Importantly, the examples show minimal efforts at recognising the importance of local peoples, which may come across as disingenuous considering the concerns expressed have not been translated into action. As such, the examples presented could be critiqued as being "reputational green-wash" (Williams & Ponsford, 2009, p. 398).

If the tourism industry were to seriously reflect on and adopt the 17 United Nations Sustainable Development Goals (SDGs) and prescribed targets, then perhaps it would be possible to contribute to sustainability, social justice and equity agendas. As an example, the growing awareness regarding climate change and strikes around the world pushing governments to do more toward the preservation of the environment, as well as the example of activist Greta Thunberg in not flying, may at least encourage sectors of the tourism industry to rethink their policies and operations and perhaps give voice to local concerns. However, and as we have advocated before, adaptation of the tourism industry to the environmental narrative may not be enough to address issues regarding inequity. Moreover, the concept of degrowth based on a 'minimalistic' view of consumption may also not be sufficient if it is not based on the concept of equity rights and local community empowerment.

Williams and Ponsford (2009) argue that inaction is attributed to diverging interests of various stakeholder groups (e.g., governments, industry, producers and consumers) as they look to each other for leadership in response to pressing sustainability. Unfortunately, those who have been left out of decision-making positions are affected the most by unsustainable tourism. Importantly, inaction can no longer be tolerated; scientists (e.g., Ripple et al., 2020) have pointed out that we are at a critical point and change is required immediately. In line with broader climate emergency activism, it is imperative that governments enforce real change in overseeing tourism corporations that are responsible for contributing to social and environmental inequalities. Corporations that have been using local resources, local images and voices but at the same time ignoring local community rights must be held accountable. As such, our proposal for an active agenda encouraging tourism corporations to engage with local communities and ensure they deliver direct benefits improving communities is imperative rather than prioritising profits. We indicated earlier there are examples of small-scale mindful businesses which are locally owned and locally responsible, and as such we do not believe this is unrealistic and indeed believe this trajectory should be mandatory. The degrowth proposition offers these organisations, as well as governments, a reconciliatory approach towards a more fair and sustainable tourism that may have a future. The failure to achieve a reconciliatory degrowth approach may mean that 'last chance tourism' will be the 'last stop' for tourism, undermining destinations, threatening vulnerable Indigenous populations, destroying flora and fauna and unbalancing essential ecosystems.

References

Adams, C. (2019, June 6). Amsterdam is giving tourists the chance to 'marry' a local for the day and explore the city on 'honeymoon'. *The Independent*. Retrieved from www.businessinsider.com/amsterdam-tourists-marry-local-for-day-honeymoon-combat-overtourism-consequences-2019–6/?r=AU&IR=T

Air New Zealand. (2019). *Air New Zealand launches 'Tiaki – care for New Zealand' inflight entertainment channel* [press release]. Retrieved from www.airnewzealand.co.nz/press-release-2019-airnz-launches-tiaki-care-for-new-zealand-inflight-entertainment-channel

Boluk, K., Cavaliere, C., & Higgins-Desbiolles, F. (2019). A critical framework for interrogating the united nations sustainable development goals 2030 Agenda in tourism. *Journal of Sustainable Tourism*, *27*(7), 847–864.

Burgen, S. (2018, June 25). 'Tourists go home, refugees welcome': Why Barcelona chose migrants over visitors. *The Guardian*. Retrieved from www.theguardian.com/cities/2018/jun/25/tourists-go-home-refugees-welcome-why-barcelona-chose-migrants-over-visitors

Burrai, E., Font, X., & Cochrane, J. (2015). Destination stakeholders' perceptions of volunteer tourism: An equity theory approach. *International Journal of Tourism Research*, *17*(5), 451–459.

Caffyn, A. (2012). Advocating and implementing slow tourism. *Tourism Recreation Research*, *37*(1), 77–80.

Cheer, J. F., Goldsworthy, K., Mathews, L., & Kanodia, S. (2017, July 19). Modern slavery and tourism: When holidays and human exploitation collide. *The Conversation*. Retrieved from http://theconversation.com/modern-slavery-and-tourism-when-holidays-and-human-exploitation-collide-78541

Dawson, J., Johnston, M. J., Stewart, E. J., Lemieux, C. J., Lemelin, R. H., Maher, P. T., & Grimwood, B. S. R. (2011). Ethical considerations of last chance tourism. *Journal of Ecotourism*, *10*(3), 250–265.

Dawson, J., Stewart, E. J., & Lemelin, R. H. (2012). Last chance tourism: Conclusion. In R. H. Lemelin, J. Dawson, & E. J. Stewart (Eds.), *Last chance tourism. Adapting tourism opportunities in a changing world* (pp. 218–229). Abingdon: Routledge.

Dawson, J., Stewart, E. J., Lemelin, R. H., & Scott, J. (2010). The carbon cost of polar bear viewing tourism in Churchill, Canada. *Journal of Sustainable Tourism*, *18*(3), 319–336.

de Grosbois, D. (2016). Corporate social responsibility reporting in the cruise tourism industry: A performance evaluation using a new institutional theory based model. *Journal of Sustainable Tourism*, *24*(2), 245–269.

Eijgelaar, E., Thaper, C., & Peeters, P. (2010). Antarctic cruise tourism: The paradoxes of ambassadorship, "last chance tourism" and greenhouse gas emissions. *Journal of Sustainable Tourism*, *18*(3), 337–354.

Fletcher, R. (2011). Sustaining tourism, sustaining capitalism? The tourism industry's role in global capitalist expansion. *Tourism Geographies*, *13*(3), 443–461.

Fletcher, R., Murray Mas, I., Blanco-Romero, A., & Blázquez-Salom, M. (2019). Tourism and degrowth: An emerging agenda for research and praxis. *Journal of Sustainable Tourism*, *27*(12), 1745–1763.

Gartry, L. (2019, March 4). Tourism Queensland chair faced perceived conflict of interest in national park tender. *ABC News Online*. Retrieved from www.abc.net.au/news/2019-03-03/queensland-tourism-perceived-conflict-of-interest-managed/10851100

Griffin, P. (2017). *The carbon majors database: CDP carbon majors report 2017*. England: CDP Worldwide. Retrieved from https://b8f65cb373b1b7b15feb-c70d8ead6ced550b4

d987d7c03fcdd1d.ssl.cf3.rackcdn.com/cms/reports/documents/000/002/327/original/Carbon-Majors-Report-2017.pdf?1499691240

Groulx, M., Boluk, K., Lemieux, C. J., & Dawson, J. (2019). Place stewardship among last chance tourists. *Annals of Tourism Research*, *75*, 202–212.

Hall, C. M. (2009a). Degrowing tourism: Decroissance, sustainable consumption and steady-state tourism. *Journal of Sustainable Tourism*, *20*(1), 46–61.

Hall, C. M. (2009b). Archetypal approaches to implementation and their implications for tourism policy. *Tourism Recreation Research*, *34*(3), 235–245.

Hall, C. M. (2013). Framing behavioural approaches to understanding and governing sustainable tourism consumption: Beyond neoliberalism, "nudging" and "green growth"? *Journal of Sustainable Tourism*, *21*(7), 1091–1109.

Henley, J. (2019, June 4). #stayontheground: Swedes turn to trains amid climate 'flight shame'. *The Guardian*. Retrieved from www.theguardian.com/world/2019/jun/04/stayontheground-swedes-turn-to-trains-amid-climate-flight-shame

Higgins-Desbiolles, F. (2006). More than an industry: Tourism as a social force. *Tourism Management*, *27*(6), 1192–1208.

Higgins-Desbiolles, F. (2008). *Capitalist globalisation, corporatized tourism and their alternatives*. New York, NY: Nova Publishers.

Higgins-Desbiolles, F. (2011). Death by a thousand cuts: Governance and environmental trade-offs in ecotourism development at Kangaroo Island, South Australia. *Journal of Sustainable Tourism*, *19*(4–5), 553–570.

Higgins-Desbiolles, F. (2018a, July 11). Why Australia might be at risk of 'overtourism'. *The Conversation*. Retrieved from https://theconversation.com/why-australia-might-be-at-risk-of-overtourism-99213

Higgins-Desbiolles, F. (2018b). Sustainable tourism: Sustaining tourism or something more? *Tourism Management Perspectives*, 25, 157–160.

Higgins-Desbiolles, F., Carnicelli, S., Krolikowshi, C., Wijesinghe, G., & Boluk, K. (2019). Degrowing tourism: Rethinking tourism. *Journal of Sustainable Tourism*, *27*(12), 1926–1944.

Inoue, Y., & Lee, S. (2011). Effects of different dimensions of corporate social responsibility on corporate financial performance in tourism-related industries. *Tourism Management*, *32*(4), 790–804.

Jack, L. (2000). *Development and application of the Kangaroo Island TOMM (tourism optimisation management model)*. Paper presented at the First National Conference on the Future of Australia's Country Towns, 29–30 June, La Trobe University Bendigo. Retrieved from www.regional.org.au/au/countrytowns/options/jack.htm

Jackson, T. (2017). *Prosperity without growth-foundations for the economy of tomorrow*. Abingdon: Routledge.

Johnston, M. E., Viken, A., & Dawson, J. (2011). Firsts and lasts in Arctic tourism: Last chance tourism and the dialectic of change. In R. H. Lemelin, J. Dawson, & E. J. Stewart (Eds.), *Last chance tourism: Adapting tourism opportunities in a changing world* (pp. 10–24). London: Routledge.

Klinsky, S., Roberts, T., Huq, S., Okereke, C., Newell, P., Dauvergne, P., . . . Keck, M. (2017). Why equity is fundamental in climate change policy research. *Global Environmental Change*, *44*, 170–173.

KLM Royal Dutch Airlines. (2019, June 28). *KLM fly responsibly* [video file]. Retrieved from www.youtube.com/watch?v=L4htp2xxhto

Lemelin, R. H., Dawson, J., & Stewart, E. J. (Eds.). (2012). *Last chance tourism: Adapting tourism opportunities in a changing world*. Abingdon: Routledge.

Lemelin, R. H., Dawson, J., Stewart, E. J., Maher, P., & Lueck, M. (2010). Last chance tourism: The doom, the gloom, and the boom of visiting vanishing destinations. *Current Issues in Tourism*, *13*(5), 477–493.

Lukacs, M. (2017, July 17). Neoliberalism has conned us into fighting climate change as individuals. *The Guardian*. Retrieved from www.theguardian.com/environment/true-north/2017/jul/17/neoliberalism-has-conned-us-into-fighting-climate-change-as-individuals

McKibben, B. (2018, October 2). The Trump administration knows the planet is going to boil. It doesn't care. *The Guardian*. Retrieved from www.theguardian.com/commentisfree/2018/oct/02/trump-administration-planet-boil-refugee-camps

Meadows, D. H., Meadows, D. L., Randers, J., & Behrens, W. W. III. (1972). *The limits to growth: A report for the Club of Rome's project on the predicament of mankind*. New York, NY: Universe Books.

Mihalic, T. (2016). Sustainable-responsible tourism discourse – Towards 'responsustable' tourism. *Journal of Cleaner Production*, *111*, 461–470.

Milano, C., Novelli, M., & Cheer, J. (2019). Overtourism and degrowth: A social movements perspective. *Journal of Sustainable Tourism*, *27*(12), 1857–1875.

Milman, O. (2019, August 28). 'Let's do it now': Greta Thunberg crosses Atlantic and calls for urgent climate action. *The Guardian*. Retrieved from www.theguardian.com/environment/2019/aug/28/greta-thunberg-arrival-in-new-york-delayed-by-rough-seas

Mosedale, J. (2016). *Neoliberalism and the political economy of tourism*. Farnham: Ashgate.

Muraca, B. (2012). Towards a fair degrowth-society: Justice and the right to a 'good life' beyond growth. *Futures*, *44*(6), 535–545.

Obrador, P. (2017). The end of sustainability? A note on the changing politics of mass tourism in the Balearic Islands. *Journal of Policy Research in Tourism, Leisure and Events*, *9*(2), 205–208.

Oklevik, O., Gössling, S., Hall, C. M., Jacobsen, J. K., Grøtte, I., & McCabe, S. (2019). Overtourism, optimisation, and destination performance indicators: A case study of activities in Fjord Norway. *Journal of Sustainable Tourism*, *27*(12), 1804–1824.

Pittis, D. (2019, September 26). Greta Thunberg was right: There is an alternative to 'eternal economic growth'. *CBC/Radio-Canada*. Retrieved from www.cbc.ca/news/business/business-climate-strike-1.5293916

Pratchett, L., Durose, C., Lowndes, V., Smith, G., Stoker, G., & Wales, C. (2009). *Empowering communities to influence local decision making: Systematic review of the evidence*. London: Department for Communities and Local Government.

Ripple, W. J., Wolf, C., Newsome, T. M., Barnard, P., & Moomaw, W. R. (2020). World scientists' warning of a climate emergency. *BioScience*, *70*(1), 8–12.

Saarinen, J. (2014). Critical sustainability: Setting the limits to growth and responsibility in tourism. *Sustainability*, *6*(1), 1–17.

Sharpley, R. (2000). Tourism and sustainable development: Exploring the theoretical divide. *Journal of Sustainable Tourism*, *8*(1), 1–19.

Sierzputowski, K. (2015, November 30). Brandalism: 82 artists install 600 fake ads across Paris to protest the COP21 climate conference. *Colossal*. Retrieved from www.thisiscolossal.com/2015/11/brandalism-fake-ads-paris/

Tourism New Zealand. (2019). *Tiaki promise – Care for New Zealand*. Retrieved from https://tiakinewzealand.com/

UNWTO. (2018). *Overtourism? Understanding and managing urban tourism growth beyond perceptions: Executive Summary*. Madrid: UNWTO.

Watts, J. (2019, March 11). Greta Thunberg, schoolgirl climate change warrior: 'Some people can let things go. I can't'. *The Guardian*. Retrieved from www.theguardian.com/

world/2019/mar/11/greta-thunberg-schoolgirl-climate-change-warrior-some-people-can-let-things-go-i-cant
Wearing, S., & Wearing, M. (2006). "Re-reading the subjugating tourist" in neoliberalism: Postcolonial otherness and the tourist experience. *Tourism Analysis*, *11*(2), 145–162.
Wheeller, B. (1993). Sustaining the ego. *The Journal of Sustainable Tourism*, *1*(2), 121–129.
Williams, P. W., & Ponsford, I. F. (2009). Confronting tourism's environmental paradox: Transitioning for sustainable tourism. *Futures*, *41*(6), 396–404.
World Commission on Environment and Development (WCED). (1987). *Our common future*. Oxford: Oxford University Press [pdf]. Retrieved from www.un-documents.net/our-common-future.pdf

11 Community-based tourism and degrowth

Esteban Ruiz-Ballesteros

Introduction

Degrowth strategies applied to tourism open up a broad perspective that enables a shift from *décrosissance du tourisme* to *tourisme de décroissance* (Bourdeau & Berthelot, 2008). With *décrosissance du tourisme*, direct action is proposed in relation to conventional tourism to transform its consumption patterns and the use of resources, essentially involving an efficiency approach combined with slow consumption (Hall, 2009). *Tourisme de décroissance*, however, emphasises the development of alternative forms of tourism, comprising new logics and practices that will have differing implications for the extent to which the natural environment is transformed. However, the environmental and economic dimensions of degrowth rely inextricably on a socio-cultural dimension that entails support for alternative economic aspirations and moderate environmental effects. In short, it is necessary to accept that organisational and relational dimensions are significant for tourism degrowth strategies.

Community-based tourism (CBT) constitutes a way of organising tourism that usually follows on from local reflection on the goals and expectations of tourism, seeking to retain local control over the activity and its social, economic and environmental effects (Ruiz-Ballesteros, 2017). CBT experiences show the key importance of the underlying social relations operating in tourism, that is, the forms of organisation, planning, distribution of profits and local participation. These relational aspects mean that an alternative form of tourism cannot emerge purely from a change in *consumption patterns*, but also needs to involve a profound transformation in the *forms of production*.

CBT is an expanding tourist modality all over the world, which has been progressively incorporated by international bodies, governments, development and international cooperation agencies, NGOs and even Indigenous organisations. CBT promotes a tourism organised from an unconventional social relations framework – but still within a market logic – that offers the potential to transform the ways of appropriation of nature and the hegemonic patterns of economisation of life. These alternative experiences of organising and producing tourism could potentially be understood as strategies for degrowth. However, this analytical perspective remains undeveloped in both the CBT and degrowth literature, which suggests the need for further research in this area to inform practice and policy more effectively.

To explore this link between CBT and degrowth, this study focuses on the organisational interfaces that translate the capitalist logic underlying all tourism ventures into a community logic that seems to fit with degrowth strategies. These interfaces can be understood as an organisational dimension of degrowth, allowing for a more informed understanding of CBT functioning and degrowth practices within tourism activities, eventually revealing CBT's potential as a degrowth strategy.

The chapter begins by presenting a general characterisation of CBT and defining the elements most likely to shape social practices relating to degrowth. Having outlined this theoretical foundation, a specific case study derived from Agua Blanca (Ecuador) and the methodology used in analysing this case study are presented. The subsequent sections set out the results and discussion, before concluding with consideration of the pertinence of a degrowth perspective for better understanding CBT and its potential (and limitations) as a strategy for degrowth.

Community-based tourism, conviviality and organisational interfaces

CBT is a way of organising tourist activity that is guided by a series of general principles: a) local ownership and control of tourist resources and infrastructures; b) local participation in the planning and development of tourist activity; c) local control over the management of tourist businesses and their profits, which must benefit the community as a whole; and d) tourism is a complement to the local and household economies and, under no circumstances, pursues capital accumulation (Dangi & Jamal, 2016; Giampiccoli & Saayman, 2018; Hiwasaki, 2006; Okazaki, 2008; Ruiz-Ballesteros, 2017).

CBT is well recognised by tourism and local development researchers and practitioners around the world. However, the channels, scope and dimensions of 'local control' over tourist activity and 'community profits' in CBT experiences are extremely variable and heterogeneous, sparking legitimate criticism and scepticism concerning the coherence and viability of CBT (Amati, 2013; Blackstock, 2005; Gascón, 2013; Mitchell & Muckosy, 2008; Taylor, 2017). Regardless, CBT goals go against current conventional practice concerning tourism and aim to reduce the dominant effect of the market within communities, through regulating tourist activity based on commonly shared criteria. Nevertheless, a question arises as to whether CBT could be considered a viable strategy for degrowth based on its more collective orientation.

Beyond productivism and consumerism, degrowth manifests through social praxis, that is, in the forms of organisation and relations operating between humans, and between humans and the environment. In seeking to understand this phenomenon, the literature on degrowth invokes conviviality (Illich, 1973) as "one of its core anthropological constructs" (Deriu, 2014, p. 82). The concept of conviviality focuses on people's capacity to control, collectively, the modern tools of industrial-capitalist society (Illich, 1973). Following Deriu (2014), an understanding of these modern tools needs to incorporate forms of social relations and institutions within which these tools are framed.

Tourism is shaped within a markedly industrial-capitalist context of activity, involving social relations, institutions and mechanisms that could be appropriated and transformed by ordinary people, thereby breaking the exclusivity that specialists and institutions such as the market have over it. Community approaches to managing and organising tourism, such as CBT, aim to turn tourism into a convivial tool insofar as these approaches generate autonomy and limit, to some extent, the monopoly of the tourist industry and its objective of accumulating capital. A convivial use of tourism would mean that a community no longer functions as a reflection of dominant capitalism, but instead contests and negotiates its logic. From this perspective, CBT could be a way of introducing conviviality and degrowth into the sphere of tourism.

This raises a question as to how conviviality would operate in practice. Its operation would depend on how tourist activity was integrated into the logic of the community and, more specifically, on ensuring that CBT became part of community-building processes. To bring this about would entail making the organisation of tourism into a community matter, framed within the context of three converging logics/practices: commonisation, collective action and reciprocity. This adaptation of tourism to the community dynamic requires the implementation of organisational interfaces that can help translate and adapt tourism to community-based logics that tie in with conviviality, thus shaping the organisational dimension of degrowth.

Commonisation is a process whereby the use of a resource becomes communal and, therefore, subject to collective institutions (Nayak & Berkes, 2011). In most cases, CBT involves developing common ownership and usage of different types of resources (landscapes, cultural elements, infrastructure) that are subject to excludability and subtractability (Ostrom, Burger, Field, Norgaard, & Policansky, 1999) and which, therefore, require collective action in order to regulate their management (decisions, supervision, initiatives, assemblies, committees). This entire process involves collective reflection and decision-making, which conditions not only the material use of resources, but also interpersonal relations and the way people relate to their environment. Although not sufficient in themselves, commonisation and collective action are necessary to generate conviviality.

In this regard, one essential element for the development of CBT is a local capacity to link the capitalist logic that underscores tourist activities (profit, growth, competition, exchange) with community-based logics rooted basically in reciprocity. Practices of reciprocity (cooperation, mutual aid, collective work) focus on social bonds (Godbout, 2013; Temple, 2003a, 2003b; Temple & Chabal, 1995), whereas practices of exchange operate outside of these social bonds, focusing on the objects that are circulated, rather than on people and their connections. Hence, exchange practices encourage accumulation over and above social integration. Reciprocity and exchange, generating two frameworks that govern both interpersonal relations and human-environment relationships, give rise to two complementary logics that coexist within communities (Temple, 2003b). However, with globalisation leading to a promotion of exchange practices as the dominant way of understanding human relations, the space for reciprocity has gradually been eroded. In many communities, the response has been to counter this growing hegemony of exchange over reciprocity through using organisational interfaces (Temple, 2003b) that are capable of turning frameworks of economistic exchange into contexts of reciprocity, thus preventing the steady erosion of the latter. It is important to identify explicitly in the context of CBT how these interfaces are manifested and through which specific practices.

Reciprocity is not an essentialist or pre-modern praxis, but rather a dynamic, fluid and eminently contextual one that involves change and adaptation. Reciprocity is embodied through multidimensional social practices that simultaneously express a symbolic content (the social bond they promote), an economic interest (calculation and profit) and an ethical meaning (obligation to reciprocate) (Walsh-Dilley, 2017). Reciprocity forms part of a moral economy and, as such, it is negotiated with the market economy. For communities, the practice of reciprocity is a pragmatic resource (Walsh-Dilley, 2013) that allows them to develop tactics (de Certau, 1990) to navigate the challenges of globalisation resiliently. The integration of communities into the global market does not necessarily need to erode reciprocity; indeed, reciprocity can be enhanced within a new hybrid economic space in which market strategies exist alongside community logics (Walsh-Dilley, 2013). This potentiality needs to be taken into account when analysing the role of reciprocity and exchange in community-based tourism.

Therefore, it appears that CBT activity, if grounded in processes of commonisation, collective action and practices of reciprocity, could become a convivial tool, developing forms of organisation that foster a strategy of degrowth, with positive economic and environmental effects on the social-ecological system. The case study examined here is intended to develop this analytical proposal further, revealing the potentialities and limits of the CBT/degrowth linkage.

Case study and method

Agua Blanca is a community (292 inhabitants, 9,000 hectares) located within the Machalilla National Park (Ecuador), in the Manabí region, one of the poorest in the country. The origin of Agua Blanca is linked to a hacienda founded at the end of the nineteenth century, but the community did not effectively emerge until the hacienda went bankrupt at the beginning of the 1970s (Ruiz-Ballesteros, 2009). The creation of the national park in 1979 brought about acute social conflict and the threat of eviction, because community subsistence was based on activities that had become illegal due to their negative environmental effects – namely, hunting, timber extraction and charcoal production.

In the mid-1980s, an archaeological project to collect artefacts left from the Manteña culture helped promote a process of cultural heritagisation (Hudson, Silva, & McEwan, 2016) that became the basis for community tourism development. Furthermore, the archaeologists acted as mediators in the conflict between the community and park officials, facilitating an agreement in which the community would gain the right to retain its territory, to develop subsistence agriculture and livestock farming, as well as CBT. In return, the community pledged to control and eradicate any activities that would have a major negative effect on conservation, such as timber extraction, charcoal production and hunting. Although there are still occasional isolated disagreements, park officials and the community work together as strategic allies, shaping a multi-level adaptive co-management approach based on a conceptualisation of the land as a commons (Ruiz-Ballesteros & Gual, 2012).

Agua Blanca has become a role model for CBT in Ecuador (Ruiz-Ballesteros & Solis, 2007). Its indigenous character (Manteño People) and its status as a community (administrative level) allow it to enjoy autonomy and decision-making capacity on its territory, as recognized by Ecuadorian law. Harnessing the tourist potential of the Machalilla National Park, the community has substantially increased its visitor numbers from 9,605 (2006) to 24,251 (2018). The local tourist product involves a guided tour lasting between two and three hours, showcasing the community's natural and archaeological heritage. The residents of Agua Blanca organise this activity through several committees that bring together 35 locals (29 men, 6 women) from 35 of the 66 households that make up the community (2018), who are subject to the authority of the communal assembly. In order to become tourism workers, members of the community must first ask the assembly. They then have to go through a training period and pass a test with experienced members of the committee. There can be no more than one member of the same family working in tourism. There is no typical profile amid tourism workers: they all have very different ages and various roles in their respective households. They work in shifts, taking care of tourists and maintaining infrastructure, including the archaeological remains, footpaths, the museum and a lagoon.

CBT accounts for more than 50% of the tourism workers' household income and is the main source of revenue for the community's council, through taxes. In addition to this collective activity, local tourism fosters revenues based on complementary family-run initiatives, involving crafts, accommodation and restaurants. All of the tourist activities carried out in the community are subject to collective regulation, and their revenues are taxed with a community levy. In addition, some local residents also work as guides outside of the community, employed by tourist agencies that operate in the park. In 2017, revenues from tourism accounted for 47% of the local monetary economy, and more than 70% of the families received some earnings from tourism. The other main pillar of the local economy is agriculture, which almost exclusively involves subsistence farming for self-consumption. Approximately 75% of the families work a communally owned vegetable plot and receive collectively managed irrigation water. Livestock (mainly goats) and employment outside of the community also provide alternative sources of income and work in an economy marked by pluri-employment. This economic model, wherein tourism is the main activity, has a positive socio-environmental effect: revenues are widely distributed in the community, which contributes to the reduction in the extraction of natural resources (coal, hunting). In this sense, it could be argued that the sustainability of the socio-ecosystem depends on tourism activity. If tourist activity fails, then it is possible to focus on livestock and vegetable plots as subsistence activities.

From an organisational perspective, two fundamental factors help understand how CBT works in Agua Blanca. First, an ongoing capacity for collective action has ensured that the tourism product itself, the consolidation of relevant infrastructure and the self-organised development of the tourist business have been sustained. Second, as is discussed in greater detail later, this collective action has been marked by an assembly strategy and a rotational logic that pursues equality of opportunities within the local community. In fact, organisational decisions are made by committees, but strategic decisions about tourism are taken by the communal assembly, where all of the households are represented, regardless of whether or not they participate in tourist activity. Their capacity for collective action hinges on a dense social network founded on kinship as well as on the necessity of working together in order to manage the commons and develop tourist activities. Rational interests and community cohesion feed collective action. However, commons and collective management do not eclipse individual interests, which are particularly characteristic of the globalised world and which are also expressed in everyday life in this community. Thus, community building, an exercise that seeks to integrate collective interests and commitments with individual desires and expectations, is in a state of permanent tension.

An exploratory and longitudinal (2006–2018) ethnographic case study (Poteete, Jansen, & Ostrom, 2010; Yin, 2009) was carried out in Agua Blanca, involving extensive and intensive (11 months) periods of fieldwork. Fieldwork was carried out between 2006–2009 and 2015–2018; the longest stay was two months long and involved only one researcher, except during 2018. Throughout the research, communication and cooperation with the community has been fluent, including the devolution of available research materials and information.

Three levels of analysis were developed, focusing on the community, the household groups and the patterns of resource extraction. The primary research methods involved participatory observation, in-depth interviews, focus groups and resource surveys.

Participant observation helped the researchers gain an in-depth understanding of everyday life in the community and how its main institutions operate. In-depth interviews enabled access to the views of 15 key informants, maintained over time, that help trace the evolution of productive activities and the community in general, the swings and shifts in tourism, and personal experiences. Focus groups were used chiefly to reconstruct the community's recent history and its most salient episodes. Resource surveys were applied to household groups in 2007 and 2018 to determine their activities, earnings and strategies, involving a collective interview with at least two of their members. Furthermore, the community's archives provided a key source of information about demographic and economic processes, as well as enabled us to reconstruct, through the minutes of the assembly meetings, the debates held and collective decisions made.

Analysis of the data obtained was based on triangulation of the different techniques used and an application of theoretical saturation, as derived from grounded theory (Charmaz, 2003; Glaser & Strauss, 1967). For each of the research topics, codifications were developed to classify the different types of data obtained. The processes studied were then analysed and described based on these codes. Our conclusions on these processes and topics were subsequently contrasted with key informants. At the end of each period of fieldwork, the results obtained were shared with the community in an assembly as a strategy to expand and contrast them.

Community-based tourism in Agua Blanca

In Agua Blanca, tourism is organised differently from conventional practice, through the development of commonisation, collective action and reciprocity. These convivial features shape CBT practices, which facilitate an understanding of the organisational dimension of degrowth and its consequences.

Commonisation and tourism

The development of CBT in Agua Blanca has occurred in conjunction with an intensive process of commonisation. Despite the inclusion of the entire territory within a protected area, which entails some forms of state control, the whole communal area is actually managed as a common property resource for harvesting, cultivation, housing and tourism. This has been possible after a long process of conflict and negotiation with the park authorities, which are far from being over. Situations of tension still arise periodically regarding management of the territory or use of resources, which are resolved through negotiation between the community and the park. Commons management is embodied practically in community institutions especially designed for the collective governance of these communal resources, in which rules and sanctions for their use are determined and applied and excluding any individual or institution outside the community as participants.

Subsistence agriculture is practised in a communal area of vegetable plots distributed among the families, involving a collective turn-based system of irrigation. Housing is only built once the assembly gives its authorisation to occupy the land and authorises timber extraction as required from the forest. There is no private property in the community and no documentation that certifies individual or family possession of any space, but only a right of usufruct, endorsed collectively.

CBT has accentuated and consolidated this commonality. The tourist infrastructure of Agua Blanca (pathways and signage, archaeological remains, the lake and adjacent installations, the crafts centre, museums and exhibits) therefore remains a collective possession financed by the community and non-governmental organisations and was built through work undertaken collectively.

This development and maintenance of communal ownership, and resistance to external interference (Ruiz-Ballesteros, 2009; Hudson et al., 2016), is a core element that needs to be grasped for understanding how this community is able to negotiate the extent of its dependence on the market and to generate relative economic autonomy, particularly in tourism. The communal ownership and use of resources strengthen the community's position on the territory: collective ownership places resources outside the market as it makes them inalienable under Ecuadorian law. It is therefore a way to protect the community.

The residents of Agua Blanca, through their common use of the territory, enjoy complete exclusivity regarding any productive activity within the community: no external agent can participate in the tourist business. This makes all tourism revenues exclusive to the community, without the need to compete with external agents, which empowers them. Revenues from CBT derive primarily from tickets sold to incoming tourists, and benefit not only the tourism workers but also the community as a whole, which receives 20% of any earnings due to the communal nature of the resources used for tourism. CBT, therefore, makes the process of commonisation explicit in the most radical way, both symbolically and economically, on account of its weight within the local economy. This commonisation fosters a very particular relationship between the people and the environment they inhabit (ranging from their homes to the cloud forest), which requires the people to act collectively in order to manage and sustain that relationship.

Collective action for tourism development

This communal system is not sustainable without an appropriate level of collective action to organise and manage it, entailing appropriate governance. Although the population barely reaches 300 inhabitants, the community of Agua Blanca is marked by clear internal diversity in relation to age, sex and incomes. The communal structure guarantees a certain subsistence minimum for individuals and access to collective resources (crops, tourism). Nevertheless, it is also possible to work outside of the community or to participate in private activities. This context helps shape a local society with a certain heterogeneity that clearly encourages differing interests, personal strategies and preferences, but at the same time, there is a clear necessity for collective management of the territory and of resource use.

Decisions are made concerning a range of different matters, and effective formulas for control and regulation have been established concerning, for example, who may build a house and how it may be built, how to regulate harvesting and gathering activities, limitations on livestock farming or the extraction of timber, and how to participate in communal or family-run tourist activities. It is also vital to consolidate perspectives at times, to provide a united front in relation to external authorities, such as the state in the form of the national park or the market in the form of external promoters seeking to access the community's resources. To operate effectively requires a substantial level of collective action, which we examined occurring through two key institutions – namely, the assembly and the tourist committee (named locally as the archaeological committee).

The assembly is the institution that formally embodies the community and provides the foundation for its collective action. In the assembly, adults of legal age to participate come together to manage issues arising in relation to any tension between individual and collective interests, and decisions are made concerning all communal issues.

In the assembly, no decisions are made without considerable debate, and such discussions frequently entail arguments and counter-arguments, comparisons and analogies, involving a very slow process of refinement. Decisions are not limited to a straightforward vote between the different choices available, nor do they occur through anonymous pronouncements. There is no majority imposed as such, nor is it necessary to achieve absolute consensus; instead, a more diffuse and intuitive logic is followed to allow for decisions: "when the matter has matured" (Raúl, president). This model of participation is particularly salient when elections for governing posts occur, which involves members nominating people, who are voted for individually by means of explicit personal declarations of support. This assembly dynamic is very different to the usual dynamic for constructing collectivist entities in Western social contexts.

The assembly is the place in which thought and action are guided, providing a tactical and ethical platform. Yet, it is not the only space in which discourses that represent reality are constructed. This also occurs in informal and smaller social contexts (e.g., households, workplace), but the assembly is the 'official' space for reality to be shaped as a community-based project, notwithstanding the conflicts and disputes evidenced in its debates. In this regard, the assembly is a kind of interface that explicitly articulates individual and collective dimensions of social reality, without negating either of these two dimensions, to help integrate their inherent contradictions (Temple, 2003a).

The evolution of this institution more recently has been particularly significant, especially concerning the presence and active participation of women. In 2006, barely one-third of women were members; by 2018, practically all women were members, and their presence in communal governance is now firmly established. Indeed, in the period 2016–2018, the vice president of the assembly was a woman.

Importantly, the assembly defines a culture of collective action through which an entire model of CBT organisation has been created and developed, setting a framework for deciding who participates in the tourist business, how participation is organised (rights and obligations), how many people can participate, what revenue collection and profit distribution model is used to apply, and even determining the extent to which tourism can complement other activities. In Agua Blanca, tour guides, according to regulations defined by the assembly, cannot engage in activities likely to have a significantly negative effect on the environment, such as hunting, timber extraction or charcoal making.

The assembly is the custodian of tourist resources, responsible for their uses and management. The tourist committee is a delegated body made up exclusively of community tour guides, who make decisions pertaining to the organisation of tourism. The assembly and the tourism committee are therefore interfaces that foster the transformation of an individual logic into a logic of collective action which enables commonisation. Community tourist activity could have been organised in other ways, but the assembly decided that it should take a collective approach under its supervision, involving community governance. Hence, CBT was set up, synonymous in this case with collective action and commonisation, and so its organisation is based on reciprocity.

Reciprocity in tourism organisation

Reciprocity in Agua Blanca is anchored in the ways in which resources are used and accessed, such as access to irrigable cropland, distribution of job opportunities and assignment of gathering quotas. For accessing CBT resources, the rationale is similar. To determine the number of people who can work as community tour guides, a minimum earnings threshold is used to determine who can give the exclusive engagement required by this tourist activity. No single family can have two members working in CBT if there are other families with no members employed in this activity. Therefore, participation in CBT cannot be concentrated within the same families, nor can it be a full-time occupation.

The practice of reciprocity in the context of CBT in Agua Blanca has two fundamental and intertwined expressions – namely, teamwork and the rotation of tasks and positions. Teamwork, more than any other activity whether ritual or festive, fosters sociability; in contrast to the individual work of the household unit, CBT activity is synonymous with 'doing together' and promoting reciprocity. Rotation is a way of allocating individual opportunities and obligations based on collective criteria (Arguedas, 1987; Michaux, Gonzáles, & Blanco, 2003), which follows the logic of distribution with a greater emphasis on equity, and it takes account of the needs of families rather than just individual needs. Hence, rotation combines social, economic, political and environmental aspects in its operation and also forges 'social bonds' (Temple, 2003a) insofar as it constructs patterns of behaviour among community members. Selective rotation within this community constitutes an expression of reciprocity, a logic that pervades community practice and, while aware of internal inequalities, this community's use of rotation seeks at least not to aggravate them.

Rotation, as a practice, can be understood as an interface that turns economistic exchange and competitiveness into a practice of reciprocity. This interface occurs because rotation is grounded in assumptions of "limiting individual advantage" (Temple, 2003b, pp. 48–49) and of a fair and equal assignation of opportunities. Under capitalism, access to opportunities is predominately regulated through processes of economistic exchange and individualistic-competitive logic. Rotation goes against this current, as does the assembly itself, and it is in the organisation of tourist activity that its most radical expression can be seen in the community.

The committee that organises CBT promotes equality of opportunities and fosters improvements in the performance and abilities of members, while profits earned on tourism activities undertaken collectively are distributed equally among all the guides. This approach embodies resistance to the assumptions of an individualist meritocracy in relation to communal organisation, as discussed by Bauman (2001). Furthermore, this approach substantially fosters cohesion, as is reported to be generally the case with any practice of reciprocity (Temple, 2003a).

This organisational model is not free from tension and conflict. In this context, any issues arising are addressed in two possible ways. First, each working group holds a brief meeting at the start and end of the day to deal with disagreements and working problems. Second, a monthly committee meeting may examine issues that require more in-depth discussion that go beyond the working group. The main problem to be solved is the linkage between collective discipline and personal interests and attitudes. A general framework of horizontal relationships and their common interest in the tourism business are the incentives that allow conflict resolution from a community logic. A "common habitus" has thus been created, which is characterised precisely by conflict resolution capacity through dialogue and agreement in assembly-type meetings.

CBT has been developed in this community to create a space for socialised learning and practice through organisational levelling, with rotation as the guiding principle, rather than encouraging specialisation and emphasising differences. Furthermore, all members of the committee receive the same monthly salary. This approach to CBT does not mean that there are no leaders, no differences and inequalities, or no people with greater skill sets compared to others. Rather, the community focuses its strategy on not aggravating or making explicit these differences: reciprocity must not yield in the face of pressure from the logic of exchange. CBT for this community involves cooperation rather than competition. This is perhaps the main organisational consequence of CBT: to enhance commonality rather than individual interest. By means of teamwork and rotation, differences and inequalities are not eliminated, but their effects are minimized. Leaders and individuals with particular skills are required, but inequalities in access to resources or benefits do not build on these differences.

This approach to CBT appears to have helped ensure the success of this tourism organisation, involving a centripetal practice of teamwork and rotation that counters, at least to some extent, the centrifugal tendencies that would otherwise be encouraged by the market through standardised tourist activity. In accordance with Temple (2003a), the case of Agua Blanca shows how tourism operating within a globalised philosophy based exclusively on exchange, comprising individualism, meritocracy, competition and increased inequality, can be turned into a practice of reciprocity based on 'limiting individual advantage', thanks to the development of interfaces such as the assembly, the tourism committee, teamwork and rotation. Hence, tourism in Agua Blanca fosters 'social bonding' over the exclusive exchange of goods (Temple & Chabal, 1995) and, therefore, lends support to the conclusion that "increasing integration into 'modern' systems and processes need not, as is often supposed, undermine the moral economy and can potentially invigorate it" (Walsh-Dilley, 2013, p. 678).

Degrowth and tourism in Agua Blanca

Agua Blanca is a relevant case study because it shows that, in the unusual context of tourism, organisational practices linked to degrowth strategies can be developed. CBT in Agua Blanca is grounded in commonisation, collective action and practices of reciprocity. Generated through different interfaces (assembly, committees, teamwork, rotation), tourism has become a convivial tool whereby the community 'controls' tourism, rather than tourism – as an overwhelming and specialised monopoly (Illich, 1973) – completely controlling the community. Within this context, it is possible to identify some CBT features and effects that foster a strategy of degrowth, with positive economic and environmental effects on the social-ecological system, as follows:

1 The manner and intensity with which tourist activity is organised strengthens *community agency* (Matarrita-Cascante, Brennan, & Luloff, 2010), facilitating resistance to market expansion. The community controls the tourist product and management of it, preventing the development of conventional tourism, unlike the case in nearby communities (Lager, 2016).
2 The goal of CBT is to offer a source of earnings to as many households as possible, preventing full-time engagement in tourism and the concentration of employment in the same families, and to the community in general through sharing profits. The *economic purpose of tourism* in Agua Blanca is not focused on individuals, but on households and the community. For example, tourism in Agua Blanca in the period from 2006 to 2017 underwent a dramatic increase in visitors (+252%) and in revenues (+356%) but, with the CBT model in place, internal socio-economic differentiation did not increase. In fact, the community Gini index was 0.223 in 2006 and 0.233 in 2017, showing an insignificant increase in inequality.
3 This form of *convivial tourism is a strategic mechanism* for controlling and regulating the extraction of environmental resources, which has allowed such resources to regenerate. The guides are expressly forbidden to hunt or to produce charcoal, the two most detrimental activities from an environmental perspective. Consequently, between 2006 and 2017, charcoal production reduced by 50%. The part-time nature of work in CBT also fosters pluri-employment. It is assumed that guides, during the ten days a month that they work exclusively in the tourist sector, earn enough to meet their family's minimum monthly monetary requirements. Their households have a vegetable plot and domestic animals that largely guarantee food self-sufficiency, and they can also complement their earnings with harvesting/gathering (ivory palm, dry timber, wild fruits) and family-oriented tourist activities (crafts, food provision, accommodation). Pluri-employment is in itself a factor promoting socio-ecological resilience.
4 In 2018, 53% of the local families participated in tourism-related committees concerned with guiding, crafts and massages. From the 1990s onwards, 30% more have participated at some point but have then left to work in other activities within or outside of the community, becoming employed as national park guides or working in construction or administration. This circumstance shows the *dynamic nature of participation* in CBT, with all former guides having participated directly in relationships of reciprocity and collective action involving greater sociability and social cohesion. None of them

has returned to hunting or charcoal making, and they and their households still maintain economic pluri-activity. As a result, tourism development, together with new patterns of reproduction (2.1 children per family in 2017), has helped to sustain the population in the territory through raising the quality of life and offering young people an alternative path to urban migration (no youth migration registered since 2006). The community does not allow in-migration, so participation in CBT is reserved only for its members. For local young people, tourism is a strategic activity to remain in the community, become independent and be able to found their own families and households, while also maintaining the current socio-ecosystem (Ruiz-Ballesteros & Ramos-Ballesteros, 2019).

CBT in Agua Blanca is a phenomenon that promotes community agency, with a household focus that supports pluri-employment and reciprocity practices, as well as having a positive effect on demographic evolution, contributing to a reduction in natural resource extraction and not increasing socio-economic inequalities. All of these features are associated with degrowth strategies. However, CBT has also contributed to the monetisation of the local economy, which has highly ambivalent effects. Although monetisation has enabled living standards to improve, it has also increased consumption (for example, in relation to energy, domestic appliances and motorbikes), community connectivity through making it easier for tourists to access the community and for young people to complete their high school education, and waste production. CBT in Agua Blanca promotes an alternative development of tourism that may resemble a *tourisme de décroissance*. However, it has other effects related to consumption and connectivity that are at least ambivalent as they intensify the use and exploitation of resources. Therefore, we always have to nuance the link between CBT and degrowth.

It seems that, from a degrowth perspective, CBT in Agua Blanca presents a certain ambivalent character, but at the same time, reveals two key features for understanding the development of degrowth strategies in tourism:

1 Its *convivial character*, as tourism has been largely run and transformed by ordinary people, generating local autonomy. Thus, through CBT, the community no longer functions as a subsidiary of a dominant capitalism, but instead contests and negotiates that logic. In Agua Blanca, commonisation, collective action and reciprocity in CBT generate community agency, a household focus and socio-economic equity, which promote its conviviality.
2 The development of interfaces has facilitated the *co-existence of market strategies with community logics*. The assembly, the committees, teamwork organisation and rotation are key CBT institutions in Agua Blanca that have promoted a shift from an individualistic capitalist rationality to a more collective community-based rationality. These CBT institutions in Agua Blanca provide the basis for understanding the socio-ecological effects of CBT for this community.

Conviviality and interfaces, as this case study showed, are analytical concepts that help elucidate a critical element for an appropriate understanding of CBT – namely, its hybrid character (between market rationality and community logic). And doing so, it allows us to focus on the 'organisational dimension of degrowth' for better approaching the potentialities of CBT as strategy for degrowth.

Agua Blanca is a very particular community, but its study reveals processes and perspectives that can be applied beyond the case study. First, it shows us that to properly understand 'tourism degrowth', it is necessary to transcend the economic domain and consider organisational aspects. Only in this way is it possible to assimilate the holistic dimension of degrowth and elucidate the basis of its effect on natural resources. While the configuration of Agua Blanca as a community – in the context of Ecuador – is very particular, and marks its organisational potential, there are analogies with other cases in America, Africa and Asia. In the Global North, rural tourism and the small localities that develop it also have analogies in terms of organisational potential. From these similar models, with community potential, we can explore multi-level adaptive co-management systems as forms of governance that could enable degrowth strategies. This is the value of exploratory case studies (Yin, 2009): their usefulness to generate theoretical-methodological perspectives applicable to other cases and contexts.

Conclusions

If CBT in the form examined in this chapter could be considered as a viable alternative form of organising tourism, perhaps as a 'post-capitalist tourism', then aspects of degrowth are likely to characterise it. Analysing CBT focused on degrowth allowed for a consideration of an aspect of CBT rarely explored in the relevant literature, that is, its hybrid character. To engage with CBT as a potentially viable strategy for degrowth, it needs to be assumed that people can "combine local, non-market and cooperative strategies with increasing integration into global markets to construct feasible livelihoods that are socially and ecologically appropriate" (Walsh-Dilley, 2013, pp. 661–662). This feature constitutes the hybrid character of CBT. In post-industrial society, as noted by Illich (1973), there are several co-existing complementary modes of production; the industrial system does not disappear, but it needs to be limited. CBT has emerged and ought to be understood in this context.

Given this context, it appears worthwhile to study how experiences of CBT express that hybrid character, for example, through interfaces and conviviality, to clarify to what extent hybridity is a consistent feature of CBT and supports its potential role as a degrowth catalyst. Therefore, as undertaken in this case study, an identification of specific interfaces and practices of conviviality that facilitate non-capitalist practices in capitalism's interstices (Fletcher, 2016), operating in the midst of the market but not through the market, and which ultimately foster processes of 'socio-ecological transition' (Martínez Alier, 2009) is likely to provide valuable insight. Through such studies, interface practices and institutions, such as assembly, rotation and teamwork as identified in this study, can become analytical tools for analysing the particular nature of this form of organising tourism and its linkage with degrowth strategies. Finally, this approach would facilitate determining in what sense, and to what extent, tourism with CBT practice can become a convivial tool (Illich, 1973), contributing to degrowth. This perspective could be useful for researchers and practitioners interested in exploring how tourism can promote degrowth from an organisational perspective. The analytical approach and the case study presented here are likely to prompt further reflection about the possibilities of a *tourisme de décroissance*.

Acknowledgements

This work was supported by the MINECO, Spain; under grant CSO2017–84893-P.

References

Amati, C. (2013). "We all voted for it": Experiences of participation in community-based ecotourism from the foothills of Mt Kilimanjaro. *Journal of Eastern African Studies*, *7*(4), 650–670.

Arguedas, J. M. (1987). *Las comunidades de España y del Perú* [The communities of Spain and Peru]. Madrid: Ministerio Agricultura.

Bauman, Z. (2001). *Community: Seeking safety in an insecure world*. London: Polity Press.

Blackstock, K. (2005). A critical look at community based tourism. *Community Development Journal*, *40*(1), 39–49.

Bourdeau, P., & Berthelot, L. (2008). Tourisme et decroissance: de la critique à l'utopie? [Tourism and degrowth: From criticism to utopia?]. In F. Flipo & F. Schneider (Eds.), *Proceedings of the first international conference on economic de-growth for ecological sustainability and social equity* (pp. 78–86). Paris: Archive ouverte en Sciences de l'Homme et de la Société.

Charmaz, K. (2003). Qualitative interviewing and grounded theory analysis. In J. A. Holstein & J. F. Gubrium (Eds.), *Inside interviewing: New lenses, new concerns* (pp. 311–330). London: Sage.

Dangi, T. B., & Jamal, T. (2016). An integrated approach to "sustainable community-based tourism". *Sustainability*, *8*(5), 475. https://doi.org/10.3390/su8050475

De Certau, M. (1990). *L'Invention du Quotidien I. Arts de Faire*. [The practice of everyday life]. Paris: Gallimard.

Deriu, M. (2014). Conviviality. In G. D'Alisa, F. Demaria, & G. Kallis (Eds.), *Degrowth: A vocabulary for a new era* (pp. 107–110). London: Routledge.

Fletcher, R. (2016). Tours caníbales puesto al día: La ecología política del turismo. [Cannibal tours updated: The political ecology of tourism]. *Ecología política*, *52*, 26–34.

Gascón, J. (2013). The limitations of community-based tourism as an instrument of development cooperation: The value of the social vocation of the territory concept. *Journal of Sustainable Tourism*, *21*(5), 716–731.

Giampiccoli, A., & Saayman, M. (2018). Community-based tourism development model and community participation. *African Journal of Hospitality, Tourism and Leisure*, *7*(4), 1–27.

Glaser, B. G., & Strauss, A. L. (1967). *The discovery of grounded theory: Strategies for qualitative research*. London: Weidenfeld and Nicolson.

Godbout, J. T. (2013). *Le don, la dette et la indentité* [Gift, debt and indentity]. Paris: Le bord de l'eau.
Hall, C. M. (2009). Degrowing tourism: Décroissance, sustainable consumption and steady-state tourism. *Anatolia*, *20*(1), 46–61.
Hiwasaki, L. (2006). Community-based tourism: A pathway to sustainability for Japan's protected areas. *Society and Natural Resources*, *19*(8), 675–692.
Hudson, C., Silva, M. I., & McEwan, C. (2016). Tourism and community: An Ecuadorian village builds on its past. *Public Archeology*, *15*(2–3), 65–86.
Illich, I. (1973). *La convivialité*. [Tools for conviviality]. Paris: Editions du Seuil.
Lager, M. T. (2016). *Montañita. Una comuna entre territorio, identidad y turismo* [Montañita. A commune between territory, identity and tourism]. Quito: Abya-Yala.
Martinez Alier, J. (2009). Socially sustainable economic de-growth. *Development and Change*, *40*(6), 1099–1119.
Matarrita-Cascante, D., Brennan, M. A., & Luloff, A. E. (2010). Community agency and sustainable tourism development: The case of La Fortuna, Costa Rica. *Journal of Sustainable Tourism*, *18*(6), 735–756.
Michaux, J., Gonzáles, M., & Blanco, E. (2003). Territorialidades andinas de reciprocidad: la comunidad. [Andean territorialities of reciprocity: The community]. In D. Temple (Ed.), *Las estructuras elementales de la reciprocidad* (pp. 75–96). La Paz: TARI/Plural Editores.
Mitchell, J., & Muckosy, P. (2008). *A misguided quest: Community-based tourism in Latin America*. London: Overseas Development Institute.
Nayak, P. K., & Berkes, F. (2011). Commonisation and decommonisation: Understanding the processes of change in the Chilika Lagoon, India. *Conservation and Society*, *9*(2), 132–145.
Okazaki, E. (2008). A community-based tourism model: Its conception and use. *Journal of Sustainable Tourism*, *16*(5), 511–529.
Ostrom, E., Burger, J., Field, C. B., Norgaard, R. B., & Policansky, D. (1999). Revisiting the commons: Local lessons, global challenges. *Science*, *284*(5412), 278–282.
Poteete, A. R., Jansen, M., & Ostrom, E. (2010). *Working together: Collective action, the commons, and multiple methods in practice*. Princeton, NJ: Pinceton University Press.
Ruiz-Ballesteros, E. (2009). *Agua Blanca. Comunidad y turismo en el Pacífico ecuatorial* [Agua Blanca. Community and tourism in the equatorial Pacific]. Quito: Abya-yala.
Ruiz-Ballesteros, E. (2017). Keys for approaching community-based tourism. *Gazeta de Antropología*, *33*(1). http://hdl.handle.net/10481/44362
Ruiz-Ballesteros, E., & Gual, M. A. (2012). The emergence of new commons. Community and multi-level governance in the Ecuadorian coast. *Human Ecology*, *40*, 847–862.
Ruiz-Ballesteros, E., & Ramos-Ballesteros, P. (2019). Social-ecological resilience as practice: A household perspective from Agua Blanca (Ecuador). *Sustainability*, *11*, 5697. https://doi:10.3390/su11205697
Ruiz-Ballesteros, E., & Solis, D. (Coords.). (2007). *Turismo comunitario en Ecuador. Desarrollo y sostenibilidad social* [Community-based tourism in Ecuador. Development and social sustainability]. Quito: Abya-Yala.
Taylor, S. R. (2017). Issues in measuring success in community-based Indigenous tourism: Elites, kin groups, social capital, gender dynamics and income flows. *Journal of Sustainable Tourism*, *25*(3), 433–449.
Temple, D. (2003a). *Teoría de la reciprocidad* [Theory of reciprocity]. La Paz: Padep-gtz.
Temple, D. (2003b). *Las estructuras elementales de la reciprocidad* [The elementary structures of reciprocity]. La Paz: TARI /Plural Editores.

Temple, D., & Chabal, M. (1995). *La Reciprocité et la Naissance des Valeurs Humaines* [Reciprocity and the birth of human values]. Paris: L'Harmattan.

Walsh-Dilley, M. (2013). Negotiating hybridity in highland Bolivia: Indigenous moral economy and the expanding market for quinoa. *Journal of Peasant Studies*, *40*(4), 659–682.

Walsh-Dilley, M. (2017). Theorizing reciprocity: Andean cooperation and the reproduction of community in highland Bolivia. *The Journal of Latin American and Caribbean Anthropology*, *22*(3), 514–535.

Yin, R. (2009). *Case study research: Design and methods* (4th ed.). Thousand Oaks, CA: Sage.

12 Don't leave town till you've seen the country

Domestic tourism as a degrowth strategy

Paul W. Ballantine

Introduction

A young man travels the world but is unable to answer questions about places in his home country of New Zealand. His standard reply becomes "Dunno, I've never been there", even when asked for information about famous New Zealand destinations from a sophisticated Frenchwoman. Produced by the New Zealand Tourist and Publicity Department in 1984, the situation described was depicted in an advertising campaign which ended with "A word of advice to all New Zealanders – don't leave town till you've seen the country" (Ngā Taonga Sound & Vision, 1984). Now considered an iconic advertisement in New Zealand, its focus was to encourage New Zealanders to explore their own country before heading off on their overseas experience.

Fast forward to early 2020, and the amount of outbound international tourism undertaken by New Zealanders has effectively reached zero, while the tourism activity of New Zealand residents is almost exclusively domestic. Of course, the reasons for this are caused by the COVID-19 global pandemic, which has led to most international borders being closed, as opposed to an unbelievably successful domestic tourism marketing campaign. COVID-19 aside, domestic tourism has become even more valuable to New Zealand than international tourism. In part, this is due to domestic tourists being able to travel at any time of year, and that they are more likely to visit those regions that tend to be overlooked by international tourists (Downes, 2019). However, a critical issue which has worked against domestic tourism within New Zealand is that airlines have made it increasingly affordable to fly to short-haul international holiday destinations, such as Australia and the South Pacific.

Building upon the themes presented elsewhere in this book, the purpose of this chapter is to explore how a renewed emphasis on domestic tourism might provide a viable strategy towards supporting degrowth and the mechanisms by which international tourism can be demarketed to New Zealand consumers. Two main approaches are discussed which might be used to lead to an outcome such as this. First, the concept of macro-social marketing is introduced as a systems framework by which international tourism might be demarketed to New Zealanders, with domestic tourism instead being promoted as a desirable alternative. Second, the consumer lifestyle of voluntary simplicity is then explored to understand what lessons the ethos of this consumption lifestyle might provide in terms of increasing the desirability of domestic tourism options. Before examining these two areas, the chapter starts by looking at the current temporal context of the global COVID-19 pandemic, as recent research has linked the potential effects of this global health crisis to the central focus of this book – that of degrowth.

The COVID-19 pandemic and tourism

COVID-19, which was declared a global pandemic on 12 March 2020, has had a significant impact on global economic, political and social-cultural systems (Sigala, 2020). Government responses to the pandemic have included requirements for social distancing, bans on travel and mobility – both domestic and international – restrictions on crowding, lockdowns which have required non-essential employees to work from home, and self- or mandatory quarantine measures. Travel and tourism activities have been halted internationally, which has devastated the tourism sector, especially in countries such as New Zealand that are highly reliant on international tourist numbers to support the industry. However, tourism is an industry vulnerable to numerous environmental, political and socio-economic risks, and the industry has proven itself to be adept at bouncing back from various crises (Novelli, Burgess, Jones, & Ritchie, 2018). Hall, Scott, and Gössling (2020) support this view, noting that while COVID-19 has led to a substantial drop in tourism-related emissions due to parked carbon assets such as aircraft and cruise ships, it is doubtful that this decrease in activity will be permanent. The reasons given for this conclusion include potential industry bailouts, cheap capital and demands for employment generation, and further deregulation to stimulate tourism activity.

Assuming this to be true, Hall, Scott, and Gössling (2020) suggest that the rebound effects of tourism activity in a post-COVID-19 world may place the industry on an even more unsustainable trajectory, with its measures of success typically being growth in visitor numbers. However, there is also the potential for COVID-19 to provide an impetus to reconsider the global tourism system in a way which is more aligned to the United Nation's Sustainable Development Goals (SDGs) (Gössling, Scott, & Hall, 2020). In tourism research, this includes questions about whether COVID-19 will support nationalism and tighter border controls; the behavioural demand responses of tourists, both short and medium term, including the impacts of the widespread adoption of videoconferencing on business travel; and the implications of financial stimulus on austerity and reducing the effects of climate change. A final area identified by the authors was the role of domestic tourism in the recovery and transformation to more resilient destinations (Gössling et al., 2020).

The COVID-19 pandemic is a crisis of those societies that are rooted in the growth paradigm (Ötsch, 2020). Sigala (2020) surmises that tourism is a result of, but also responsible for a) an interconnected and global world; b) the negative effects of pollution, waste and climate change; c) global, national and regional economic development and growth; and d) the superiority of capitalist values in both decision-making and policy formulation. To address these issues, it has been suggested that tourism research in a post-COVID-19 era should challenge traditional growth-oriented paradigms and assumptions, and instead focus on reimagining and resetting tourism (Gössling et al., 2020; Hall et al., 2020; Sigala, 2020).

At a macro-level, Sigala (2020) argues that COVID-19 tourism research should address mindsets which include globalisation as an unstoppable force; neoliberal capitalism as the best system and decision-making tool for organising and allocating resources; and growth as the sole way for developing and measuring success. At a micro-level, she suggests that COVID-19 tourism research "should question and reset why tourism is viewed, practiced and managed as a way to 'escape', 'relax', 'socialise', 'construct identities/status', 'learn' and reward themselves from a routine, unpleasant and meaningless life" (Sigala, 2020, p. 314). This should include why tourism is viewed as an escape from a mundane and boring life, why people should travel long distances to be away from home to learn and grow, and why companies have led to communities and people being commoditised in order to drive economic development.

While there is evidence to suggest that the tourism industry post-COVID-19 might rebound in a way that is even less sustainable, there is also an opportunity to use this crisis to challenge many of the assumptions and practices grounded in a neoliberal worldview. This would be confronting for many of those involved with tourism research given that Hall (2011, pp. 298–299) suggests that

> the tourism academy could be loosely described as a bunch of relatively time and money rich people trying to find ways of getting other relatively time and money rich people to travel and travel more, sometimes with a good cause in mind like conserving heritage or creating jobs for the poor, but it's still about encouraging consumption.

In this regard, it is critical to consider how the degrowth discourse might be used to inform the approaches taken to encourage consumers to consider domestic tourism as a preferable and more sustainable alternative to their current international travel practices.

The degrowth discourse in tourism

Hall (2009) was one of the first authors to link the concept of degrowth to the sustainability of tourism. However, Fletcher (2019) notes that the origins of degrowth can be traced back to the 1960s and 1970s from authors who published in disciplines such as political and social ecology and ecological economics. Underlying the concept of degrowth is the central thesis that "growth is uneconomic and unjust, that it is ecologically unsustainable and that it will never be enough" (Kallis, Demaria, & D'Alisa, 2015, p. 6). Latouche (2003, pp. 3–4) contributed to the initial degrowth literature, defining it as being "a necessity, not a principle, an ideal, but the objective of a post-development society". This idea was ground in the notion of achieving social prosperity without the need for infinite growth. Latouche (2006, 2007) also defines what degrowth does not include. Specifically, degrowth does not condemn poor countries to poverty, nor does it mean zero growth or a return to the past. Moreover, it does not suggest a need for a new social order, such as authoritarianism, which is incompatible with democracy.

Degrowth can be understood in different ways, from a narrow economic transformation to a much broader cultural paradigm shift (Fletcher, 2019). Central to most discussions of degrowth is the need "to understand critically and undo the phenomenon of growth – a material, ecological, historical, discursive and institutional phenomenon that is at the heart of the Western imaginary and its colonial dominance – and to propose alternatives to it" (Kallis, 2018, p. 9). This can include both macro- and micro-level initiatives. Examples of specific macro-level initiatives include resource and emission caps, extraction limits, work-sharing (reduced work hours), basic income and income caps, consumption and resource taxes with affordability safeguards, supporting innovative models of local living, ethical banking and green investments. Micro-level initiatives to support degrowth include cycling, sharing, reusing and repurposing, vegetarianism or veganism, co-housing and consumer cooperatives.

A common misunderstanding equates degrowth with an economic decline or recession. However, as Kallis et al. (2018, p. 294) explain, "involuntary declines are not degrowth in themselves, and countries in recession or depression are not degrowth experiments". Thus, moments and places of crisis, burst financial bubbles, natural disasters and similar events are not degrowth. Instead, degrowth can be conceptualised as a "radical political and economic reorganization leading to drastically reduced resource and energy throughput" (Kallis et al., 2018, p. 291).

Given the negative environmental impacts directly attributable to tourism, it is considered a critical sector for implementing degrowth strategies (Hall, 2009). Tourism is enabled by a consumer dynamic which is never satisfied and where tourists continually seek out newer and more novel destinations and experiences (Higgins-Desbiolles, 2010). The desire of sensation seeking through travel continues to go unchallenged, further perpetuating the continuous growth agenda of the tourism industry and supporting the ability for tourism corporates to sustain their profit agendas (Higgins-Desbiolles, Carnicelli, Krolikowski, Wijesinghe, & Boluk, 2019; see also Chapter 10, this volume). Moreover, while the neoliberal growth agenda is not unique to the tourism industry, the negative outcomes of pursuing such a strategy has led to a growing interest in degrowth in tourism research.

Tourists are actively targeted in the competitive tourism marketplace to drive the endless growth that is a requirement of the contemporary political economy in many countries. Higgins-Desbiolles et al. (2019) argue that this illustrates the discriminatory nature of mobility in this era; refugees are not considered welcome, whereas tourism (for the privileged) is developed without consideration of its scale or overall impact on destinations, many of which are already overdeveloped. Andriotis (2018) discusses the potential for tourists to travel in a manner that reduces resource use and other negative impacts. This might include, for example, backpackers who move slowly between destinations and who spend considerable time in one place trying to minimise consumption. Tourism degrowth also converges with several other campaigns which have sought to mitigate the negative impacts of tourism growth (Fletcher, 2019), including the advocacy of 'slow' (Fullagar, Markwell, & Wilson, 2012), 'responsible' (Spenceley, 2012) and 'pro-poor' (Scheyvens, 2009).

Hall (2011) observes that travel for the sake of travel, and the taken-for-granted nature of leisure and business mobility, are a sign of consumerism, a form of capitalism which has been prevalent over recent decades. Notions of unfettered growth in tourism also only serve to reinforce the neoliberal agenda present in modern economies. If promoting domestic, as opposed to international, tourism presents a viable approach to supporting degrowth, one way to achieve this might be to target the system which enables the undesired behaviour. Here, macro-social marketing provides a framework by which this change might occur.

Tourism degrowth through macro-social marketing

Macro-social marketing is defined as the application of social marketing to target societal change, as opposed to change at the individual level (Kennedy & Parsons, 2012). Taking a macro-social marketing approach, public policy or other initiatives can be employed to sabotage the marketing mix of undesirable products (i.e., international tourism) in order to promote desirable alternatives (Kennedy, 2016; Lefebvre & Kotler, 2011), in this case, greater levels of domestic tourism. Macro-social marketing can be used to address the 'wicked' problems which are perpetuated by marketing systems (Kennedy, 2016). These wicked problems often result due to the use (or non-use) of products, which can accumulate into a widespread systematic macro-level problem (Kemper & Ballantine, 2017). While climate change is a common focus of the negative outcomes of international tourism, other effects include the introduction of invasive species, biodiversity loss, land use change, pollution and water consumption (Hall, 2011).

Macro-social marketing attempts to change institutional norms in order to affect systemic change (Kennedy & Parsons, 2012). In this vein, change in social institutions is bought about by the process of institutionalisation and is based on systems thinking that aims to create change in interconnected levels of society (Kennedy, 2016). Solutions to wicked problems such as climate change thus require multi-level initiatives which target 'people' and 'place', or "more specifically, individual as well as economic, social and structural barriers" (Kemper & Ballantine, 2019, p. 39). In this instance, the 1984 "don't leave town till you've seen the country" campaign took a social marketing approach, where the message was targeted at consumers (i.e., potential tourists). A macro-social marketing approach would be required to go beyond this, targeting the public and private organisations that enable and support the tourism industry, in addition to the socio-cultural environment which drives tourist behaviour.

The socio-cultural environment includes societal beliefs, attitudes, values and norms (Swinburn, Egger, & Raza, 1999). The cultural environment can be more micro focused, such as the culture of a household or peer group, or macro focused, such as the mass media (Swinburn et al., 1999). Moreover, cultural messages which target individual beliefs or knowledge are considered an individual approach, while the use of media to influence norms is considered a structural intervention (Cohen, Scribner, & Farley, 2000). At the macro-level, mass media is an influential source of socio-cultural information (Swinburn et al., 1999). Media can be unidirectional in the case of programmatic content, advertisements and public relations campaigns (Hovell, Wahlgren, & Gehrman, 2002), which provides a basis for some behaviours, norms and expectations. However, media can also be the news media, journalism and reporting. Here, how language is used to convey a message (what is said vs. what is not said) influences formal (e.g., political) and informal (e.g., social media, conversational) discourse (Kemper & Ballantine, 2019).

A systems approach to macro-social marketing initiatives was introduced by Kemper and Ballantine (2017), who outlined how the multi-level perspective (MLP) on socio-technical transitions can be used to frame, analyse and address wicked problems. The MLP is a systems theory that organises socio-technical systems into niches, regimes and landscapes (Kemper & Ballantine, 2017), which refers to the micro (individual), meso (community) and macro (societal) levels of innovation systems (Geels, 2010). Systematic change can be achieved through regime shifts, which involves interacting processes both within and between these levels (Geels, 2010; Kemper & Ballantine, 2017).

Society consists of several socio-technical systems which involve technologies, actors and institutions, including cultural practices, cultural meanings and knowledge (Geels, 2004). Examples of socio-technical systems include those associated with energy, food, communications, leisure and mobility (Kemper & Ballantine, 2017). Addressing why desirable innovations, products or processes are not adopted requires an understanding of the overarching structures of markets, institutional and regulatory systems, inadequate infrastructures for change, and patterns of consumer demand (Smith, Stirling & Berkhout, 2005). For example, these specifically relate to sunk investments, vested interests, and favourable subsidies and regulations (Unruh, 2000). Within a tourism context, the infrastructure which has been built to support mass international tourism represents significant sunk costs, while large-scale tourism businesses have a vested interest in maintaining the status quo. Moreover, even prior to the outbreak of COVID-19, many international airlines, for example, also received significant subsidies to ensure their survival, yet alone growth.

Kemper and Ballantine (2017) argue that macro-social marketing should develop a more top-down approach which addresses the landscape, where influencing public opinion and changing policy and the structure and mechanisms of markets induces change at the micro (niche) and meso (regime) levels. This can be achieved through developing systematic instruments and process strategies, both of which include participant selection, framing the problem or transition message, and using an array of policy instruments (Loorbach & Rotmans, 2010). Within this, it is critical that macro-social marketers identify and understand the normative rules which influence behaviour. These rules include values, norms, role expectations, duties, rights and responsibilities – all of which are internalised through the socialisation process (Geels, 2004). Furthermore, given that macro-social marketing is intertwined with the political sphere, consumer acceptance of public policy becomes a critical part of systematic change (Kemper & Ballantine, 2017). Moreover, political and corporate resistance are significant barriers towards the successful implementation of macro-social marketing initiatives.

To encourage greater adoption of domestic tourism alternatives, changing public perception is critical. Within New Zealand, and as highlighted at the start of this chapter, many people often have limited travel experience within their own country, instead preferring to travel overseas. If greater adoption of domestic tourism alternatives might achieve degrowth objectives through macro-social marketing, it is necessary to frame international tourism alternatives as being environmentally unsound and to promote more local and slower alternatives. The regulatory environment also needs to be addressed, for example, removing subsidies from the airline industry and increasing direct and indirect taxes on the tourism activities which are known to have negative short- and long-term environmental effects. Similarly, the media also needs to be harnessed to change the narratives around international tourism being desirable, when it is already known that the quality of the touristic experience does not have to be related to the distance travelled or the number of locations visited in a short period of time.

Kemper and Ballantine (2017, p. 389) argue that in order to address wicked problems, "we must break down the specific behaviors and associated institutions (including social and cultural practices) which promote and inhibit that (un) desirable behavior, or more specifically, collective behavior". In attempting to reduce the negative impacts of international tourism by promoting domestic tourism alternatives, there is a need to break down the causes as to why New Zealand residents choose to travel overseas instead of choosing to explore their own country. The non-sustainability of certain tourism practices relates to issues of availability (including pricing, distribution and promotion) and socio-cultural practices and norms, and these types of issues need to be addressed through macro-social marketing.

Lessons from some social movements may also be useful in understanding how domestic tourism can be promoted to support a degrowth agenda. One specific consumer movement is voluntary simplicity (or voluntary simplifiers, as a group), and this is where we turn our attention to next.

Can voluntary simplicity inform a tourism degrowth agenda?

The idea of voluntary simplicity was first introduced by Gregg (1936), who took his inspiration from the great spiritual leaders of history, who he believed practiced the lifestyle. When describing the term, Gregg (1936) identifies concepts such as 'singleness of purpose', 'sincerity and honesty within' and the 'avoidance of exterior clutter', as being the central tenets of voluntary simplification. Etzioni (1998, p. 620) defines the practice as "the choice out of free will . . . to limit expenditures on consumer goods and services, and to cultivate non-materialistic sources of satisfaction and meaning". Further to this point, Alexander and Ussher (2012) describe voluntary simplicity as a diverse social movement comprising people who resist high-consumption lifestyles and who instead are seeking a lower consumption but higher quality of life alternative. All of these definitions emphasise the voluntary nature of the lifestyle, in that it should not occur through either coercion by authorities or as a means of budgeting through tough economic times (Elgin, 1981; Leonard-Barton, 1981).

Etzioni (1998) argues that consumers adopt the lifestyle on a sliding scale of involvement, identifying three levels of intensity: downshifters, strong simplifiers and holistic simplifiers. Downshifters are moderate simplifiers, forgoing some consumer goods while maintaining the majority of their consumer lifestyle. Bekin, Carrigan, and Szmigin (2005) liken this approach to conspicuous simplicity, where people's old possessions are replaced with items that symbolise their new, simplified lifestyle. Shaw and Newholm (2002) describe downshifters as exhibiting a mostly self-centred response to the perception of a rushed and mediocre lifestyle present in contemporary society. Taylor-Gooby (1998) goes further, stating that downshifters may be in contradiction to the lifestyle as their affluence is retained, thus contradicting the ideology.

The symbolism associated with downshifters can be illustrated by a consumer who chooses to buy a hybrid or electric automobile; where individual ownership is still considered important, but so too is the motivation to make a public statement about their green values. This is similar to the problem of "anti-consumption consumption in the tourism industry" (Hall, 2011, p. 300), where green travel experiences to remote destinations are actively promoted with little consideration of the carbon emissions which might result. However, there is potential to successfully promote domestic tourism alternates to downshifters, as there is evidence to suggest that those consumers who make decisions based on environmental and social concerns are often interested in transferring these values to a tourism context (Miller, 2003).

Strong simplifiers are those who give up high-paying jobs in order to live off less income, restricting their consumption accordingly (Ballantine & Creery, 2010). These simplifiers also include those who are motivated to reduce the time they spend at work, in essence pursuing a lower-stress life. Etzioni (1998) posits that those people who voluntarily reduce their income tend to be stronger simplifiers than those who only moderate their lifestyle, the rationale for this being that a significant reduction in income leads to more comprehensive lifestyle simplification than does selective downshifting behaviour.

Holistic simplifiers are similar to strong simplifiers, but adjust their entire lifestyle to fit the voluntary simplicity ethic (Ballantine & Creery, 2010); this can involve moving house to a less affluent area (perhaps rural) with the intention of leading a simpler life (Etzioni, 1998). McGouran and Prothero (2016) consider holistic simplifiers to be the most dedicated of all voluntary simplifier groups. Elgin and Mitchell (1977) view this as 'full' voluntary simplicity, as simplifiers in this group wholeheartedly embrace the ethos of voluntary simplicity.

For either strong or holistic simplifiers, Hall (2011) argues that voluntary simplicity has arguably had some impact with the development of staycations and slow tourism, options which may be seen as desirable for both of these simplifier types. Here, as opposed to travel by airplane or private car to chosen destinations, the tourism experience might focus on day trips away from their normal home or a prolonged period of time at a single destination. Moreover, the application of voluntary simplicity and sustainable consumption to tourism suggests the quality of the experience should be seen as equally important to the quantity. While tourism can be a meaningful activity for people, there is no reason to suggest that happiness is a correlate of travel frequency and distance (Hall, 2011). Indeed, "the most authentic tourists of all may be those wanting to visit friends and relations because of the connectedness it provides" (Hall, 2007, p. 1140).

The main themes which emerge from the voluntary simplicity literature relevant to degrowth centre around the changes in consumption behaviour which occur as part of adopting the lifestyle, and can be viewed in terms of reducing consumption, ethical consumption and sustainable consumption. For voluntary simplifiers, reducing consumption involves limiting consumption through activities such as sharing, buying second-hand and eliminating clutter (e.g., Bekin, Carrigan, & Szmigin, 2007; Craig-Lees & Hill, 2002; Huneke, 2005; Shaw & Newholm, 2002), while ethical consumption involves a person considering the environmental and social impacts of their consumption choices, which may result in activities like buying fair-trade and/or environmentally friendly products (e.g., McDonald, Oates, Young, & Hwang, 2006; Shaw & Newholm, 2002). Voluntary simplifiers may also undertake sustainable consumption, where activities such as recycling and composting can occur (e.g., Bekin et al., 2007; Huneke, 2005). Each of these reduced consumption patterns also has relevance to degrowth within tourism. A focus on domestic tourism addresses aspects of reducing the consumption required for tourism to occur, while local and slower tourism practices address aspects of ethical consumption, where such tourists are more sensitive to the host community than mass tourists, who may be more interested in the 'quantity' of experience. Issues of sustainable consumption are addressed with domestic tourism having a lesser effect on the environment compared to international tourism alternatives.

A critical issue to consider when applying aspects of voluntary simplicity to degrowth in tourism is that much of the extant voluntary simplifier literature focuses on those people who have already fully adopted the lifestyle (Ballantine, Arbouw, & Ozanne, 2011). McDonald et al. (2006) recognise the importance of a large consumer segment that displays voluntary simplifier behaviour but are not committed or fully converted to the voluntary simplifier lifestyle, terming these people beginner voluntary simplifiers. Within the beginner voluntary simplifier segment, they distinguish among three types of early simplifiers: a) apprentice simplifiers, who are voluntary simplifiers in the making; b) partial simplifiers, who settle for lifestyles with some beginner simplifier features; and c) accidental simplifiers, who simplify involuntarily because of economic reasons and are often not concerned with the ethical or environmental aspects of consumption.

Although not explicitly focusing on the beginner voluntary simplifier group, Ballantine and Creery (2010) explored the disposition activities of voluntary simplifiers, noting the difficulty associated with the early stages of adopting the lifestyle and its role in identity construction. The authors also suggested a vague endpoint to the journey of becoming a voluntary simplifier, with some of their participants professing a desire to live 'off the grid' and others indicating that they will always have some reliance on the market for what they need. In this respect, the journey of becoming a voluntary simplifier can be viewed as a liminal transition (Noble & Walker, 1997), where an individual separates themselves from an existing role (i.e., a mainstream consumer) and undergoes a transitional process (the liminal period) before fully incorporating their new role as a voluntary simplifier into their daily lives.

This perspective is consistent with the idea that consumption and anti-consumption, and the attitudes individuals hold about their choices, can act as a form of identity creation and expression (Zavestoski, 2002), with other authors observing that identity creation can also occur through tourism (Hall, 2011). Furthermore, anti-consumption can be practised in many ways and does not have to mean non-consumption; it ranges from resistance to the rejection of consumption, and has varying degrees of visibility (Hogg, Banister, & Stephenson, 2009). However, irrespective of what form of anti-consumption behaviour occurs, the literature suggests that these behaviours are context dependent and that the behaviours, challenges and attitudes associated with anti-consumption are highly affected by social, cultural and market influences (Zavestoski, 2002). In this respect, while extant research on voluntary simplicity has focused on the individual, any consideration of tourism degrowth requires a wider systematic perspective – one which can be informed by macro-social marketing should widespread societal change be desired. However, an understanding of voluntary simplification is essential to inform the messaging which would need to occur as part of a macro-social marketing strategy.

Support for this is provided by Alexander and Ussher (2012), who observe that the voluntary simplifier movement needs to radicalise and expand into the social, economic and political mainstream if it is to respond to issues such as unsustainable consumption practices and contribute to a degrowth society. Alexander (2013, p. 300) goes further, stating that

> simple living is unlikely to be an effective response to the ecological problems of overconsumption in the absence of structural change. Accordingly, there is little doubt that structural change by way of legal, political and economic reform is a necessary part of any transition beyond growth capitalism.

In short, "with respect to the affluent societies, at least, degrowth depends on voluntary simplicity" (Alexander, 2013, p. 300).

Conclusion

On some levels the "don't leave town till you've seen the country" slogan was prophetic for the New Zealand tourism industry, despite its 1984 origins. All research takes place in a temporal context, and at the time this chapter was being finalised, the world was in the grip of COVID-19. In many respects, COVID-19 has done more to reduce international tourism activity than what any degrowth advocate could wish for. However, the tourism industry is vulnerable to many risks and exogenous shocks, and the expectation is that it has the potential to bounce back from the effects of COVID-19. That said, the advertising campaign outlined at the start of this chapter predated greater awareness of the negative consequences of mass tourism and its growth-oriented paradigms and assumptions, but there is the potential for COVID-19 to provide an impetus to reconsider the global tourism system. This chapter sought to understand how the two areas of macro-social marketing and voluntary simplicity might help with this endeavour.

At its core, the purpose of degrowth is to critically understand and undo the phenomenon of growth. While this chapter touched on some macro- and micro-level initiatives that can be enacted to support degrowth, many of these initiatives overlap with ideas that have been discussed in the macro-social marketing literature to address wicked problems or in the voluntary simplifier literature to explore how consumers are able to reduce their consumption activities. Degrowth is not an economic decline or recession, as has been caused by COVID-19, but it does provide a backdrop to consider what theories or frameworks might be used to more permanently degrow mass international tourism in favour of domestic alternatives.

Macro-social marketing has the potential to address how public policy or other initiatives can be used to sabotage the marketing of international tourism. The global tourism industry is a complex and intertwined system, so it is therefore critical for macro-social marketers to identify and understand the normative rules which influence tourist behaviour in order for change to occur. Current tourism practices need to be framed as being environmentally unsound, the regulatory environment should be targeted for change, and the media needs to be harnessed to promote local and slower tourism alternatives. Here, activists can play a role by "promoting post growth models of progress both to governments and the constituencies upon whose mandate democratic governments depend" (Alexander, 2012, p. 13).

To inform the macro-social marketing approach that might be taken to pursue a degrowth agenda, the literature on voluntary simplicity provides some insight as to how the various levels of society and social norms might be targeted for change to occur. While voluntary simplicity is a lifestyle which should not occur by coercion, the ethos which underpins the practice can be conveyed through various media, consistent with the approaches suggested by the macro-social marketing literature. In many respects, forms of anti-consumption have become increasingly popularised through the Marie Kondo 'KonMari' method (Kondo, 2014) (which was turned into a television show) or documentaries such as *Minimalism: A Documentary about the Important Things* (https://minimalismfilm.com/). The same could also occur in a tourism context.

Arnould (2007) argues that the whole idea of anti-consumption reinforces social classes, because only the wealthy can afford to resist the market by adopting voluntary simplicity lifestyles. Similarly, Taylor-Gooby (1998) suggests that voluntary simplicity is a luxury for rich and secure minorities, arguing that the 'greening' of capitalism should be a worldwide project. If the degrowth of international tourism is to be achieved, a critical starting point is to target those consumers who travel most. Understanding what is known about different voluntary simplifier types, the consumption themes that underpin the behaviour and the experiences of beginner voluntary simplifiers provides some initial direction about how degrowth in tourism might be messaged to consumers who would otherwise be able to continue their wasteful tourism practices. However, to support this, Alexander (2013, p. 13) suggests that until "a government seriously embraces a post growth model of progress – either voluntarily or by force of ecological or financial necessity – will a top down politics of voluntary simplicity be taken seriously".

The need to promote a message similar to that of the 1984 "don't leave town till you've seen the country" campaign is even more critical in today's world than it was more than 35 years ago. Compared to 1984, international tourism is now cheaper, more available and more desirable to those consumers who are able to afford it. At the same time, the negative impacts of unfettered growth in international tourism have also become more prevalent. It is hoped that this chapter has given some insight into how degrowth in international tourism might be achieved through a renewed focus on domestic tourism.

References

Alexander, S. (2012). *Degrowth implies voluntary simplicity: Overcoming barriers to sustainable consumption*. SSRN 2009698. http://dx.doi.org/10.2139/ssrn.2009698

Alexander, S. (2013). Voluntary simplicity and the social reconstruction of law: Degrowth from the grassroots up. *Environmental Values*, *22*(2), 287–308.

Alexander, S., & Ussher, S. (2012). The voluntary simplicity movement: A multi-national survey analysis in theoretical context. *Journal of Consumer Culture*, *12*(1), 66–86.

Andriotis, K. (2018). *Degrowth in tourism: Conceptual, theoretical and philosophical issues*. Wallingford: CABI.

Arnould, E. J. (2007). Should consumer citizens escape the market? *The Annals of the American Academy of Political and Social Science*, *611*(1), 96–111.

Ballantine, P. W., Arbouw, P., & Ozanne, L. K. (2011). Learning to resist: The challenges faced by beginner voluntary simplifiers. In R. Ahluwalia, T. L. Chartrand, & R. K. Ratner (Eds.), *Advances in consumer research* (Vol. 39, pp. 404–408). Duluth, MN: Association for Consumer Research.

Ballantine, P. W., & Creery, S. (2010). The consumption and disposition behaviour of voluntary simplifiers. *Journal of Consumer Behaviour*, *9*(1), 45–56.

Bekin, C., Carrigan, M., & Szmigin, I. (2005). Defying marketing sovereignty: Voluntary simplicity at new consumption communities. *Qualitative Market Research*, *8*(4), 413–429.

Bekin, C., Carrigan, M., & Szmigin, I. (2007). Beyond recycling: 'Commons-friendly' waste reduction at new consumption communities. *Journal of Consumer Behaviour*, *6*(5), 271–286.

Cohen, D. A., Scribner, R. A., & Farley, T. A. (2000). A structural model of health behavior: A pragmatic approach to explain and influence health behaviors at the population level. *Preventive Medicine*, *30*(2), 146–154.

Craig-Lees, M., & Hill, C. (2002). Understanding voluntary simplifiers. *Psychology & Marketing*, *19*(2), 187–210.
Downes, S. (2019, June 20). Kiwis who have never been to the other island. *Stuff*. Retrieved from www.stuff.co.nz/ travel/kiwi-traveller/113630581/kiwis-who-have-never-been-to-the-other-island
Elgin, D. (1981). *Voluntary simplicity: Toward a way of life that is outwardly simple*. New York, NY: Morrow.
Elgin, D., & Mitchell, A. (1977). Voluntary simplicity. *The Co-Evolution Quarterly*, *3*(1), 4–19.
Etzioni, A. (1998). Voluntary simplicity: Characterization, select psychological implications, and societal consequences. *Journal of Economic Psychology*, *19*(5), 619–643.
Fletcher, R. (2019). Ecotourism after nature: Anthropocene tourism as a new capitalist 'fix'. *Journal of Sustainable Tourism*, *27*(4), 522–535.
Fullagar, S., Markwell, K., & Wilson, E. (Eds.). (2012). *Slow tourism: Experiences and mobilities*. Bristol: Channel View Publications.
Geels, F. W. (2004). From sectoral systems of innovation to socio-technical systems: Insights about dynamics and change from sociology and institutional theory. *Research Policy*, *33*(6–7), 897–920.
Geels, F. W. (2010). Ontologies, socio-technical transitions (to sustainability), and the multi-level perspective. *Research Policy*, *39*(4), 495–510.
Gössling, S., Scott, D., & Hall, C. M. (2020). Pandemics, tourism and global change: A rapid assessment of COVID-19. *Journal of Sustainable Tourism*. doi:10.1080/09669582.2020.1758708
Gregg, R. (1936). *The value of voluntary simplicity*. Wallingford, PA: Pendle Hill.
Hall, C. M. (2007). Response to Yeoman et al: The fakery of 'The authentic tourist'. *Tourism Management*, *28*(4), 1139–1140.
Hall, C. M. (2009). Degrowing tourism: Décroissance, sustainable consumption and steady-state tourism. *Anatolia*, *20*(1), 46–61.
Hall, C. M. (2011). Consumerism, tourism and voluntary simplicity: We all have to consume, but do we really have to travel so much to be happy? *Tourism Recreation Research*, *36*(3), 298–303.
Hall, C. M., Scott, D., & Gössling, S. (2020). Pandemics, transformations and tourism: Be careful what you wish for. *Tourism Geographies*. doi:10.1080/14616688.2020.1759131
Higgins-Desbiolles, F. (2010). The elusiveness of sustainability in tourism: The culture-ideology of consumerism and its implications. *Tourism and Hospitality Research*, *10*(2), 116–129.
Higgins-Desbiolles, F., Carnicelli, S., Krolikowski, C., Wijesinghe, G., & Boluk, K. (2019). Degrowing tourism: Rethinking tourism. *Journal of Sustainable Tourism*, *27*(12), 1–19.
Hogg, M. K., Banister, E. N., & Stephenson, C. A. (2009). Mapping symbolic (anti-) consumption. *Journal of Business Research*, *62*(2), 148–159.
Hovell, M. F., Wahlgren, D. R., & Gehrman, C. A. (2002). The behavioral ecological model. In R. J. DiClemente, R. A. Crosby, & M. C. Kegler (Eds.), *Emerging theories in health promotion practice and research: Strategies for improving public health* (pp. 347–385). San Francisco: John Wiley & Sons.
Huneke, M. E. (2005). The face of the un-consumer: An empirical examination of the practice of voluntary simplicity in the United States. *Psychology & Marketing*, *22*(7), 527–550.
Kallis, G. (2018). *Degrowth (The economy: Key ideas)*. New York, NY: Agenda Publishing.
Kallis, G., Demaria, F., & D'Alisa, G. (2015). Introduction: Degrowth. In G. D'Alisa, F. Demaria, & G. Kallis (Eds.), *Degrowth: A vocabulary for a new era* (pp. 1–18). New York, NY: Routledge.

Kallis, G., Kostakis, V., Lange, S., Muraca, S., Paulson, S., & Schmelzer, M. (2018). Research on degrowth. *Annual Review of Environment and Resources*, *43*(1), 291–316.

Kemper, J. A., & Ballantine, P. W. (2017). Socio-technical transitions and institutional change: Addressing obesity through macro-social marketing. *Journal of Macromarketing*, *37*(4), 381–392.

Kemper, J. A., & Ballantine, P. W. (2019). Targeting the structural environment at multiple social levels for systemic change. *Journal of Social Marketing*, *10*(1), 38–53.

Kennedy, A. M. (2016). Macro-social marketing. *Journal of Macromarketing*, *36*(3), 354–365.

Kennedy, A. M., & Parsons, A. (2012). Macro-social marketing and social engineering: A systems approach. *Journal of Social Marketing*, *2*(1), 37–51.

Kondo, M. (2014). *The life-changing magic of tidying up: The Japanese art of decluttering and organizing*. New York, NY: Ten Speed Press.

Latouche, S. (2003). Por una sociedad del decrecimiento. *Le Monde Diplomatique*, 97. Edición Española.

Latouche, S. (2006). *Le Pari de la décroissance*. Paris: Fayard.

Latouche, S. (2007). *Petit traité de la décroissance sereine*. Paris: Mille et Une Nuits.

Lefebvre, R. C., & Kotler, P. (2011). Design thinking, demarketing and behavioral economics: Fostering interdisciplinary growth in social marketing. In G. Hastings, K. Angus, & C. Bryant (Eds.), *The Sage handbook of social marketing* (pp. 80–94). London: Sage.

Leonard-Barton, D. (1981). Voluntary simplicity lifestyles and energy conservation. *Journal of Consumer Research*, *8*(3), 243–252.

Loorbach, D., & Rotmans, J. (2010). The practice of transition management: Examples and lessons from four distinct cases. *Futures*, *42*(3), 237–246.

McDonald, S., Oates, C. J., Young, C. W., & Hwang, K. (2006). Toward sustainable consumption: Researching voluntary simplifiers. *Psychology & Marketing*, *23*(6), 515–534.

McGouran, C., & Prothero, A. (2016). Enacted voluntary simplicity – Exploring the consequences of requesting consumers to intentionally consume less. *European Journal of Marketing*, *50*(1/2), 189–212.

Miller, G. A. (2003). Consumerism in sustainable tourism: A survey of UK consumers. *Journal of Sustainable Tourism*, *11*(1), 17–39.

Ngā Taonga Sound & Vision. (1984). *NZ tourism – Don't leave town till you've seen the country*. Retrieved from www.ngataonga.org.nz/set/item/173

Noble, C. H., & Walker, B. A. (1997). Exploring the relationships among liminal transitions, symbolic consumption, and the extended self. *Psychology & Marketing*, *14*(1), 29–47.

Novelli, M., Burgess, L. G., Jones, A., & Ritchie, B. W. (2018). No Ebola . . . still doomed' – The Ebola-induced tourism crisis. *Annals of Tourism Research*, *70*, 76–87.

Ötsch, W. (2020). *What type of crisis is this? The coronavirus crisis as a crisis of the economicised society*. Working Paper Series 57. Bernkastel-Kues: Institute für Ökonomie & Philosophie, Cusanus Hochschule für Gesellschaftsgestaltung.

Scheyvens, R. (2009). Pro-poor tourism: Is there value beyond the rhetoric? *Tourism Recreation Research*, *34*(2), 191–196.

Shaw, D., & Newholm, T. (2002). Voluntary simplicity and the ethics of consumption. *Psychology & Marketing*, *19*(2), 167–185.

Sigala, M. (2020). Tourism and COVID-19: Impacts and implications for advancing and resetting industry and research. *Journal of Business Research*, *117*, 312–321.

Smith, A., Stirling, A., & Berkhout, F. (2005). The governance of sustainable socio-technical transitions. *Research Policy*, *34*(10), 1491–1510.

Spenceley, A. (2012). *Responsible tourism: Critical issues for conservation and development*. London: Routledge.

Swinburn, B., Egger, G., & Raza, F. (1999). Dissecting obesogenic environments: The development and application of a framework for identifying and prioritizing environmental interventions for obesity. *Preventive Medicine*, *29*, 563–570.

Taylor-Gooby, P. (1998). Comments on Amitai Etzioni: Voluntary simplicity: Characterization, select psychological implications, and societal consequences. *Journal of Economic Psychology*, *19*(5), 645–650.

Unruh, G. C. (2000). Understanding carbon lock-in. *Energy Policy*, *28*(12), 817–830.

Zavestoski, S. (2002). The social-psychological bases of anticonsumption attitudes. *Psychology & Marketing*, *19*(2), 149–165.

13 Degrowing tourism

Can grassroots form the norm?

O. Cenk Demiroglu and Ethemcan Turhan

Introduction

Tourism and degrowth are two notions that are being increasingly discussed together in terms of the wider debate on the need for limiting economic production and consumption for the sake of restored and improved well-being of socio-natures (Andriotis, 2018; Fletcher, Mas, Romero, & Blázquez-Salom, 2020). Socio-natures in this formulation refer to the intrinsically social characteristics of environmental changes across different scales, which are determined by power and inequality as products of historical-geographical processes (Castree, 2001). These power differentials and inequalities are also the determinants of winners and losers from the rapid growth of global tourism. Lenzen et al. (2018) estimated that the global tourism industry's carbon footprint is expected to grow at 4% per annum. Yet the global tourism industry, the source of millions of jobs and related contribution to GDP, is itself a highly vulnerable sector to the impacts of climate change (Scott, Hall, & Gössling, 2019). This chapter approaches this issue by means of a more bottom-up focus that investigates winter tourism, related patterns of consumption and the interaction of these with the ongoing climate crisis. Degrowth is emerging as a popular pathway to the mitigation of the recreational greenhouse gas emissions that contribute to global warming. At the center of attention here is the climate activist and lobbyist groups within Nordic Europe's ski tourism community, and the main focus in this chapter is on the contents and the efficacy of their discourses on climate action. In particular, what and how are these groups communicating about climate change mitigation with what relevance to the idea of degrowth, and can the forms of action of such grassroots movements prove to be mainstream or not?

Degrowth and tourism through a climate change lens

The contemporary quest for reaching a balanced sustainability (see Chapters 1 and 14, this volume) among its three pillars of economy, nature and society has become even more pronounced by the emergency of the climate crisis, and has led to the emergence of alternative discourses to the dominant economic growth paradigm mainly marked by neoliberalism and economism. The transition to "green", "bio-" or "circular" economies, as well as a "steady-state economy" and "degrowth", have become a significant part of academic, business and political agendas to alleviate the chronic socio-natural problems of the status quo (D'Amato et al., 2017; D'Amato, Droste, Winkler, & Toppinen, 2019; Eaton & Sheng, 2019).

From a mainstream perspective the green economy has been defined as "one that results in improved human well-being and social equity, while significantly reducing environmental risks and ecological scarcities" (UNEP, 2011, p. 2), the circular economy as "turning goods that are at the end of their service life into resources for others; closing loops in industrial ecosystems and minimizing waste through material reuse, extended product life, repair, re-manufacturing or upgrading", and the bioeconomy as "where the use of biological resources (plants, animals, microorganisms) plays a leading role and biotechnology has an important impact on economic output" (Eaton & Sheng, 2019, p. 1.15). A steady-state economy is construed as "constant stocks of physical wealth (artifacts) and constant population, each maintained at some chosen, desirable level by a low rate throughput" (Daly, 1974, p. 15), while degrowth is defined as "a process of political and social transformation that reduces a society's throughput ['the energy and resource flows in and out of an economy'] while improving the quality of life" (Kallis et al., 2018, p. 4.2) as well as "a downscaling of production and consumption that increases human well-being and enhances ecological conditions and equity on the planet" (Research and Degrowth (R & D), 2020). In this context, the degrowth concept is seen by some (Hall, 2009; Kerschner, 2010; Washington & Twomey, 2016) as a means to achieve (or recover) the steady states of economic, natural and social equilibria.

Both steady-state economy and degrowth deviate from the green, bio- and circular economies that partly fail to reject the growth imperative in their essences. This is especially so with the concept of the green economy, which has increasingly been adopted or promoted by high-level organizations within the past decade (UNEP, 2011; OECD, 2011), which also has strong links to the notion of the circular economy in the case of the EU (European Commission, 2019). The idea of the green and circular economies have been contested for their relatively weak normative justification, especially in terms of environmental sustainability (Sandberg, Klockars, & Wilén, 2019), and even criticized for greenwashing as neoliberal business-as-usual (Turhan & Gündoğan, 2017).

Based on the perspectives of some of the leading researchers on sustainability transition, D'Amato et al. (2019, p. 460) have found that the most favored synergic remedies to growth paradigm would be "circular solutions towards economic-environmental decoupling in a degrowth perspective", followed by "a mix of circular and green economy solutions" and "a green economy perspective, with an emphasis on natural capital and ecosystem services, and critical towards growth". Greater collaboration between the proponents of circularity and degrowth for a mutually beneficial synthesis was also encouraged by Schröder et al. (2019), drawing on, among others, opportunities for degrowth principles to guide the circular economy's goal of driving creative industries to replace extractive ones. A number of other researchers (Hobson & Lynch, 2016; Valenzuela & Böhm, 2017; Corvellec, Böhm, Stowell, & Valenzuela, 2020) also maintain critiques against the circular economy concept for its economic growth orientation, technocentrism, rebound effects and over-ambitious waste management proposals. Degrowth, however, is based on sound philosophical underpinnings but is still and, perhaps even meant to be, short of practical tests (Burkhart, Schmelzer, & Treu, 2020) and proposals consistent with its manifesting bottom-up approaches (Kerschner, 2010; Cosme, Santos, & O'Neill, 2017). Instead, it has been cast more as a potential gateway than a pathway for the sustainability transition efforts. The latter contradiction is even more relevant for this chapter as we deal with the engagement of grassroots movements with the degrowth discourse for the decarbonization of tourism.

While the introductory and concluding chapters of this book give full accounts of the present and anticipated degrowth and tourism debates, here we briefly re-introduce the subject from a climate change perspective to set the framework for understanding the engagement and efficacy of grassroots movements. To our knowledge, the relatively recent literature on tourism and degrowth (see Chapter 1, this volume) has so far not directly studied the issue within an explicit climate change context, attracting the foci of only a handful of studies (Hall, 2009, 2010, 2011, 2015; Hollenhorst, Houge-Mackenzie, & Ostergren, 2014; Gascon, 2019; Menton et al., 2020; Sun, Lin, & Higham, 2020), as well as Prideaux (this volume), who foresees the challenges of degrowth as a potential adaptation means for destinations negatively impacted by climate change. Among the earlier studies, only Gascon (2019) has approached the issue from a demand perspective, which is also the scope of this chapter, criticizing efforts of the World Tourism Organization towards positioning tourism mobilities as a human right, as acknowledged in Article 7 of the Global Code for Ethics in Tourism (UNWTO, 2001), and thereby implicitly setting the stage to passivize radical mitigation and calling for sustainability such as those of the degrowth discourse.

Tourism is one of the largest sectors of the global economy, constituting roughly 10% of the Gross World Product and employment, when its indirect and induced effects are taken into account (WTTC, 2020). In doing so, however, it also makes up at least 5–8% of global emissions (UNEP & UNWTO, 2005; Lenzen et al., 2018). From degrowth and other sustainability transition perspectives, where less working time is advocated both to downscale production and to improve life quality (see also Chapter 4, this volume), the tourism and recreation industries become even more critical areas of analyses as they directly relate to the consequences for increased leisure time, where "man the worker", *Homo Faber*, would be inevitably faced by the rise of "man the player", *Homo Ludens* (Huizinga, 1949), who is more associated with the creative destructions needed for the sustainable transition (Rojek, 2005).

The complexities of the tourism-climate-degrowth relationship extend to a framework (Figure 13.1) that goes beyond the impact-emissions reciprocality and the tourism system's adaptation and mitigation responses, in addition to the aforementioned lesser-known leisure dimensions of degrowth. By indicating a

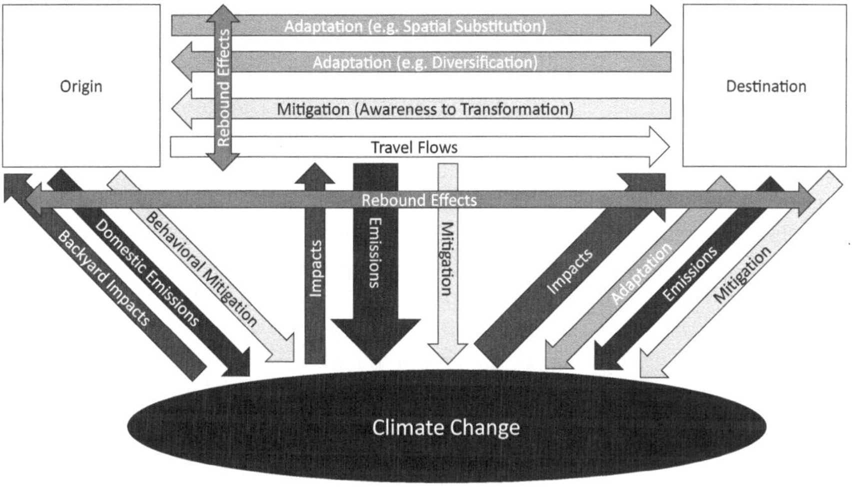

Figure 13.1 An extended framework of tourism system and climate change inter-relationships

Source: Authors

rebound effects context that recognizes "maladaptation" and "malmitigation" consequences as well (Aall, Hall, & Groven, 2016), sustainability transition alternatives can be better exemplified. The tourism system pollutes the climate through emissions, especially in its transit (e.g. fossil fuel consumption of air and car travel) and destination (energy use and loss of sinks for accommodation, shopping, entertainment, food and beverage) spaces. However, what is usually missing in this equation regarding a net forcing effect is the compensation of domestic emissions at the origins, i.e. the carbon footprints of work and alternative backyard leisure and recreation options.

The application of sustainability transitions concepts is most relevant to the mitigation responses in Figure 13.1 but is also highly related to impact-adaptation cycles due to inter- and cross- rebound effects between adaptation and mitigation (Aall et al., 2016). For example, trends such as "transformative tourism" claim increased environmental awareness (Walter, 2016), but fall short of anticipating what happens next with their visitors – do they stop travelling back to the "transformative" destination and even spread the word? What then are the socio-economic consequences for the destination and even its competitors? How careful are visitors with their emissions and other environmental footprints when engaged with staycations or other proximity tourism options?

Even without positioning themselves as transforming spaces, destinations' direct mitigation efforts such as greening practices, e.g. switching to renewable energies and recycling, are also precarious, as greener destinations may be more attractive to some customer segments but then increased travel emissions would lead to net increased emissions at the systematic scale – should travel to/from these destinations still be based on fossil fuels. Technological and policy interventions to such travel may either complicate the issue further or not be a viable option. For instance, introducing a carbon tax on air travel may hinder long-haul trips to exotic destinations, among which the least emitting but also the most vulnerable tourism nations exist. Gössling and Higham (2020) argue that air passenger duties may potentially reduce excessive growth rates while increasing lengths of stay, perhaps even contributing to slow tourism. On the technological side, efficiency improvements providing less per capita emissions do not necessarily translate into aggregate reductions in sectors such as tourism, where projected growth rates outscore efficiencies (Hall, 2011), which is just another example of why decoupling ideas with green and circular ambitions may end up as wishful thinking that may never deliver the much-needed absolute and fast decoupling on a global scale (Hickel & Kallis, 2020; Vadén et al., 2020). Moreover, Hickel (2019) asserts that technocentrism, a major critique of bio-, green and circular economies, around yet-to-be-tested "miraculous" solutions such as Bio-Energy with Carbon Capture and Storage (BECCS) may increase the risks of the climate crisis, pushing reliance away from more radical political and behavioral transitions that are much needed in the short window of response to abrupt climate change.

The Intergovernmental Panel on Climate Change (IPCC; 2018, p. 470) acknowledges that strategies to reconcile low-carbon trajectories with sustainable development and ecological sustainability, including those of green growth and degrowth, show low levels of agreement among research. That being said, both the Low Energy Demand (LED), used in the *Global Warming of 1.5°C* special report (IPCC, 2018), and SSP1 (van Vuuren et al., 2017) of the next-generation Shared Socio-economic Pathway (SSP) scenario family (Riahi et al., 2017), to be used in the IPCC Sixth Assessment Report, do simulate green economy strategies with reduced energy consumptions but sustained economic growths (Hickel, 2019). Tourism is seen and proposed, at least by leading international bodies such as the UNEP and UNWTO (2012), as one of the critical sectors to drive green growth and lead the way to a green economy-based sustainability transition. However, many experts (Gössling, Scott, & Hall, 2013; Hall, 2013, 2015; Stroebel, 2015) object to this reasoning, especially from a climate change perspective:

> The green economy framing sets the tourism industry on a controversial path towards future growth by promoting rules and norms of climate change mitigation that reflect an adjustment of neoliberal capitalism to the environmental crisis that has been criticised more broadly in terms of the green economy. The framing does not question the existing business model that causes environmental harm. Rather, it provides the legitimacy for tourism to grow by presenting tourism as a strategy for poverty alleviation, employment and income generation for communities in developing countries, and by highlighting the potential to reduce the negative impacts of tourism activity. However, if the tourism industry fails to contribute to development and emissions increase, a direction that the researchers referred to in this paper expect

> to be likely as a result of growing demand, the impacts on the environment and societies in destinations could be detrimental.
>
> (Stroebel, 2015, p. 2237)

Instead of the green economy-based approach, a more radical transformation such as those guided by degrowth principles may be required (Hall, 2011, 2015; Demiroglu, 2017) towards a steady-state economy with an environmentalist primacy that perceives the economy as a "subsystem of the earth ecosystem" (Daly, 1993, p. 811). At this stage, the engagement of and advocacy by grassroots movements towards degrowth and other sustainability transition alternatives for reasons of mitigation play a potentially central role in understanding the future pathways of decarbonization in tourism. In the remainder of the chapter, we present our initial findings on the climate change mitigation discourses of a grassroots organization that manifests itself at the intersections of environmentalism and winter sports recreation and tourism.

Case study: Protect Our Winters (POW)

Protect Our Winters (POW) is a grassroots movement with its initial foundation in 2007 in the USA. Its history (see following quote) reminds us about the "victim" status of tourism and/or outdoor recreation and how it may motivate individuals to act on the climate problem, even if it would require considering tourism as both "part of the problem" and "part of the solution" (Aall, 2014).

> In 2007, pro snowboarder Jeremy Jones saw that more and more resorts he'd always counted on for good riding were closed due to lack of snow. Something was clearly going on, and he felt the need to act. But he couldn't find any organizations focused on mobilizing the snowsports community on climate – there was a gap between the impacts that climate change was already having on our great escapes, and organized action to address it.
>
> So Jeremy founded Protect Our Winters.
>
> (POW, 2020a)

Ski tourism is the first and the most climate change–affected tourism type (Scott, Hall, & Gössling, 2012), and a plethora of studies (Demiroglu, Dannevig, & Aall, 2013; Steiger, Scott, Abegg, Pons, & Aall, 2019) have assessed the mostly negative impacts on ski destinations and areas around the world. To no surprise, a high-risk perception and actual observation, and perhaps high expectations on efficacy (Roser-Renouf, Maibach, Leiserowitz, & Zhao, 2014), have triggered climate change activism among the ski tourism industry and the community. Adaptation efforts, such as snowmaking, have already been put in place by the industry, but also collective, albeit voluntary, mitigation action has become common among members of the National Ski Areas Association in the USA. Likewise, similar initiatives have come into existence in the European Alps, especially through leadership of the International Commission for the Protection of the Alps

(CIPRA) (Demiroglu & Sahin, 2015), and further projects, such as Smart Altitude supported by the European Regional Development Fund (Polderman et al., 2020), are promoting energy efficiency and transition (but being limited to ski area operations are missing the major emitter – the transit space).

Nordic Europe is one cradle of skiing with traditions and artefacts dating back millennia. For centuries, skiing has been a major object of transport, combat and hunting, but also a leisure activity for sports and recreation, especially from the nineteenth century onwards. Accelerated by post-WWII developments, skiing has become one of the most iconic objects of mass tourism consumption in both its Nordic roots and elsewhere, including but not limited to established destinations such as in the Alps, North America and Japan, and emerging ones in Russia, China and Turkey (Demiroglu, 2014; Demiroglu, Lundmark, & Strömgren, 2019). Vanat (2020) reports that around 135 million people engage in snow sports worldwide. Sweden, Norway and Finland are home to more than 500 downhill ski areas, and national participation in skiing is among the highest in the world. Iceland and Greenland also offer few local and exotic ski areas, whereas Denmark is mainly a skier generator to its Nordic neighbors and elsewhere (Demiroglu, 2014; Demiroglu et al., 2019, 2020; Vanat, 2020).

Vanat (2020) notes that 20% of some 2,000 ski resorts in the world attract at least 100,000 skier visits per year each and thereby represent 80% of the market in total. Such industrial clusters receive criticism for monopolization and urbanization of mountain landscapes to the "Aspen effect", where the local and working populations become displaced due to increasing prices, further contributing to transport congestion (Clifford, 2002; Stoddart, 2012; Lasanta, Beltran, & Vaccaro, 2013; Park & Pellow, 2013). The documentary film *Resorting to Madness* (Coldstream Creative, 2006) tells of the negative societal and environmental impacts of North America's large ski corporations' business models based on aggressive, continuous growth goals. Somewhat similar corporatization and resortification patterns are also found in Nordic Europe, especially following the growing dominance of some companies in Swedish and Norwegian markets since the 1990s (Demiroglu et al., 2019, 2020).

Contrary to neoliberal business practices, environmental and climate activism from the ski community has also been on the rise in Nordic Europe (Demiroglu & Sahin, 2015). During 2008–2012, the Norwegian Ski and Biathlon Federations as well as the Ski Association ran the "White Winter" program to communicate their environmental and climatic guidelines with their 216,000 members registered in some 1,200 clubs. While its outcomes were meant to institute permanent measures, they have rather been put on the shelf, but other initiatives such as supporting research and development for sustainable snowmaking have been on the agenda (Ski Sport, 2019). In Sweden, the Save Our Snow (SOS) organization has initiated climate awareness and called for increased activism in the industry and among skiers. The SOS, founded in 2014, was also the antecedent of POW's Sweden branch founded in 2016 (POW Sweden, 2020a). In fact, the international extension of the US-based POW (2020b) to Nordic Europe included Norway in 2013 (POW Norge, 2020a) and Finland in 2015 (POW Finland, 2020a).

As of mid-2020, the US-based parent accounts of POW on Facebook and Instagram had 111,500 and 209,000 followers, respectively. Along with its chapters in Canada, the Alps (France-Switzerland-Austria-Germany), the UK, Australia, New Zealand, Japan and Nordic Europe, POW now can claim to be in the midst of a spectrum between the locally engaged grassroots activism organizations and the globally institutionalized mainstream (Salazar, 1996). However, for POW, holding on to its grassroots character is critical as, at different regional scales, it needs to deal with different perspectives and challenges. For example, while in the USA the main agenda may be geared towards lobbying against a federal government in denial of anthropogenic climate change, in the Nordics and other regions where the climate crisis is relatively much better acknowledged at the political levels, the main concern could be about the "how" of transformation. We briefly examine the plans and activities of POW's Nordic chapters, based on their web and social media communications and in a Nordic context of traditional values connected to environmentalism, sports (the "*idrett/idrott*" philosophy) (Lund, 2007) and outdoor life, i.e. "*friluftsliv*" (Aall, 2014), and any efficacy attributed to or anticipated for such activist and lobbyist efforts. In doing so, we also try to identify what sustainability transition discourses the Nordic POW narratives are heading.

In 2019, all of the eight European chapters of POW assembled in POW Europe to develop a common strategic plan (POW Europe, 2020a), similar to but differentiated from its US version (POW, 2019a). This four-year roadmap (2020–2023) identified the three main goals as "inspiration and engagement", "mobilization" and "leverage". These goals are connected to different strategies, tactics and actions, with some country-specific tasks, and depart from POW's main framework called the "Theory of Change". We briefly summarize these concepts and practices as follows, taking account of the similarities and differences among the Nordic and other contexts.

POW as the catalyst

Within just more than a decade, the POW movement has grown from a handful of concerned skiers to an international network supported and followed by thousands ready to change themselves and others. POW believes that the potential of snow sports and outdoor enthusiasts is large enough to deliver a global impact. While participants in mountain recreation and tourism add up to 36 million in the USA (POW, 2019a) and exceed 60 million in Europe (POW Europe, 2020a), even a small proportion of this extensive community is considered enough to enable desired change.

In extending their base, POW utilizes a diverse set of tools, sometimes using narratives that move the audiences out of their comfort zones with hard facts. While carefully avoiding any doomsaying that may lead to inaction, such communications do end up amplifying awareness. *The Economic Contributions of Winter Sports in a Changing Climate* report (Hagenstad, Burakowski, & Hill, 2018), for instance, already had 11 million reads in its first year (POW, 2019b). Likewise, the websites of the Nordic chapters greet their visitors with the same awful truth

(POW Finland, 2020b). Yet in the Nordic context, the emphasis shifts a little bit away from the economic concerns. For the Norwegian, climate change threat is about winter sports, "thus lifestyle, but also the heritage connected to snow play and ski culture" (translated from POW Norge, 2020b). In Sweden, climate change targets "Swedish people's soul" since "the snow sports give a playful joy to children of all ages, keep everyone strong and healthy, and provide a much-needed closeness to and understanding of nature" (translated from POW Sweden, 2020b). Actual efficacy, however, would eventually need to come with reduced emissions, rather than increased reach or empathies.

POW Europe's primary concern is to ensure a resilient policy towards carbon neutrality with net zero emissions by 2050, thereby limiting warming to 1.5°C (note the different target setting at 2°C by POW USA in their *2020 Strategic Plan*), but they also highlight their goal as "to build the national and European political will to enable meaningful climate policies, and shift cultural norms to make those policies durable" (POW Europe, 2020a, p. 13). According to their "Theory of Change" (POW, 2019a, p. 4; POW Europe, 2020a, p. 5), such resilience is achieved at the intersections of supporting (or electing, in the case of the USA) climate champions and influencing climate policy but also through changing cultural norms. After all, those norms will foster the adoption of policies and innovations, or even the opposite, in which cultural resilience resists socio-political change (Norgaard, 2011; Hulme, 2011). Following POW's style and example, "so it becomes cool, conscious and smart to take the train or other public transport, car share or e-bike" (POW Europe, 2020a, p. 9).

Informing or reminding snow sports enthusiasts that they are both the victims and the causes but also the solutions of climate crisis (Aall, 2014), coupled with strong lifestyle values towards sports and nature as in the Nordic context, can help POW institute norms of climate-friendliness (see Stern, Dietz, Abel, Guagnano, & Kalof, 1999 for a wider discussion). However, removing the gaps and even contradictions between attitudes and behavior still poses major challenges. This may even be true for those of us who scholarly travel to disseminate knowledge on climate change and tourism (Høyer, 2009). In the case of an unwillingness to fly less even among the most environmentally aware tourists, Holden (2016, p. 76) explains that "by engaging in pro-environmental behaviour in our everyday lives we may gain a 'right' to behave for a time in the way we wish to without a consideration of the environmental consequences", in a way, extending Gascon's (2019) criticism towards the UN's frivolous claiming of tourism as a human right. Juvan, Ring, Leisch, and Dolnicar (2016) empirically segment tourists with pro-environmental beliefs but no sustainable vacation behaviors as "government blamers", "struggling seekers" (for sustainable vacation options) and "impact neglecters" (due to limited environmental literacy).

Whereas adaptation rewards mostly locally and in nearer terms, mitigation is for the global good and takes a long time to be effective on limiting warming, thus perhaps, finding the commons in altruistic values. Therefore, in addition to the norm setting, motivation and knowledge play important roles to spark that extra for mitigation (Ortega-Egea, García-de-Frutos, & Antolín-López, 2014). In the case of POW, both information dissemination and engagement activities make top of the agenda. The three Nordic chapters carry out communications and events for these purposes and work with scientists, businesses and athletes (POW Finland, 2020c, 2020d; POW Norge, 2020c; POW Sweden, 2020c, 2020d). These collaborations are critical for the intended behavioral transitions, as research shows that both content credibility and endorser profiles are major determinants for promoting environmental behavior (Elgaaied-Gambier, Monnot, & Reniou, 2018).

The web pages of the three chapters recommend some standard rules for becoming climate-friendly, which can be combined as: a) smart travelling, b) green eating, c) simple living, d) repairing, reusing, recycling, e) dollar voting, f) continuous learning, and g) role modeling (POW Finland, 2020e; POW Norge, 2020d; POW Sweden, 2020e). Moreover, POW Sweden (2019) has also developed the application "POW Habits" to engage winter sports enthusiasts with carbon accounting principles and carbon footprint calculations in an interactive game fashion. Such applied education with a purpose is also critical, as otherwise, in this example, the recreationists may end up using calculator tools for offsetting reasons rather than any actual proactive mitigation (Hall, 2013). But, perhaps, one of the most practical forms of facilitation by POW towards self-decarbonization is the introduction of the POW Pass as a major action regarding the cultural change focus of the 2020 Strategic Plan (POW Europe, 2020a). Half to three-quarters of tourism's 5–8% global carbon footprint share results from travel to/from the destination, air and car travel in particular (UNEP & UNWTO, 2005; Lenzen et al., 2018), and this is where the major paradoxes arise regarding climate-friendliness versus (de)carbonization of tourism (Hall, 2013; Holden, 2016; Juvan et al., 2016). According to POW Europe (2020a, p. 4), the emissions share of extra-destination travel in mountain tourism remains at 60% of the total. With the launch of the POW Pass, in partnership with major rail operators, the facilitation of smart travel is encouraged by providing outdoor enthusiasts with a practical alternative and an economic incentive. POW Norway also takes on the extra responsibility to collaborate with its partner resorts towards promoting bus services to destinations not covered by the rail network.

POW as the activist

At a time when a single teenage activist, Greta Thunberg, can be misperceived as a villain rather than as a heroine for promoting her sustainable alternative to transatlantic travel in 2019 (Mkono, Hughes, & Echentille, 2020), POW's mobilization and lobbying activities will need to be carefully designed to deliver the best reach and efficacy. So far, POW has been involved in numerous actions from encouraging independent advocacy to leading campaigns of its own or joining other movements. As Stoddart (2011) puts it in a nutshell, referring to the early days of POW and other organizations, "slogans like 'global cooling', 'save our snow', or 'protect our winters' are empty without addressing the ironic relationship between skiing, mobility and climate change" (Stoddart, 2011, p. 27) and "responses to ecological irony are limited . . . insofar as they individualize responsibility for political action on climate change" (Stoddart, 2011, p. 26). He further notes that individual or household lifestyle changes, e.g. downshifting or simple living, are not enough by themselves to accomplish the much-needed socio-political change to reverse the environmental crisis. Roser-Renouf et al. (2014) suggest that climate communicators should prioritize building a sense of *collective* efficacy among their audiences, even before illustrating the hard facts on negative climate change impacts, to avoid any inaction. Such efforts may also engage the so-called "government blamers" (Juvan et al., 2016) in the intended pro-environmental behavior by communicating with them the state political agendas and actions for a sustainability transition.

POW has a double role in ensuring a resilient climate policy. POW encourages its audiences to pledge for voting and to vote for the right parties who will deliver the best practices in combatting the climate crisis. Combined with the climate-friendliness norms instituted among the public, these practices are expected to be sustainable for bottom-up adoption. To educate the voters in their critical governmental choices, objective, non-partisan guidance is provided based on a filter of parties' relevant pledges (POW Sweden, 2018; POW, 2019a). However, POW applies direct pressure on or support for policy-makers to ensure regulatory frameworks and decisions for mitigation. In the three Nordic countries, where carbon-neutrality targets are agreed upon, popular government support is high on the agenda, especially in Finland (POW Europe, 2020a). Sweden, home to Greta Thunberg and the "flight shame" trend, is also facing controversial developments from a climate perspective, such as the Scandinavian Mountains Airport opened in 2019 to cater mainly for ski tourists to Swedish and Norwegian resorts operated by Scandinavia's leading ski corporation, and heliskiing investments infamous for their emissions per visitor. Such attempts are under watch by POW Sweden, with some field protests and demonstrations already held in 2019. In Norway, current debates and proposals center around carbon pricing, renewable energies, climate-friendly transportation and protection of public lands (POW Norge, 2020d, 2020e, 2020f, 2020g). While all of these topics have their specific agendas, they all commonly relate to Norway's biggest dilemma in the face of climate change – preparing to become a carbon-neutral country while accumulating an enormous wealth from exporting fossil fuels and their exogenous consumption emissions, which are technically not accounted for by Norway, but which still drive warming at the global scale, including Norway.

Towards a new path

In 2014, 63% of the keen summer skiers in Norway had not heard about POW, including its parent chapter (Demiroglu, Dannevig, & Aall, 2018). In just less than a decade, POW has grown into an international network beyond its winter sports sphere, doing its best to positively influence individuals, businesses and politics for the sake of a solution to the climate crisis. In doing so, its discourse seems to have focused on a green economy with hints of circular practice and technocentrism. While this implication is limited to the documents examined in this chapter, it is also evident in POW's future vision, where green policies and technological changes play central roles to keeping up with a leisure-as-usual approach (POW Europe, 2020a, p. 3):

> In the future we envision, the world embraces renewable energy, zero carbon transport, innovative technologies and market instruments to achieve carbon neutrality by 2050. In the process, we do not sacrifice, but instead turbo-charge, joy. We ride fast and efficient trains or drive zero emissions vehicles to ski areas, climbing spots and trails; individuals and industries pay a fair price for their carbon footprint; and elected officials at all levels of government make addressing climate a top priority. We recognise our actions here as a model for the world. Healthy, strong, wildly adventurous, we still never forget our privilege, knowing that our battle against climate change is not only for ourselves, but for billions facing profound suffering around the world.

As observed in many other arenas, the COVID-19 pandemic has triggered discussions and reflections in the POW community that post-pandemic tourism habits and geographies will not be the same (Prideaux, Thompson, & Pabel, 2020; see also Chapter 14, this volume). A major fruit of this nerve-wracking period, in addition to the undesired outcomes such as cancellation of Eurostar's "ski train" running between the UK and the Alps, was the decision to expand into a "new path" (POW Europe, 2020b), where POW collects public opinions, ideas and stories regarding (forced) adaptation and challenges during the mobility-restricted times of COVID-19, to be submitted to the consultation forum of the EU regarding their climate action within the European Green Deal (European Commission, 2019). Already, some feedback has been documented, with hints towards degrowth, as evident in the words of Heidi Kalmari when answering what she wants to maintain from the pandemic experience (translated from POW Finland, 2020f):

> A lot of people are now thinking about the deeper meanings of things and also enjoying it as the unnecessary hustle and bustle has diminished. I would like the "less is more" mindset to be maintained in the future – in everything, at work, in everyday life and in society.

Economic growth orientation and technological reliance, especially regarding decarbonization, of the green economy discourse and the subsequent overconfidence in decoupling make proposals like the European Green Deal potentially more responsive to the now chronic environmental alarms such as the climate crisis. Sooner or later, but sooner, more radical alternatives will need to be tested once such roadmaps fail to deliver on their aspirations. In this sense, degrowth, or syntheses founded on degrowth ideals (D'Amato et al., 2019; Schröder et al., 2019), should be on the highest-level agendas, not least through the advocacy of bottom-up organizations.

A great unknown in the implications of degrowth, especially from a climate crisis perspective, is the complexities of leisure. While increased leisure is key to the life quality goal of the degrowth movement, it can also bring along consumptive consequences such as through increased ecosystem services, as well as challenges with reinventing livelihoods in established destinations, at a time when staycations, backyard recreation and proximity tourism have already started to emerge as corona-safe and sustainable alternatives, and destinations like the Alps discuss paradigmatic shifts in their tourism character due to climatic and demographic changes (Bausch & Gartner, 2020). Perhaps, it is already time for proponents of degrowth to think more thoroughly about the linkages between leisure and decarbonization and inform and engage concerned communities like POW on these matters.

This chapter does not conclude here. It has rather served as a departure for transdisciplinary research on climate change, tourism/recreation and degrowth. For some, skiing may sound like the toy of the elite, and their activism like a "trustafarian" trend (Kennedy, 2019). Yet those skiers, along with their fellow snowboarders, have been first on the field, just like many farmers, to witness the early signs of climate change. Some have seized the day by either ignoring the issue and blaming it on the system or by simply reorganizing their leisure plans in terms of time and space. Others, like Jeremy Jones, have considered this to be an issue of mitigation, rather than adaptation, and led the way towards mobilizing transformation. Now the question is moving away from "if" to "how" we should change – even if that means an urgent need for a systemic revolution.

References

Aall, C. (2014). Sustainable tourism in practice: Promoting or perverting the quest for a sustainable development? *Sustainability*, *6*(5), 2562–2583.

Aall, C., Hall, C. M., & Groven, K. (2016). Tourism: Applying rebound theories and mechanisms to climate change mitigation and adaptation. In T. Santarius, H. J. Walnum, & C. Aall (Eds.), *Rethinking climate and energy policies: New perspectives on the rebound phenomenon* (pp. 209–225). Cham, Switzerland: Springer.

Andriotis, K. (2018). *Degrowth in tourism: Conceptual, theoretical and philosophical issues*. Wallingford: CABI.

Bausch, T., & Gartner, W. C. (2020). Winter tourism in the European Alps: Is a new paradigm needed? *Journal of Outdoor Recreation and Tourism*, *31*. https://doi.org/10.1016/j.jort.2020.100297

Burkhart, C., Schmelzer, M., & Treu, N. (Eds.). (2020). *Degrowth in movement(s): Exploring pathways for transformation*. Alresford, UK: Zero Books.

Castree, N. (2001). Socializing nature: Theory, practice, and politics. In N. Castree & B. Braun (Eds.), *Social nature: Theory, practice, and politics* (pp. 1–21). Malden, MA: Wiley-Blackwell.

Clifford, H. (2002). *Downhill slide: Why the corporate ski industry is bad for skiing, ski towns, and the environment*. San Francisco, CA: Sierra Club Books.

Coldstream Creative. (2006). *Resorting to madness: Taking back our mountain communities*. Retrieved from https://vimeo.com/86809444

Corvellec, H., Böhm, S., Stowell, A., & Valenzuela, F. (2020). Introduction to the special issue on the contested realities of the circular economy. *Culture and Organization*, *26*(2), 97–102.

Cosme, I., Santos, R., & O'Neill, D. W. (2017). Assessing the degrowth discourse: A review and analysis of academic degrowth policy proposals. *Journal of Cleaner Production, 149*, 321–334.

D'Amato, D., Droste, N., Allen, B., Kettunen, M., Lähtinen, K., Korhonen, J., . . . Toppinen, A. (2017). Green, circular, bio economy: A comparative analysis of sustainability avenues. *Journal of Cleaner Production, 168*(1), 716–734.

D'Amato, D., Droste, N., Winkler, K. J., & Toppinen, A. (2019). Thinking green, circular or bio: Eliciting researchers' perspectives on a sustainable economy with Q method. *Journal of Cleaner Production, 230*(1), 460–476.

Daly, H. (1974). The economics of the steady state. *The American Economic Review, 64*(2), 15–21.

Daly, H. (1993). Steady-state economics: A new paradigm. *New Literary History, 24*(4), 811–816.

Demiroglu, O. C. (2014). *Kış Turizmi*. Ankara: Detay.

Demiroglu, O. C. (2017). Turizm, Yeşil Büyüme ve İklim Değişikliği. In M. Ucal (Ed.), *İklim Değişikliği ve Yeşil Boyut: Yeşil Ekonomi, Yeşil Büyüme* (pp. 146–157). Istanbul: H. Böll Stiftung.

Demiroglu, O. C., Dannevig, H., & Aall, C. (2013). The multidisciplinary literature of ski tourism and climate change. In M. Kozak & N. Kozak (Eds.), *Tourism research: An interdisciplinary perspective* (pp. 223–237). Cambridge, UK: Cambridge Scholars.

Demiroglu, O. C., Dannevig, H., & Aall, C. (2018). Climate change acknowledgement and responses of summer (glacier) ski visitors in Norway. *Scandinavian Journal of Hospitality and Tourism, 18*(4), 419–438.

Demiroglu, O. C., Lundmark, L., Saarinen, J., & Müller, D. K. (2020). The last resort? Ski tourism and climate change in Arctic Sweden. *Journal of Tourism Futures, 6*(1), 91–101.

Demiroglu, O. C., Lundmark, L., & Strömgren, M. (2019). Development of downhill skiing tourism in Sweden: Past, present, future. In U. Pröbstl, H. Richins, & S. Türk (Eds.), *Winter tourism: Trends and challenges* (pp. 305–323). Wallingford: CABI.

Demiroglu, O. C., & Sahin, U. (2015). *Ski community activism on the mitigation of climate change* [pdf]. Retrieved from https://ipc.sabanciuniv.edu/Content/Images/CKeditor Images/20200323-14030356.pdf

Eaton, D., & Sheng, F. (Eds.). (2019). *Inclusive green economy: Policies and practice*. Shanghai: Zayed International Foundation for the Environment, and Tongji University.

Elgaaied-Gambier, L., Monnot, E., & Reniou, F. (2018). Using descriptive norm appeals effectively to promote green behavior. *Journal of Business Research, 82*, 179–191.

European Commission. (2019). *A European green deal: Striving to be the first climate-neutral continent*. Retrieved from https://ec.europa.eu/info/strategy/priorities-2019-2024/european-green-deal_en

Fletcher, R., Mas, I. M., Romero, A. B., & Blázquez-Salom, M. (Eds.). (2020). *Tourism and degrowth: Towards a truly sustainable tourism*. Abingdon: Routledge.

Gascon, J. (2019). Tourism as a right: A "frivolous claim" against degrowth? *Journal of Sustainable Tourism, 27*(12), 1825–1838.

Gössling, S., & Higham, J. (2020). The low-carbon imperative: Destination management under urgent climate change. *Journal of Travel Research*. https://doi.org/10.1177/0047287520933679

Gössling, S., Scott, D., & Hall, C. M. (2013). Challenges of tourism in a low-carbon economy. *WIREs Climate Change, 4*(6), 525–538.

Hagenstad, M., Burakowski, E. A., & Hill, R. (2018). *Economic contributions of winter sports in a changing climate*. Boulder, CO: Protect Our Winters.

Hall, C. M. (2009). Degrowing tourism: Décroissance, sustainable consumption and steady-state tourism. *Anatolia*, *20*(1), 46–61.

Hall, C. M. (2010). Changing paradigms and global change: From sustainable to steady-state tourism. *Tourism Recreation Research*, *35*(2), 131–143.

Hall, C. M. (2011). Policy learning and policy failure in sustainable tourism governance: From first- and second-order to third-order change? *Journal of Sustainable Tourism*, *19*(4–5), 649–671.

Hall, C. M. (2013). Framing behavioural approaches to understanding and governing sustainable tourism consumption: Beyond neoliberalism, "nudging" and "green growth"? *Journal of Sustainable Tourism*, *21*(7), 1091–1109.

Hall, C. M. (2015). Economic greenwash: On the absurdity of tourism and green growth. In M. Reddy & K. Wilkes (Eds.), *Tourism in the green economy* (pp. 361–380). London: Earthscan by Routledge.

Hickel, J. (2019). Degrowth: A theory of radical abundance. *Real-World Economics Review*, *87*(19), 54–68.

Hickel, J., & Kallis, G. (2020). Is green growth possible? *New Political Economy*, *25*(4), 469–486.

Hobson, K., & Lynch, N. (2016). Diversifying and de-growing the circular economy: Radical social transformation in a resource-scarce world. *Futures*, *82*, 15–25.

Holden, A. (2016). *Environment and tourism*. Abingdon: Routledge.

Hollenhorst, S. J., Houge-Mackenzie, S., & Ostergren, D. M. (2014). The trouble with tourism. *Tourism Recreation Research*, *39*(3), 305–319.

Høyer, K. G. (2009). A conference tourist and his confessions: An essay on a life with conference tourism, aeromobility and ecological crisis. *Tourism and Hospitality Planning & Development*, *6*(1), 53–68.

Huizinga, J. (1949). *Homo ludens*. London: Routledge & Kegan Paul.

Hulme, M. (2011). A town called Bygdaby. *Nature Climate Change*, *1*, 83.

IPCC. (2018). *Global warming of 1.5°C*. Geneva: IPCC. In Press.

Juvan, E., Ring, A., Leisch, F., & Dolnicar, S. (2016). Tourist segments' justifications for behaving in an environmentally unsustainable way. *Journal of Sustainable Tourism*, *24*(11), 1506–1522.

Kallis, G., Kostakis, V., Lange, S., Muraca, B., Paulson, S., & Schmelzer, M. (2018). Research on degrowth. *Annual Review of Environment and Resources*, *43*(1), 4.1–4.26.

Kennedy, T. (2019). Your rich friend who goes on ski holidays could save the planet. *Vice Media Group*. Retrieved from www.vice.com/en_in/article/d3n3bm/ski-holiday-melting-snow-climate-change

Kerschner, C. (2010). Economic de-growth vs. Steady-state economy. *Journal of Cleaner Production*, *18*(6), 544–551.

Lasanta, T., Beltran, O., & Vaccaro, I. (2013). Socioeconomic and territorial impact of the ski industry in the Spanish Pyrenees: Mountain development and leisure-induced urbanization. *Pirineos*, *168*, 103–128.

Lenzen, M., Sun, Y.-Y., Faturay, F., Ting, Y.-P., Geschke, A., & Malik, A. (2018). The carbon footprint of global tourism. *Nature Climate Change*, *8*, 522–528.

Lund, M. (2007). Norway: How it all started. *Skiing Heritage*, *19*(3), 8–13.

Menton, M., Larrea, C., Latorre, S., Martinez-Alier, J., Peck, M., Temper, L., & Walter, M. (2020). Environmental justice and the SDGs: From synergies to gaps and contradictions. *Sustainability Science*. https://doi.org/10.1007/s11625-020-00789-8

Mkono, M., Hughes, K., & Echentille, S. (2020). Hero or villain? Responses to Greta Thunberg's activism and the implications for travel and tourism. *Journal of Sustainable Tourism*. https://doi.org/10.1080/09669582.2020.1789157

Norgaard, K. M. (2011). *Living in denial: Climate change, emotions, and everyday life*. Boston, MA: MIT Press.

OECD. (2011). *Towards green growth: A summary for policy makers* [pdf]. Retrieved from www.oecd.org/greengrowth/48012345.pdf

Ortega-Egea, J. M., García-de-Frutos, N., & Antolín-López, R. (2014). Why do some people do "more" to mitigate climate change than others? Exploring heterogeneity in psycho-social associations. *PLoS One*, *9*(9), e106645.

Park, L. S. H., & Pellow, D. N. (2013). *The slums of Aspen: Immigrants vs. the environment in America's Eden*. New York, NY: NYU Press.

Polderman, A., Haller, A., Viesi, D., Tabin, X., Sala, S., Giorgi, A., . . . Bidault, Y. (2020). How can ski resorts get smart? Transdisciplinary approaches to sustainable winter tourism in the European Alps. *Sustainability*, *12*(14), 5593.

Prideaux, B., Thompson, M., & Pabel, A. (2020). Lessons from COVID-19 can prepare global tourism for the economic transformation needed to combat climate change. *Tourism Geographies*. https://doi.org/10.1080/14616688.2020.1762117

Protect Our Winters (POW). (2019a). *2020 strategic plan* [pdf]. Retrieved from https://protectourwinters.org/wp-content/uploads/2019/12/POW-2020-strategic_plan.pdf

Protect Our Winters (POW). (2019b). *2018 annual report* [pdf]. Retrieved from https://protectourwinters.org/wp-content/uploads/2019/12/R3-POW-0001-AnnualReport-digital.pdf

Protect Our Winters (POW). (2020a). *Our work*. Retrieved from https://protectourwinters.org/our-work/

Protect Our Winters (POW). (2020b). *POW international*. Retrieved from https://protectourwinters.org/pow-international/

Protect Our Winters (POW) Europe. (2020a). *2020 strategic plan Europe* [pdf]. Retrieved from https://protectourwinters.eu/wp-content/uploads/2020/05/strat_doc_v8_EU.pdf

Protect Our Winters (POW) Europe. (2020b). *Forging a new path*. Retrieved from https://protectourwinters.eu/forging-a-new-path

Protect Our Winters (POW) Finland. (2020a). *POW Finland*. Retrieved from www.protectourwinters.fi/pow-finland/

Protect Our Winters (POW) Finland. (2020b). *Ilmasto*. Retrieved from www.protectourwinters.fi/ilmasto/

Protect Our Winters (POW) Finland. (2020c). *Lähettiläät*. Retrieved from www.protectourwinters.fi/lahettilaat/

Protect Our Winters (POW) Finland. (2020d). *Yhteistyökumppanit*. Retrieved from www.protectourwinters.fi/yhteistyossa/

Protect Our Winters (POW) Finland. (2020e). *POWin polku-talvien-pelastamiseen*. Retrieved from; www.protectourwinters.fi/polku-talvien-pelastamiseen/

Protect Our Winters (POW) Finland. (2020f). *Uudelle polulle, kohti kestävää tulevaisuutta*. Retrieved from www.protectourwinters.fi/uudelle-polulle-kohti-kestavaa-tulevaisuutta/

Protect Our Winters (POW) Norge. (2020a). *Om oss*. Retrieved from www.protectourwinters.no/protect-our-winters-norge/

Protect Our Winters (POW) Norge. (2020b). *Langrennsmiljøet våkner til klimakrisen*. Retrieved from. www.protectourwinters.no/2020/02/20/langrennsmiljoet-vakner-til-klimakrisen/

Protect Our Winters (POW) Norge. (2020c). *Riders Alliance Norge*. Retrieved from www.protectourwinters.no/riders-alliance-norge/

Protect Our Winters (POW) Norge. (2020d). *Sett en pris på karbon!* Retrieved from www.protectourwinters.no/2020/04/02/sett-en-pris-pa-karbon/

Protect Our Winters (POW) Norge. (2020e). *Skap en økonomi som støtter fornybar energi!* Retrieved from www.protectourwinters.no/2020/04/07/skap-en-okonomi-som-stotter-fornybar-energi/

Protect Our Winters (POW) Norge. (2020f). *Heia elektrifisering og innovasjon av transport!* Retrieved from www.protectourwinters.no/2020/04/04/heia-elektrifisering-og-innovasjon-av-transport/

Protect Our Winters (POW) Norge. (2020g). *Beskytte offentlig land!* Retrieved from www.protectourwinters.no/2020/04/10/beskytte-offentlig-land/

Protect Our Winters (POW) Sweden. (2018). *En guide till hur du väljer vintern i valet 9 september 2018* [pdf]. Retrieved from http://xn – vljvintern-q5a.se/POW_valjvintern_guide_180722.pdf

Protect Our Winters (POW) Sweden. (2019). *POW habits*. Retrieved from www.powhabits.se/

Protect Our Winters (POW) Sweden. (2020a). *Kort om oss*. Retrieved from www.protectourwinters.se/kort-om-oss/

Protect Our Winters (POW) Sweden. (2020b). *Varför vi fnns*. Retrieved from www.protectourwinters.se/varfor-finns-vi/

Protect Our Winters (POW) Sweden. (2020c). *Ambassadörer*. Retrieved from www.protectourwinters.se/ambassadorer/

Protect Our Winters (POW) Sweden. (2020d). *Våra vänner*. Retrieved from www.protectourwinters.se/vanner/

Protect Our Winters (POW) Sweden. (2020e). *Vad du kan göra*. Retrieved from www.protectourwinters.se/vad-du-kan-gora/

Research and Degrowth (R & D). (2020). *Definition*. Retrieved from https://degrowth.org/definition-2/

Riahi, K., Van Vuuren, D. P., Kriegler, E., Edmonds, J., O'Neill, B. C., Fujimori, S., . . . Lutz, W. (2017). The shared socioeconomic pathways and their energy, land use, and greenhouse gas emissions implications: An overview. *Global Environmental Change*, *42*, 153–168.

Rojek, C. (2005). *Leisure theory: Principles and practice*. Basingstoke: Palgrave Macmillan.

Roser-Renouf, C., Maibach, E. W., Leiserowitz, A., & Zhao, X. (2014). The genesis of climate change activism: From key beliefs to political action. *Climatic Change*, *125*, 163–178.

Salazar, D. (1996). The mainstream-grassroots divide in the environmental movement: Environmental groups in Washington State. *Social Science Quarterly*, *77*(3), 626–643.

Sandberg, M., Klockars, K., & Wilén, K. B. (2019). Green growth or degrowth? Assessing the normative justifications for environmental sustainability and economic growth through critical social theory. *Journal of Cleaner Production*, *206*, 133–141.

Schröder, P., Bengtsson, M., Cohen, M., Dewick, P., Hofstetter, J., & Sarkis, J. (2019). Degrowth within – Aligning circular economy and strong sustainability narratives. *Resources, Conservation and Recycling*, *146*, 190–191.

Scott, D., Hall, C. M., & Gössling, S. (2012). *Tourism and climate change: Impacts, adaptation and mitigation*. London: Routledge.

Scott, D., Hall, C. M., & Gössling, S. (2019). Global tourism vulnerability to climate change. *Annals of Tourism Research*, *77*, 49–61.

Ski Sport. (2019). *Prosjekt hvoit vinter 2.0*. Retrieved from https://skisport.no/nyheter/prosjekt-hvit-vinter-2-0-2/

Steiger, R., Scott, D., Abegg, B., Pons, M., & Aall, C. (2019). A critical review of climate change risk for ski tourism. *Current Issues in Tourism*, *22*(11), 1343–1379.

Stern, P. C., Dietz, T., Abel, T. D., Guagnano, G., & Kalof, L. (1999). A value-belief-norm theory of support for social movements: The case of environmentalism. *Human Ecology Review*, *6*(2), 81–97.

Stoddart, M. C. J. (2011). "If we wanted to be environmentally sustainable, we'd take the bus": Skiing, mobility and the irony of climate change. *Human Ecology Review*, *18*(1), 19–29.

Stoddart, M. C. J. (2012). *Making meaning out of mountains: The political ecology of skiing*. Vancouver: UBC Press.

Stroebel, M. (2015). Tourism and the green economy: Inspiring or averting change? *Third World Quarterly*, *36*(12), 2225–2243.

Sun, Y. Y., Lin, P. C., & Higham, J. (2020). Managing tourism emissions through optimizing the tourism demand mix: Concept and analysis. *Tourism Management*, *81*, doi:10.1016/j.tourman.2020.104161

Turhan, E., & Gündoğan, A. C. (2017). The post-politics of the green economy in Turkey: Re-claiming the future? *Journal of Political Ecology*, *24*, 277–295.

United Nations Environment Programme (UNEP). (2011). *Towards a green economy: Pathways to sustainable development and poverty eradication*. Paris: UNEP.

United Nations Environment Programme (UNEP) & United Nations World Tourism Organization (UNWTO). (2005). *Making tourism more sustainable: A guide for policy makers*. Paris: UNEP & UNWTO.

United Nations Environment Programme (UNEP) & United Nations World Tourism Organization (UNWTO). (2012). *Tourism in the green economy: Background report*. Madrid: UNEP & UNWTO.

United Nations World Tourism Organization (UNWTO). (2001). *Global code of ethics for tourism*. Retrieved from www.unwto.org/global-code-of-ethics-for-tourism#:~:text=As%20a%20fundamental%20frame%20of,key%2Dplayers%20in%20tourism%20development.

Vadén, T., Lähde, V., Majava, A., Järvensivu, P., Toivanen, T., Hakala, E., & Eronen, J. T. (2020). Decoupling for ecological sustainability: A categorisation and review of research literature. *Environmental Science & Policy*, *112*, 236–244.

Valenzuela, F., & Böhm, S. (2017). Against wasted politics: A critique of the circular economy. *Ephemera: Theory & Politics in Organization*, *17*(1), 23–60.

van Vuuren, D. P., Stehfest, E., Gernaat, D. E., Doelman, J. C., Van den Berg, M., Harmsen, M., . . . Girod, B. (2017). Energy, land-use and greenhouse gas emissions trajectories under a green growth paradigm. *Global Environmental Change*, *42*, 237–250.

Vanat, L. (2020). *2020 international report on snow & mountain tourism: Overview of the key industry figures for ski resorts* [pdf]. Retrieved from https://vanat.ch/RM-world-report-2020.pdf

Walter, P. G. (2016). Catalysts for transformative learning in community-based ecotourism. *Current Issues in Tourism*, *19*(13), 1356–1371.

Washington, H., & Twomey, P. (2016). *A future beyond growth: Towards a steady state economy*. Abingdon: Routledge.

World Travel and Tourism Council (WTTC). (2020). *Global economic impacts & trends 2020*. Retrieved from https://wttc.org/Research/Economic-Impact/moduleId/1445/itemId/91/controller/DownloadRequest/action/QuickDownload

14 COVID-19 pandemic, tourism and degrowth

C. Michael Hall and Siamak Seyfi

Introduction

According to many media and academic commentators, the COVID-19 pandemic has provided a space for the potential transformations of the tourism industry, as well as the context in which it operates, and from being a force of exploitative, capitalist globalization and neoliberal injustices (Higgins-Desbiolles, 2020) towards a more resilient, responsible and sustainable tourism (Cheer, 2020; Galvani, Lew, & Perez, 2020; Ioannides & Gyimóthy, 2020; Kwok & Koh, 2020; Rowen, 2020). Niewiadomski (2020) argues that one of the fundamental effects of COVID-19 is that the world is experiencing a temporary de-globalization. He further states that failure and incapability of the market in resolving the crisis, with a consequent strengthening and revival of the nation-state, can be seen in the border control and travel restrictions taken by government and the range of governmental recovery packages that have been implemented. Undoubtedly, the limitations on global mobility as a consequence of COVID-19 is significant, but this does represent only one dimension of globalization. Other key dimensions of globalization, such as the global financial system and the mobility of capital and the global ICT network, continue to operate and, as the emergence of new platforms, such as Zoom and other web-based video conferencing tools, indicates, may have even been strengthened in some aspects.

Brouder (2020) argues that the COVID-19 crisis has been different from other crises, such as 11 September 2001 or the global financial crisis of 2008, and believes that the former provides a unique opportunity for transformative change in tourism. Brouder (2020) further proposes a matrix of potential evolutionary pathways towards tourism transformation, which he claims largely relies on sufficient institutional innovation occurring from both the tourists and the hosting destinations as well as emerging new paths. The inherent uncertainty surrounding the pandemic and its magnitude, volatility and profile suggests that its character is very different compared with previous ones (UNWTO, 2020a) and has challenged any prediction and conceptualization of the transformative paths. As a result, the coronavirus pandemic is so far-reaching in terms of its potential economic, social, environmental and political impacts and has grown so rapidly and abruptly that making forecasts has become quite challenging.

For many concerned with the effects of growthism in tourism and in the wider ecological economy, COVID-19 is regarded as potentially enabling a sustainability or socio-technical transition (Cohen, 2020; Sarkis, Cohen, Dewick, & Schröder, 2020; Wells, Abouarghoub, Pettit, & Beresford, 2020). From a tourism perspective, the need for an effective sustainability transition is regarded as critical (Hall, Prayag, & Amore, 2017; Higgins-Desbiolles, Carnicelli, Krolikowski, Wijesinghe, & Boluk, 2019; Hall, Prayag, Fieger, & Dyason, 2020a). As Fletcher, Murray, Blázquez-Salom, and Asunció (2020) suggest:

> even if the COVID-19 crisis ends relatively soon, we cannot afford to return to levels of travel experienced previously, particularly by the wealthiest segment of the world's population. This is not only because of the social unrest overtourism provoked, but also because the industry's environmental damages (including climate change as well as pollution and resource depletion) which were already beyond unsustainable.

Since the advent of the COVID-19 pandemic, many believe that the mobility restrictions across the globe have been able to curb some forms of pollution and have contributed to the actions against climate change. For instance, a report showed that the crisis temporarily cut CO_2 emissions in China by 25% in the aftermath of the lockdown measures in the country (Myllyvirta, 2020), while air quality improved considerably in many parts of the world during lockdowns (Baldasano, 2020; Li et al., 2020; Mahato, Pal, & Ghosh, 2020; Sharma, Zhang, Gao, Zhang, & Kota, 2020). Even though the air quality effects experienced by people during lockdowns were substantial, the actual long-term impact is potentially negligible. Forster et al. (2020) estimate that global NOx emissions declined by as much as 30% in April 2020, contributing a short-term cooling effect. However, this cooling trend was offset by a ~20% reduction in global SO_2 emissions that weakens the aerosol cooling effect, causing short-term warming. As a result, they estimate that the direct effect of the pandemic-driven response represents a cooling of only around $0.01 \pm 0.005°C$ by 2030 compared to a baseline scenario that follows current national policies (Forster et al., 2020). Their conclusion being that substantial cuts in greenhouse gas emissions are still necessary.

COVID-19 is both a public health crisis and a real-time experiment in downsizing the consumer economy; many suggest that this pandemic may represent the beginning of a sustainable consumption transition (Goffman, 2020; Ateljevic, 2020; Brouder, 2020) and offers an opportunity to reset and reshape tourism in a more sustainable way (Hall et al., 2020a; Cheer, 2020; Higgins-Desbiolles, 2020). This is also commented on by major global institutions:

> The pandemic is an unprecedented wake-up call, laying bare deep inequalities and exposing precisely the failures that are addressed in the 2030 Agenda for Sustainable Development and the Paris Agreement on climate change.
>
> (United Nations, 2020a)

This outbreak has challenged the SDGs which were supposed to be attained by 2030, and many of the goals enshrined in the SDGs are facing severe setbacks as a result of the economic and social consequences of COVID-19 (Barbier & Burgess, 2020; Leal Filho, Brandli, Lange Salvia, Rayman-Bacchus, & Platje, 2020). Caught in a vicious negative feedback development loop, the lack of progress towards achieving the SDGs has actually made many countries, and especially developing countries, much more vulnerable to the pandemic than they should be (UN, 2020a), with the impacts on the global economy making it extremely difficult to generate "financial and technical support for the poorest and most vulnerable people and countries hardest hit" (UN, 2020a, p. 1). According to Sumner, Hoy, and Ortiz-Juarez (2020), as a result of COVID-19 global poverty could increase for the first time since 1990 and could represent a reversal of approximately a decade of progress in poverty reduction.

> In some regions the adverse impacts could result in poverty levels similar to those recorded 30 years ago. Under the most extreme scenario of a 20 per cent income or consumption contraction, the number of people living in poverty could increase by 420–580 million, relative to the latest official recorded figures for 2018.
>
> (Sumner et al., 2020, p. 2)

Table 14.1 shows the various effects of this pandemic on each goal of the SDGs (UN, 2020b) and the responses that are required to be taken for pandemic recovery. As Barbier and Burgess (2020, p. 3) observe:

> The COVID-19 pandemic is causing a growing financial burden on all countries, disrupting economies and causing hundreds of thousands of deaths globally. Low and middle-income economies will additionally suffer from the lack of international funding available for achieving the 17 Sustainable Development Goals (SDGs), climate change mitigation and adaptation, and biodiversity conservation. The pandemic is likely to further undermine progress towards the SDGs by 2030, which was already faltering even before the outbreak.

Undoubtedly, the COVID-19 pandemic has slowed down the "runaway consumption train" (United Nations Development Program (UNDP), 1998). However, it has also both directly and indirectly caused incredible turbulence within the global economic system as a result of its effects on supply chains, the capacity to engage in international trade, international mobility, finance, cash flows, retail and consumer behaviour (Hall, Prayag, Fieger, & Dyason, 2020b; Shafi, Liu, & Ren, 2020), and especially tourism (Gössling, Scott, & Hall, 2020; Hall et al., 2020a). A clear outcome of the impacts of COVID-19 and the closing of many international borders during lockdowns is that the economic and employment effects

Table 14.1 COVID-19 and the SDGs

SDG	*COVID-19 effects*	*COVID-19 responses*
Goal 1: No poverty	Intensifying the global poverty rate; loss of income and increase in poverty rate, especially for vulnerable groups of society	Supporting low- and middle-income countries and vulnerable groups such as women and children
Goal 2: Zero hunger	Disruption in food production and distribution	Ensuring prompt measures needed for viable food supply chains to mitigate the risk of large shocks; boosting social protection programmes; keeping global food trade and domestic supply chain going; supporting smallholder farmers' ability to increase food production.
Goal 3: Good health and well-being	Devastating effects on health outcomes	Coordinating support required from the international community; accelerating research and development of a vaccine and treatments; providing guidance and advice for people to look after their mental health
Goal 4: Equality education	Closure of schools; lack of proper and efficient remote learning; acceptability of remote learning for all	Helping countries in mobilizing resources and implementing innovative and context-appropriate solutions to provide education remotely, leveraging high-tech, low-tech and no-tech approaches; seeking equitable solutions and universal access .
Goal 5: Gender equality	Women's economic gain at risk; increased levels of violence against women; increased exposure to COVID-19 among health and social care women workers	Mitigating gender-based violence, including domestic violence; providing social protection and economic stimulus packages for women and girls; coordinating mechanisms that include gender perspectives
Goal 6: Clean water and sanitation	Supply disruptions and inadequate access to clean water, sanitation	Funding and support to reach more girls and boys with basic water, sanitation and hygiene facilities in remote areas and slump; continuing support to affected, at-risk, low-capacity and fragile countries
Goal 7: Affordable and clean energy	Supply and personnel shortage and disruption in access to electricity; further weakening health system response and capacity	Prioritizing energy solutions to power health clinics and first responders; keeping vulnerable consumers connected; increasing reliable, uninterrupted, and sufficient energy production in preparation for a more sustainable economic recovery

(Continued)

Table 14.1 (Continued)

SDG	*COVID-19 effects*	*COVID-19 responses*
Goal 8: Decent work and economic growth	Suspension of economic activities; lower income; less work time; massive unemployment	Ensuring the availability of essential health services and protection of health systems; helping people cope with adversity, through social protection and basic services; protecting jobs, supporting SMEs and informal sector workers through economic response and recovery programmes; promoting social cohesion and investing in community-led resilience and response systems
Goal 9: Industry, innovation and infrastructure	Substantial reduction in global manufacturing growth; disruptions in global value chains and the supply of products	Providing and making digital technologies accessible for all; investments in infrastructure to accelerate economic recovery
Goal 10: Reduced inequalities	Deepened existing inequalities, especially most vulnerable groups and communities; economic inequalities and fragile social safety	Scaling up international support and political commitment aiming to support low- and middle-income countries and vulnerable groups; investing in policies and institutions that can turn the tide on inequality
Goal 11: Sustainable cities and communities	Population living in slumps face higher risk of exposure to COVID-19 for high population density and poor sanitation conditions	Supporting local governments and community-driven solutions in informal settlements; providing urban data, evidence-based mapping and knowledge for informed decision; mitigating economic impact and initiate recovery
Goal 12: Responsible consumption and production	Offered an opportunity to build recovery plans and responsible and sustainable consumption and production patterns	Making a profound and systemic shift to a more sustainable economy that works for both people and the planet and social change
Goal 13: Climate action	Reduced commitment to climate action; less environmental footprints	Creating a green transition; adding green jobs and sustainable and inclusive growth, green economy; investing in sustainable solutions; confronting all climate risks; and cooperating
Goal 14: Life below water	Temporary shutdown of activities as well as reduced human mobility and resource demands	Devising long-term solutions for the health of our planet as a whole; reviving the ocean and building a sustainable ocean economy
Goal 15: Life on land	Highlighted the need to address threats to ecosystems and wildlife	Helping nations manage COVID-19 waste; delivering a transformational change for nature and people; working to ensure economic recovery packages to create resilience for future crises; and modernizing global environmental governance
Goal 16: Peace, justice and strong institutions	Increased geopolitical conflicts; increased level of risk for conflict areas	Respecting human rights; calling for a global ceasefire; protecting UN personnel and their capacity to continue critical operations; supporting refugees and displaced people
Goal 17: Partnerships for the goals	Aggregate backlash against globalization; massive recession of global economy	Increasing international cooperation on public health through global solidarity mainly for developing countries and the most vulnerable populations, including refugees and internally displaced persons

have led to calls for economic stimulus packages from governments (International Labour Organisation (ILO), 2020; International Monetary Fund (IMF), 2020; OECD, 2020)). However, the nature of any stimulus package and its connection, if any, to decarbonization, environmental and welfare policies will clearly have long-term implications for the nature of growth and sustainability. Perhaps somewhat optimistically, Newell and Dale (2020) state:

> Returning to 'business as usual' will not happen, and we are entering a period of 'new normal'. COVID-19 has exposed vulnerabilities that extend beyond pandemic issues, necessitating thinking beyond solely pandemic responses and addressing broader resilience to a range of disturbances. . . . Economic recovery after such a global pause can be accelerated by a green new deal leading to a post-pandemic carbon-neutral economy. However, economic stimulation needs to embrace diversification and avoid focusing investment in single industries or sub-industries, regardless of whether these are 'green' (i.e., eco-tourism, solar).

The need for an appropriate response, including degrowth strategies, to the global heating crisis has been long-standing. However, to what extent does the COVID-19 pandemic provide a realistic basis for a sustainable transition? This chapter discusses the impact of COVID-19 on tourism and the role of degrowth as a response. It first examines why the COVID-19 pandemic provides a potential opportunity for slowing growth and enabling sustainable consumption before looking at COVID-19 and its effects on tourism. The chapter then concludes by noting that many of the government responses to COVID-19 are not geared towards green responses but are instead meant to reinforce business as usual or worse. As a result, the prospects for a global sustainable transition appear limited.

Slowing growth

Changing consumption and concomitant lifestyles is a socio-political issue, not just an economic and environmental one, especially with respect to factoring in equity within and between societies. There can be no presumption that growth alone increases welfare, but rather welfare is an issue of distribution of wealth. If progressive taxes and appropriate regulation and state intervention were necessary for the functioning of the welfare state as a response to the socio-economic shocks of World War II and the preceding Depression, then similar socio-technical system change is surely required for the current COVID-19 and environmental shocks. This is particularly important because of the limited capacities for changing individual behaviours via social marketing interventions or nudging in the time period available to avoid disastrous environmental change arising from current patterns of material consumption (Hall, 2013, 2014, 2016). As Vermeulen (2009, p. 25) argues, the focus of responses to overconsumption needs to be on

> structures as a whole, rather than their individual actions. Short-term solutions may rely on improving efficiencies within existing modes of production

and consumption (reformist changes). In the longer term, however, what is needed is a rethink of how and what we consume (transformist changes).

Hall (2011, 2015) essentially identifies three different approaches to the growth crisis: 1) a business as usual (BAU) approach that, if anything, may only intensify existing market-oriented approaches to the problem of growth – what in policy terms may be described as a first order change; 2) a green growth approach that still utilizes quantitative notions of output but which puts a strong focus on technological solutions and greater efficiency in material/resource/energy (MRE) use; and 3) a degrowth approach that seeks to integrate efficiency and sufficiency approaches.

Given the role of rebound effects and the interconnectedness of growth and MRE consumption, "Energy-efficient technological improvements as the solution for the world's energy and environmental problems will not work. Rather energy-efficient technology improvements are counter-productive, promoting energy consumption. Yet energy efficiency improvements continue to be promoted as a panacea" (Polimeni, Mayumi, Giampietro, & Alcott, 2008, p. 169). Nevertheless, it is important to emphasize that this does not mean that MRE-efficient technologies should not be promoted. Instead, it depends on their context and the overall nature of consumption, not only within tourism but also in the transfer of consumption between tourism and other aspects of what individuals consume within specific socio-technical systems. As Polimeni et al. (2008, p. 169) note: "If individual energy consumption behaviours are significantly altered to reduce consumption and this behaviour is unwavering, then energy efficient technologies can further reduce energy consumption".

The third approach reflects what has been suggested in the debate over growth and the environment since the 1960s (Boulding, 1966; Daly, 1991; Latouche, 2009; Polimeni et al., 2008; Hall, 2015), which is that a sufficiency approach is required that limits consumption patterns in relation to bio-physical constraints. As Czech (2006, p. 1653) comments, so long as economic growth is the focus, whether 'green' or not, "technological progress will not result in biodiversity conservation; rather, an expansion of the human niche and the consumption of more natural resources will result".

Hall (2009, 2015) sought to develop a framework for efficiency (green growth) and sufficiency (degrowth) in sustainable tourism in order to respond to the problematic fixation with visitor growth and economic growth, what Georgescu-Roegen (1977) termed "growth mania" (Figure 14.1). Hall (2009, 2010) integrated Daly's notion of a steady-state economy with that of degrowth (Kerschner, 2010; Buch-Hansen, 2014) and suggested the notion of steady-state tourism, defining it as a tourism system (whether destination, regional or global) that encourages qualitative development, with a focus on quality of life and social and ecological well-being measures, but not aggregate quantitative growth to the detriment of natural capital. Arguing that the problem with tourism is that the larger something has grown, the greater, *ceteris paribus*, are its maintenance costs (Hall, 2011, 2015). One of the significant aspects of Hall's model is that he argued that the lower rates of maintenance throughput in a tourism system that arise from a degrowth/steady-state approach

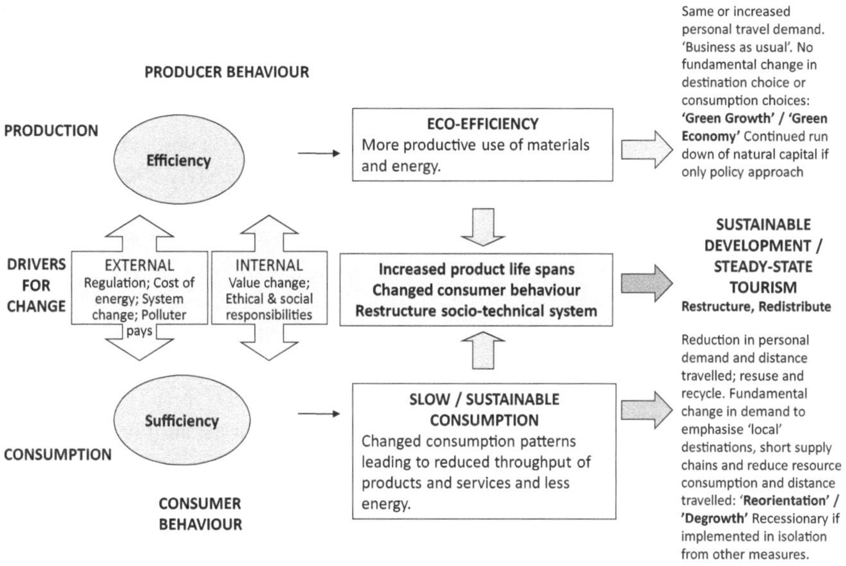

Figure 14.1 Efficiency and sufficiency as sustainable tourism degrowth strategies

would have to be carefully managed, because an emphasis on short supply chains, local destinations and reduced resource consumption and distance travelled would be recessionary if implemented in isolation from other measures (Hall, 2015). This is therefore something that can be examined in the 'natural experiment' that arises from the impact of the COVID-19 pandemic on tourism.

COVID-19

The novel coronavirus COVID-19 spread swiftly around the world from the end of 2019. As of mid-August 2020, nearly 21 million cases have been reported and over 750,000 people had died of this virus worldwide (World Health Organization (WHO), 2020). Quarantines, mobility restrictions and strict social distancing measures implemented in response to the pandemic led to an unprecedented downturn in the global economy, disrupted supply chains, and sharply reduced production and consumption, leading to dramatic declines in employment and the GDP (Gössling et al., 2020). The World Bank's (2020) global economic baseline forecast suggested a 5.2% contraction in global GDP in 2020, triggering the deepest global recession in decades and meaning that per capita incomes in most emerging and developing economies will shrink (World Bank, 2020). However, the long-term economic impact will depend on the combination of: a) the medical capacity to restrict the effects of the coronavirus; b) the measures put in place to restrict transmission; c) the characteristics of government interventions to boost the economy; and d) business and consumer behaviour.

Tourism, which is often regarded as one of the most labour-intensive sectors of the economy, has been dramatically affected by the pandemic because of the impact on international aviation and travel (Hall et al., 2020a). The July 2020 International Air Transport Association (IATA) forecast is for global enplanements to fall 55% in 2020 compared to 2019. Passenger numbers are expected to rise 62% in 2021 compared to 2020, but still down almost 30% compared to 2019. A full recovery to 2019 levels is not expected until 2023 (IATA, 2020). According to the UNWTO (2020b), the near-complete lockdown imposed by countries around the world in response to the pandemic in May 2020 led to a 98% fall in international tourist numbers in that month when compared to 2019. The UNWTO also reported a fall of 300 million international tourists, representing a 56% year-on-year drop in tourist arrivals between January and May 2020. This translates into a loss of US$320 billion in international tourism receipts – more than three times the amount during the global economic crisis of 2009 (UNWTO, 2020b).

The economies of emerging destinations and destinations largely dependent on tourism, such as Small Islands Development States (SIDS), are being hit hard by the collapse of the tourism sector. For instance, tourism – which accounts for 50–90% of GDP and employment in some countries in the Caribbean – has largely been economically crippled by the pandemic, with millions of jobs at risk (Srinivasan, Muñoz, & Chensavasdijai, 2020). The virus also affected the world's top tourist destinations. For instance, Europe, which was the most visited continent by international tourists and welcomed 672 million tourists in 2019, nearly half of the international arrivals in the world, is estimated to experience a financial loss of roughly €1 billion per month (Brzozowski, 2020). Despite the effects of the pandemic on the global tourism industry, many believe that the COVID-19 pandemic has provided a rare opportunity for tourism degrowth and transformation and questioning the meaning of globalized tourism in an attempt to encourage a more responsible and/or sustainable tourism (Hall et al., 2020a; Ateljevic, 2020; Brouder, 2020; Cheer, 2020; Higgins-Desbiolles, 2020; Gössling et al., 2020; Prayag, 2020). However, this transformation is neither guaranteed nor necessarily the most appropriate response with respect to tourism from a degrowth perspective.

COVID-19 and slowing down consumption

As discussed earlier, the COVID-19 pandemic slowed down the global economy, leading to a massive recession and dramatically changed consumer spending. Consumer spending is one of the most important driving forces for global economic growth (Baker, Farrokhnia, Meyer, Pagel, & Yannelis, 2020; Jones, 2020), and the COVID-19 pandemic has triggered a massive transformation in customer purchasing patterns, with many customers grappling with uncertainty (Hall, 2020b; Jones, 2020). The empirical study of Andersen, Hansen, Johannesen, and Sheridan (2020) shows the effect of social distancing laws on consumer spending in Scandinavia. They reported that aggregate spending dropped by around 25% in Sweden and by four additional percentage points in Denmark. Similarly, the study of Chen, Qian, and Wen (2020) on the impact of COVID-19 on consumption after China's outbreak in late January 2020 shows that spending on goods and services were significantly affected, with a decline of 33% and 34%, respectively; with dining and entertainment and travel declining by 64% and 59%.

The COVID-19 pandemic has caused significant disruptions in the global tourism industry (Hall et al., 2020a, 2020b; Gössling et al., 2020). The COVID-19 pandemic brought international travel to an abrupt halt and therefore greatly affected global mobility (UNCTAD, 2020). Remarkably, the UNWTO reported in May 2020 that 100% of global destinations had introduced some form of travel restrictions, which represents the most severe restriction on international travel in history (UNWTO, 2020a). Due to the resulting travel restrictions, as well as the slump in demand among travelers, the COVID-19 pandemic has had significant impacts on lodging, car rental, cruise lines and airlines, and the broader aviation and travel sector. Significant reductions in passenger numbers culminated in the cancellation of flights, which drastically decreased airline profits and caused many of them to lay off staff or declare bankruptcy. The IATA reported that the airline industry lost $314 billion in revenues due to the sharp reduction in the number of passengers. They also indicated that airlines would require $200 billion in bailouts to survive the crisis (Jasper & Philip, 2020). Similar to the aviation sector, the cruise industry and lodging sector also witnessed sharp declines (Cheer, 2020) and a collapse in share prices. For example, Carnival's stock dropped by nearly 60% while Royal Caribbean and Norwegian have lost more than 70% of their value (Businesswire, 2020). This was an especially important issue for many small island nations that depend heavily on the employment and cash flow that cruise ships generate. Other sectors within the tourism and hospitality industry were also substantially crippled. For instance, short-term rental platforms like Airbnb were affected (Kuhzady, Seyfi, & Béal, 2020). Although P2P accommodation and Airbnb in particular were increasing in popularity prior to the COVID-19 epidemic, major shifts have been seen in the demand for short-term rentals (DuBois, 2020). For instance, Airbnb's revenue in 2020 is expected to be less than half of what it was in 2019 (Kuhzady et al., 2020).

The pandemic, albeit temporarily, did reduce 'over-tourism', with many local residents in tourist destinations unhappy with the overwhelming number of tourists which they believed were disrupting their life, a phenomenon which gained popular interest among tourism academics and practitioners as well as wider media coverage (Milano, Cheer, & Novelli, 2019; Phi, 2019; Sæþórsdóttir, Hall, & Wendt, 2020). However, in light of the COVID-19 outbreak and growing mobility restrictions and restricted lockdowns and quarantine in nearly all tourism destinations in response to the pandemic, the concern for overtourism was replaced by a newfound worry of 'undertourism', or the complete absence of tourism, which was previously only a marginal issue for major tourist destinations (Romagosa, 2020; Sæþórsdóttir et al., 2020).

Yet the overtourism/undertourism issue clearly restates the classic issue of finding the appropriate balance between tourism and destinations. Importantly, this was being articulated before COVID-19. In the case of coastal areas, for example,

Hjalager (2020) was suggesting that undertourism existed in many areas where there appeared to be no other development opportunities apart from tourism. Similarly, Haywood (2020, p. 605) commented that "in reality, the vast majority of communities-as-destinations suffer from under-tourism – a problem that is bound to become even more serious as economies tank and poverty levels ratchet up [as a result of COVID-19]". Such a position reflects the seemingly automatic position of many tourism researchers, as well as the tourism industry, that tourism should be used as a form of economic development, which stands in contrast to a broader perspective that the best form of tourism may well be little or no tourism at all (Hall, 2008). Just as significantly, the undertourism/overtourism question of balance also returns us to the key question of balance in the understanding of what sustainability means in a tourism context.

For example, the UNEP and UNWTO (2005) publication *Making Tourism More Sustainable: Guide for Policy Makers*, described by Eugenio Yunis, UNWTO Head of sustainable development of tourism as "applicable world-wide . . . a 'bible' for all decision-makers who are encouraged to be actively involved in the development of an environmentally and socially responsible tourism which creates long term economic benefits for the businesses and destinations" (Yunis, 2006, p. 2), argues that sustainable tourism is based on the three pillars (economic, social, environmental) of sustainable development and that "delivering sustainable development *means striking a balance between them*" (UNEP & UNWTO, 2005, p. 9; emphasis added).

This notion of balance is one of the cornerstones of the so-called sustainable tourism paradigm that has come to be normalized by the tourism industry, policymakers and the majority of the tourism academy (Hall, 1994, 2011). To illustrate this point, Hall (2011) provided the example of an inquiry undertaken by the British Independent Television Commission in 1998 with respect to an attack on the environmental movement in a Channel Four television programme "Against Nature", first shown in the United Kingdom in November and December 1997. In the programme's defence, Michael Jackson, Chief Executive of Channel Four, wrote: "The small but significant group of people who hold views opposed to the environmental lobby have rarely been seen on British television" (quoted in Edwards, 1998, p. 201). In response an editorial in the *Ecologist* magazine stated: "Jackson's view is the norm for a culture in which business dominance is so total, so normal, that any challenges to that domination are seen as 'biased' and 'strange', requiring immediate balance" (Edwards, 1998, p. 201). Similarly, in an academic vein, Edgell (2006, p. 24) states that, "For sustainable tourism to be successful, long-term policies that balance environmental, social, and economic issues must be fashioned", with the preface to the first edition of the book noting that it

> stresses that positive sustainable tourism development is dependent on forward looking policies and new management philosophies that seek harmonious relations between local communities, the private sector, not-for-profit organizations, academic institutions, and governments at all levels to develop

> practices that protect natural, built, and cultural environments *in a way compatible with economic growth.*
>
> (Edgell, 2006, p. xiii; emphasis added)

The preface to the third edition states: "Sustainable tourism is part of an overall shift that recognizes that orderly economic growth combined with concerns for the environment and quality-of-life social values, will be the driving force for long-term progress in tourism development and policies" (Edgell, 2020). Nevertheless, the argument is somewhat circular, as the book also states that "orderly economic growth" is "part of the goal of sustainable tourism" (Edgell, 2020). Although economic growth has become an undefined 'orderly economic growth' in the book's new edition, it is still economic growth.

In addition to the fetishization of economic growth, tourism stakeholders, even given the issues of overtourism, fail to address the fundamental growth problem. Instead, the ongoing message of international tourism bodies in relation to tourism and sustainability is that a continued focus on improved competitiveness, efficiency, the market and growth is the answer, even though it must be done "better" (Zurab Pololikashvili, Secretary-General of the UNWTO, in UNWTO, 2018).

> "Tourism's sustained growth brings immense opportunities for economic welfare and development", said the UNWTO Secretary-General, while warning at the same time that it also brings in many challenges. "Adapting to the challenges of safety and security, constant market changes, digitalization and the limits of our natural resources should be priorities in our common action". . . . The UNWTO Secretary-General stressed education and job creation, innovation and technology, safety and security; and sustainability and climate change as the priorities for the sector to consolidate its contribution to sustainable development and the 2030 Agenda, against the backdrop of its expansion in all world regions and the socio-economic impact this entails. To address these issues, Mr. Pololikashvili concluded that "public/private cooperation as well as public/public coordination must be strengthened, in order to translate tourism growth into more investment, more jobs and better livelihoods".
>
> (UNWTO, 2018)

Sustainability itself is strongly positioned by the UNWTO as an economic and competitive value rather than as an ethical or environmental one. For example, in its focus on the SDGs, it comments: "many companies already seem to acknowledge that their contribution [to the SDGs] should be integrated into core business and form an inherent part of the creation of value to succeed on today's markets" (UNWTO & UNDP, 2017, p. 41). Hall (2019) argues that the UNWTO, like all major tourism institutions, treats sustainability and overtourism as managerial issues, i.e. that all that is needed to solve them is to improve management practices, together with better information and greater efficiency. For example, the UNWTO and UNDP (2017, p. 14) state: "Both countries and companies lack frameworks to capture, aggregate and report on the full economic, social and environmental impacts of tourism [with respect to] Improving performance by measuring impact and sharing knowledge". And they go on to define sustainable tourism as "tourism that takes full account of its current and future economic, social and environmental impacts" (UNWTO & UNDP, 2017, p. 17). Although, of course, account for impact is not the same as changing practices, measurement, surveillance, control and regulation lie at the core of managerialist values developed in an economic and philosophical context where process is subordinated to output (and profit) (Lynch, 2014).

The conclusion to the foreword of the WTTC and McKinsey & Company (2017, p. 5) report on *Managing Overtourism in Tourism Destinations* states:

> To solve this challenge [of overtourism], leaders must be willing to identify and address the barriers (including beliefs, norms, and structures) that are holding us back from effectively managing overcrowding. And they must look for ways to compromise: when overcrowding goes too far, the repercussions are difficult to reverse.

However, cuts in visitor numbers is rarely the strategy adopted to manage tourism. Instead, the managerial focus is on shifting demand in space and time and searching for greater efficiencies. At the macro-level, economic growth and its relationship to visitor growth are not fundamentally questioned. Such perspectives also constrain responses to COVID-19. Demands from industry to open up travel bridges and bubbles for economic reasons have been widespread and, in many countries, have arguably outweighed health concerns, leading to further waves of COVID-19 cases (Australian Leisure Management, 2020; BBC, 2020; McIlroy & Cranston, 2020; Sullivan, 2020). The active promotion of domestic tourism in many countries as a result of holiday-makers not being able to travel internationally has also highlighted the issues involved in shifting tourist consumers from one location to another. For example, record numbers of visitors to the UK countryside have caused huge problems, with complaints "that a new generation of holidaymakers are treating the countryside like a festival site, leaving behind tents, chairs and excrement, as well as endangering rare habitats and wildlife" (Barkham, 2020). The problem of overtourism has therefore simply shifted in time and space. Encouraging people to travel domestically may therefore reduce the amount of emissions, but the UK experience shows that other environmental impacts have occurred instead. Fundamentally, there is therefore a need for changed behaviours and strategies that better acknowledge the environmental and social dimensions of tourism as well as the economic ones. As the COVID-19 experience shows, slowing tourism-related consumption does have recessionary impacts, but restarting tourism as part of a sustainable transition also means that tourism needs to be reimagined in order to contribute to sustainability.

Conclusion

The lack of tourism as the result of COVID-19 health concerns and associated travel restrictions has significantly affected the socio-economic condition of destinations worldwide (Sæþórsdóttir et al., 2020). However, the future of post-viral tourism is largely dependent on different issues such as the duration of the pandemic, the severity and stringency of social distancing interventions on business, the magnitude and effectiveness of government stimulus packages, and business and consumer behaviours (Hall et al., 2020a). As this chapter has observed, a number of commentators and academics have suggested that the coronavirus crisis may help move the world toward responsible, sustainable and socially innovative tourism (Galvani et al., 2020; Romagosa, 2020). Such optimistic perspectives have, however, tended not to fundamentally challenge treating issues of growth beyond managerialist options.

As Hall (2015) has suggested, changing consumption and concomitant lifestyles is a socio-political issue, not just an economic and environmental one, factoring in equity within and between societies in particular (Khor, 2011). There can be no presumption that growth alone increases welfare, but rather that welfare is an issue of distribution of wealth. If progressive taxes and appropriate regulation and state intervention were necessary for the functioning of the welfare state as a response to the socio-economic shocks of World War II and the preceding Depression, then similar socio-technical system change is surely required for the current shocks (Hall, 2015). Vermeulen (2009, p. 25) argues that the focus of responses to overconsumption needs to be on

> structures as a whole, rather than their individual actions. Short-term solutions may rely on improving efficiencies within existing modes of production and consumption (reformist changes). In the longer term, however, what is needed is a rethink of how and what we consume (transformist changes).

Issues of growth, including the response to COVID-19, sustainable tourism and overtourism are framed as managerial problems by government, industry and most tourism researchers. Yet, as Hall (2011, p. 661) pointed out, the continuing contribution of a growing tourism industry to unsustainable global change "raises a clear question as to whether sustainable tourism can actually be achieved via a so-called 'balanced' approach that seeks to continue to promote economic growth". Green growth and the management of visitor growth as part of the response to overtourism are little more than a marginal reform of a socio-economic system that is unsustainably geared towards economic growth. They are not major shifts in the tourism policy paradigm (Hall, 2011). It is not just a case of tourism getting more efficient, as important as that is. Tourism consumption needs to be spatially and temporally shifted in order to reduce its overall emissions and MRE consumption. But a degrowth response to tourism also needs to go further than managerial and technological responses and deal with the nature of consumptive behaviour; otherwise, destination managers, mediated by state agencies, corporate interests and economic rationality, may 'manage' resources into oblivion – what Hall (2019) described as "Brundtland-as-usual". Instead, tourism and travel behaviours need to shift from being focused on efficiency and management to traveling within the environmental, social and economic limits of the ecosystems of which humanity is a part.

References

Andersen, A. L., Hansen, E. T., Johannesen, N., & Sheridan, A. (2020). *Pandemic, shutdown and consumer spending: Lessons from Scandinavian policy responses to COVID-19*. arXiv preprint arXiv:2005.04630.

Ateljevic, I. (2020). Transforming the (tourism) world for good and (re) generating the potential 'new normal'. *Tourism Geographies*, *22*(3), 467–475.

Australian Leisure Management. (2020, August 16). WTTC calls for leadership from Australian government to save the travel & tourism sector. *Australian Leisure Management*. Retrieved from www.ausleisure.com.au/news/wttc-calls-for-leadership-from-australian-government-to-save-the-travel-tourism-sector/

Baker, S. R., Farrokhnia, R. A., Meyer, S., Pagel, M., & Yannelis, C. (2020). *How does household spending respond to an epidemic? Consumption during the 2020 covid-19 pandemic* (No. w26949). Washington, DC: National Bureau of Economic Research.

Baldasano, J. M. (2020). COVID-19 lockdown effects on air quality by NO_2 in the cities of Barcelona and Madrid (Spain). *Science of the Total Environment*, *741*, 140353.

Barbier, E. B., & Burgess, J. C. (2020). Sustainability and development after COVID-19. *World Development*, *135*, 105082.

Barkham, P. (2020, August 14). 'The worst of human nature': UK staycationers' trail of destruction. *The Guardian*. Retrieved from www.theguardian.com/environment/2020/aug/14/uk-staycations-countryside-coast-visitors-trail-of-destruction

BBC. (2020, June 10). Coronavirus: Restart tourism to beat virus, says United Nations. *BBC News*. Retrieved from www.bbc.com/news/business-52997197

Boulding, K. E. (1966). The economics of the coming spaceship earth. In H. Jarrett (Ed.), *Environmental quality in a growing economy* (pp. 3–14). Baltimore: John Hopkins University and Resources for the Future.

Brouder, P. (2020). Reset redux: Possible evolutionary pathways towards the transformation of tourism in a COVID-19 world. *Tourism Geographies*, *22*(3), 484–490.

Brzozowski, A. (2020). *European tourism sector urges measures to mitigate COVID-19 impact*. Retrieved from www.euractiv.com/section/coronavirus/news/european-tourism-sector-urges-measures-to-mitigate-covid-19-impact/

Buch-Hansen, H. (2014). Capitalist diversity and de-growth trajectories to steady-state economies. *Ecological Economics*, *106*, 167–173.

Businesswire. (2020). *Impact of COVID-19 on the global cruise industry: Compare key industry players' actions*. Retrieved from www.businesswire.com/news/home/20200417005169/en/Impact-COVID-19-Global-Cruise-industry-Compare-Key

Cheer, J. M. (2020). Human flourishing, tourism transformation and COVID-19: A conceptual touchstone. *Tourism Geographies*, *22*(3), 514–524.

Chen, H., Qian, W., & Wen, Q. (2020). *The impact of the COVID-19 pandemic on consumption: Learning from high frequency transaction data*. SSRN 3568574.

Cohen, M. J. (2020). Does the COVID-19 outbreak mark the onset of a sustainable consumption transition? *Sustainability: Science, Practice and Policy*, *16*(1), 1–3.

Czech, B. (2006). If Rome is burning, why are we fiddling? *Conservation Biology*, *20*(6), 1563–1565.

Daly, H. E. (1991). *Steady-state economics* (2nd ed.). Washington, DC: Island Press.

DuBois, D. (2020). *Impact of the coronavirus on global short-term rental markets*. Retrieved from www.airdna.co/blog/coronavirus-impact-on-global-short-term-rental-markets

Edgell, D. L., Sr. (2006). *Managing sustainable tourism: A legacy for the future*. Binghamton: Haworth Press.

Edgell, D. L., Sr. (2020). *Managing sustainable tourism: A legacy for the future* (3rd ed.). Abingdon: Routledge.

Edwards, D. (1998). Learning to think the right thoughts: "ChannelOne". *The Ecologist, 28*(4), 201–203.

Fletcher, R., Murray, I. M., Blázquez-Salom, M., & Asunción, B. R. (2020, March 24). Tourism, degrowth, and the COVID-19 Crisis. *POLLEN Ecology Network*. Retrieved from https://politicalecologynetwork.org/2020/03/24/tourism-degrowth-and-the-covid-19-crisis/

Forster, P. M., Forster, H. I., Evans, M. J., Gidden, M. J., Jones, C. D., Keller, C. A., . . . Turnock, S. T. (2020). Current and future global climate impacts resulting from COVID-19. *Nature Climate Change*. doi:10.1038/s41558-020-0883-0.

Galvani, A., Lew, A. A., & Perez, M. S. (2020). COVID-19 is expanding global consciousness and the sustainability of travel and tourism. *Tourism Geographies, 22*(3), 567–576.

Georgescu-Roegen, N. (1977). The steady state and ecological salvation: A thermodynamic analysis. *Bioscience, 27*(4), 266–270.

Goffman, E. (2020). In the wake of COVID-19, is glocalization our sustainability future? *Sustainability: Science, Practice and Policy, 16*(1), 48–52.

Gössling, S., Scott, D., & Hall, C. M. (2020). Pandemics, tourism and global change: A rapid assessment of COVID-19. *Journal of Sustainable Tourism*, 1–20. https://doi.org/10.1080/09669582.2020.1758708

Hall, C. M. (1994). *The politics of tourism*. Chichester: John Wiley & Sons.

Hall, C. M. (2008). *Tourism planning* (2nd ed.). Harlow: Pearson.

Hall, C. M. (2009). Degrowing tourism: Décroissance, sustainable consumption and steady-state tourism. *Anatolia, 20*(1), 46–61.

Hall, C. M. (2010). Changing paradigms and global change: From sustainable to steady-state tourism. *Tourism Recreation Research, 35*(2), 131–143.

Hall, C. M. (2011). Policy learning and policy failure in sustainable tourism governance: From first- and second-order to third-order change? *Journal of Sustainable Tourism, 19*(4–5), 649–671.

Hall, C. M. (2013). Framing behavioural approaches to understanding and governing sustainable tourism consumption: Beyond neoliberalism, "nudging" and "green growth"? *Journal of Sustainable Tourism, 21*(7), 1091–1109.

Hall, C. M. (2014). *Tourism and social marketing*. Abingdon: Routledge.

Hall, C. M. (2015). Economic greenwash: On the absurdity of tourism and green growth. In V. Reddy & K. Wilkes (Eds.), *Tourism in the green economy* (pp. 339–358). London: Earthscan.

Hall, C. M. (2016). Intervening in academic interventions: Framing social marketing's potential for successful sustainable tourism behavioural change. *Journal of Sustainable Tourism, 24*(3), 350–375.

Hall, C. M. (2019). Constructing sustainable tourism development: The 2030 agenda and the managerial ecology of sustainable tourism. *Journal of Sustainable Tourism, 27*(7), 1044–1060.

Hall, C. M., Prayag, G., & Amore, A. (2017). *Tourism and resilience: Individual, organisational and destination perspectives*. Bristol: Channel View.

Hall, C. M., Prayag, G., Fieger, P., & Dyason, D. (2020b). Beyond panic buying: Consumption displacement and COVID-19. *Journal of Service Management*. doi:10.1108/JOSM-05-2020-0151

Hall, C. M., Scott, D., & Gössling, S. (2020a). Pandemics, transformations and tourism: Be careful what you wish for. *Tourism Geographies, 22*(3), 577–598.

Haywood, K. M. (2020). A post-COVID future: Tourism community re-imagined and enabled. *Tourism Geographies, 22*(3), 599–609.

Higgins-Desbiolles, F. (2020). Socialising tourism for social and ecological justice after COVID-19. *Tourism Geographies*, *22*(3), 610–623.

Higgins-Desbiolles, F., Carnicelli, S., Krolikowski, C., Wijesinghe, G., & Boluk, K. (2019). Degrowing tourism: Rethinking tourism. *Journal of Sustainable Tourism*, *27*(12), 1926–1944.

High, E. (2020). *Outcomes from European sustainable development report webinar*. Retrieved from www.unsdsn.org/outcomes-from-european-sustainable-development-report

Hjalager, A. M. (2020). Land-use conflicts in coastal tourism and the quest for governance innovations. *Land Use Policy*, *94*, 104566.

International Air Transport Association (IATA). (2020, July 28). Recovery delayed as international travel remains locked down. *IATA Pressroom. Press Release No: 63*. Retrieved from www.iata.org/en/pressroom/pr/2020-07-28-02/

International Labour Organisation (ILO). (2020). *COVID-19 has exposed the fragility of our economies*. Retrieved from www.ilo.org/global/about-the-ilo/newsroom/news/WCMS_739961/lang – en/index.htm

International Monetary Fund (IMF). (2020). *Policy responses to COVID-19*. Retrieved from www.imf.org/en/Topics/imf-and-covid19/Policy-Responses-to-COVID-19

Ioannides, D., & Gyimóthy, S. (2020). The COVID-19 crisis as an opportunity for escaping the unsustainable global tourism path. *Tourism Geographies*, *22*(3), 624–632.

Jasper, C., & Philip, S. (2020, March 18). Airlines need up to $200 billion to survive virus, IATA says. *Bloomberg*. Retrieved from www.bloomberg.com/news/articles/2020-03-17/airlines-need-up-to-200-bailout-to-survive-virus-iata-warns

Jones, K. (2020). These charts show how COVID-19 has changed consumer spending around the world. *World Economic Forum*. Retrieved from www.weforum.org/agenda/2020/05/coronavirus-covid19-consumers-shopping-goods-economics-industry

Kerschner, C. (2010). Economic de-growth vs. steady-state economy. *Journal of Cleaner Production*, *18*(6), 544–551.

Khor, M. (2011). *Risks and uses of the green economy concept of sustainable development, poverty and equity*. Research Paper No. 40. Geneva: South Centre.

Kuhzady, S., Seyfi, S., & Béal, L. (2020). Peer-to-peer (P2P) accommodation in the sharing economy: A review. *Current Issues in Tourism*. doi:10.1080/13683500.2020.1786505

Kwok, A. O., & Koh, S. G. (2020). COVID-19 and extended reality (XR). *Current Issues in Tourism*. doi:10.1080/13683500.2020.1798896

Latouche, S. (2009). *Farewell to growth*. Cambridge: Polity Press.

Leal Filho, W., Brandli, L. L., Lange Salvia, A., Rayman-Bacchus, L., & Platje, J. (2020). COVID-19 and the UN sustainable development goals: Threat to solidarity or an opportunity? *Sustainability*, *12*(13), 5343. https://doi.org/10.3390/su12135343

Li, L., Li, Q., Huang, L., Wang, Q., Zhu, A., Xu, J., . . . Azari, M. (2020). Air quality changes during the COVID-19 lockdown over the Yangtze River Delta Region: An insight into the impact of human activity pattern changes on air pollution variation. *Science of The Total Environment*, *732*, 139282. https://doi.org/10.1016/j.scitotenv.2020.139282

Lynch, K. (2014). New managerialism, neoliberalism and ranking. *Ethics in Science and Environmental Politics*, *13*(2), 141–153.

Mahato, S., Pal, S., & Ghosh, K. G. (2020). Effect of lockdown amid COVID-19 pandemic on air quality of the megacity Delhi, India. *Science of the Total Environment*, *730*, 139086. https://doi.org/10.1016/j.scitotenv.2020.139086

McIlroy, T., & Cranston, M. (2020, June 8). Pressure builds for Australia to reopen. *Financial Review*. Retrieved from www.afr.com/politics/federal/pressure-builds-for-australia-to-reopen-20200608-p550hk

Milano, C., Cheer, J. M., & Novelli, M. (Eds.). (2019). *Overtourism: Excesses, discontents and measures in travel and tourism*. Wallingford: CABI.

Myllyvirta, L. (2020). Analysis: Coronavirus temporarily reduced China's CO_2 emissions by a quarter. *Carbon Brief*. Retrieved from www.carbonbrief.org/analysis-coronavirus-has-temporarily-reduced-chinas-co2-emissions-by-a-quarter

Newell, R., & Dale, A. (2020). COVID-19 and climate change: An integrated perspective. *Cities & Health*. doi:10.1080/23748834.2020.1778844

Niewiadomski, P. (2020). COVID-19: From temporary de-globalisation to a re-discovery of tourism? *Tourism Geographies*, *22*(3), 651–656.

Organisation for Economic Co-operation and Development (OECD). (2020). *Tourism policy responses to the coronavirus (COVID-19)*. Retrieved from www.oecd.org/coronavirus/policy-responses/tourism-policy-responses-to-the-coronavirus-covid-19-6466aa20/

Phi, G. T. (2019). Framing overtourism: A critical news media analysis. *Current Issues in Tourism*, 1–5. https://doi.org/10.1080/13683500.2019.1618249

Polimeni, J. M., Mayumi, K., Giampietro, M., & Alcott, B. (2008). *The Jevons Paradox and the myth of resource efficiency improvements*. London: Earthscan.

Prayag, G. (2020). Time for reset? COVID-19 and tourism resilience. *Tourism Review International*, *24*(2–3), 179–184.

Romagosa, F. (2020). The COVID-19 crisis: Opportunities for sustainable and proximity tourism. *Tourism Geographies*, *22*(3), 690–694.

Rowen, I. (2020). The transformational festival as a subversive toolbox for a transformed tourism: Lessons from Burning Man for a COVID-19 world. *Tourism Geographies*, *22*(3), 695–702.

Sarkis, J., Cohen, M. J., Dewick, P., & Schröder, P. (2020). A brave new world: Lessons from the COVID-19 pandemic for transitioning to sustainable supply and production. *Resources, Conservation and Recycling*, *159*, 104894. doi:10.1016/j.resconrec.2020.104894

Shafi, M., Liu, J., & Ren, W. (2020). Impact of COVID-19 pandemic on micro, small, and medium-sized enterprises operating in Pakistan. *Research in Globalization*, *2*, 100018. https://doi.org/10.1016/j.resglo.2020.100018

Sharma, S., Zhang, M., Gao, J., Zhang, H., & Kota, S. H. (2020). Effect of restricted emissions during COVID-19 on air quality in India. *Science of the Total Environment*, *728*, 138878. https://doi.org/10.1016/j.scitotenv.2020.138878

Srinivasan, K., Muñoz, S., & Chensavasdijai, V. (2020). *COVID-19 pandemic and the Caribbean: Navigating uncharted waters*. Retrieved from https://blogs.imf.org/2020/04/29/covid-19-pandemic-and-the-caribbean-navigating-uncharted-waters/

Sullivan, A. (2020, June 5). Coronavirus quarantine plans raise serious questions for tourism in Ireland and Britain. *DW*. Retrieved from www.dw.com/en/coronavirus-quarantine-plans-raise-serious-questions-for-tourism-in-ireland-and-britain/a-53690865

Sumner, A., Hoy, C., & Ortiz-Juarez, E. (2020). *Estimates of the impact of COVID-19 on global poverty*. Working Paper 2020/43. Helsinki: UNU-WIDER United Nations University World Institute for Development Economics Research.

Sæþórsdóttir, A. D., Hall, C. M., & Wendt, W. (2020). From boiling to frozen? The rise and fall of international tourism to Iceland in the era of overtourism. *Environments*, *7*(8), 59. doi:10.3390/environments7080059

United Nations (UN). (2020a). *Shared responsibility, global solidarity: Responding to the socio-economic impacts of COVID-19*. New York, NY: UN Secretary General [pdf]. Retrieved from www.un.org/sites/un2.un.org/files/sg_report_socio-economic_impact_of_covid19.pdf

United Nations (UN). (2020b). *The sustainable development goals: Our framework for COVID-19 recovery*. Retrieved from www.un.org/sustainabledevelopment/sdgs-framework-for-covid-19-recovery/

UNCTAD. (2020). *COVID-19 and tourism: Assessing the economic consequences* [pdf]. Retrieved from https://unctad.org/en/PublicationsLibrary/ditcinf2020d3_en.pdf

United Nations Development Program (UNDP). (1998). *Human development report 1998*. New York, NY: UNDP.

United Nations Environmental Programme and the World Tourism Organization (UNEP & UNWTO). (2005). *Making tourism more sustainable: A guide for policy makers*. Paris: UNEP.

UNWTO. (2018). *Press release: Tourism can and should lead sustainable development: UNWTO Secretary-General opens ITB 2018*. PR 18020, 06 Mar 18. Madrid: UNWTO.

UNWTO. (2020a). *100% of global destinations now have covid-19 travel restrictions*. Retrieved from www.unwto.org/news/covid-19-travel-restrictions

UNWTO. (2020b, July 28). *Impact of COVID-19 on global tourism made clear as UNWTO counts the cost of standstill*. Retrieved from www.unwto.org/news/impact-of-covid-19-on-global-tourism-made-clear-as-unwto-counts-the-cost-of-standstill

UNWTO & UNDP. (2017). *Tourism and the sustainable development goals – Journey to 2030*. Madrid: UNWTO.

Vermeulen, S. J. (2009, September). Sustainable consumption: A fairer deal for poor consumers. *Environment and Poverty Times*. No. 6. Arendal: UNEP & GRID.

Wells, P., Abouarghoub, W., Pettit, S., & Beresford, A. (2020). A socio-technical transitions perspective for assessing future sustainability following the COVID-19 pandemic. *Sustainability: Science, Practice and Policy*, *16*(1), 29–36.

World Bank. (2020). *Pandemic, recession: The global economy in crisis*. Retrieved from www.worldbank.org/en/publication/global-economic-prospects

World Health Organization (WHO). (2020). *Coronavirus disease (COVID-19) pandemic*. Retrieved from www.who.int/emergencies/diseases/novel-coronavirus-2019

World Travel & Tourism Council (WTTC) and McKinsey & Company. (2017). *Coping with success: Managing overcrowding in tourism destinations*. London: WTTC.

Yunis, E. (2006). *12 aims of a sustainable tourism*. Paper presented at EcoTrans European network for sustainable tourism development. Making tourism more sustainable – helpful instruments and examples of good practice in Europe, 3 February. Reisepavillon fairgrounds, Hanover, Germany [pdf]. Retrieved from www.ecotrans.org/docs/1_hamele_Yunis_intro_aims.pdf

15 Conclusions – degrowing tourism

Can tourism move beyond BAU (Brundtland-as-Usual)?

C. Michael Hall, Linda Lundmark and Jundan Jasmine Zhang

Introduction

This book on the potential value of degrowth strategies in tourism is being concluded in the midst of the COVID-19 pandemic. Yet again, tourism is facing another crisis in which the economic contribution of tourism, and international tourism in particular, has failed to materialise. However, as noted throughout this book, many contributors and commentators feel that, somehow, this time things are different for trying to achieve a more sustainable form of tourism.

COVID-19, as significant as it is, is only the latest in a series of challenges to better understand the embeddedness of assumptions of growth in tourism practices and the paradigms and institutional structures that surround them, and their implications. It has exposed numerous vulnerabilities in the global tourism system as well as in the economies, societies and environments that tourism contributes to. Whether we have moved to "a new normal" or continue with business-as-usual in some ways is a moot point. Regardless of how interpreted, COVID-19 has exposed systemic vulnerabilities that extend beyond pandemic issues, that necessitate thinking beyond solely pandemic responses in a narrow tourism business context, as significant as that might be, and addressing broader resilience to change and the range of disturbances that affect the tourism system and destinations in particular. Clearly, global environmental change and global heating are at the forefront of such disturbances (IPCC, 2018) but, as Newell and Dale (2020) note, the COVID-19 outbreak has also illustrated the vulnerability of global supply chains and transport networks to crises, with a renewed focus on the importance of local economy and production, economic diversification and social connectivity. For many researchers, including even some in tourism, such vulnerabilities are nothing new and reflect the risks of growth fetishism without an appreciation of its consequent effects and dangers. In briefly summarising some of the main contributions of this book and in providing some pointers for future work, it is therefore helpful to revisit the various challenges to growthism that have emerged since the 1960s.

Been there, done that

The first challenges to economic growth emerged in the 1960s at a time of increased awareness of environmental problems (Carson, 1962) and the threat of overpopulation (Ehrlich, 1968), and were clearly connected to some of the concerns of the modern environmental movement. Such concerns were very much restricted to the developed world, and economic doubts about the feasibility or desirability of economic growth were initially few. However, the influential work of Boulding (1966) on the economics of "spaceship earth" and Mishan (1967) started to open up initial spaces of critique and also contributed to the emergence of important work by Ayres and Kneese (1969) on the negative externalities of production and consumption and Odum (1971) that led to the development of industrial ecology.

Also of significance from this period was Georgescu-Roegen's (1971) research on entropy, economic processes and energy/matter transformation that served as one of the foundations for ecological economics as well as concepts of degrowth, and Daly's work on the stationary state economy (Pigou, 1943), which developed into steady-state economics (Daly, 1972, 1974, 1991) and a subsequent influence on tourism thinking. Broader awareness of the work that challenged growth was, as noted by several chapters in the present volume, raised by the publication of the Club of Rome's Limits to Growth (LTG) report (Meadows, Meadows, Randers, & Behrens, 1972).

In each of the LTG scenarios that Meadows et al. (1972) ran, population collapsed during the twenty-first century due to ever-increasing pollution and food shortages, along with other factors such as soil erosion. The study also coincided with the hosting of the first United Nations Conference on the Human Environment held in Stockholm, the regular hosting of which laid the subsequent foundations for sustainable development and the SDGs. However, the LTG's main policy recommendation of stabilisation, which was similar to Daly's (1972) work on steady-state economics, was generally dismissed by politicians and policy-makers, as well as by the vast majority of academics. Those who argue for sustainable and steady-state development solutions, incuding degrowth and the right-sizing of economic production and consumption so as to fit within the earth's biocapacity were, and still are, very much in the minority (Hall, 2011a).

In a foretaste of the more recent debates over responsibilities for and limitations on carbon emissions (IPCC, 2018), delegates at the Stockholm conference made it clear that they were not going to accept policies arising from resource limits that would hamper what they regarded as their future form of development (Beckerman, 1972). This is perhaps not surprising given the dismissal of LTG by the vast majority of economists. For example, Beckerman (1972) suggested:

> that the problem of environmental pollution is a simple matter of correcting a minor resource misallocation by means of pollution charges, and that most of the common objections to such a policy can be demolished with the aid of no more economics than that which is the stock-in-trade of any second year economics student. . . . [LTG] was such a brazen, impudent piece of nonsense that nobody could possibly take it seriously so that it would be a waste of time talking about it.
>
> (Beckerman, 1972, p. 327)

As Perez-Carmona (2013) noted, again anticipating much of the contemporary debate over response to climate change adaptation, 'The common argumentative line was that technological progress and the market mechanism could prevent scarcity and pollution from constituting a substantial limitation on long-term economic growth' (2013, p. 91). Indeed, it is important to note that growth in tourism and aviation and its consequences were also emerging as an issue. For example, Mishan (1970) concluded his evaluation of the Commission on the Third London Airport by noting that:

> equity is wholly ignored. If indeed, the business tycoons and the Mallorca holiday-makers are shown to benefit, after paying their fares, to such an extent that they could more than compensate the victims of aircraft spillover, the cost-benefit criterion is met. But compensation is not paid. The former continue to enjoy the profit and the pleasure; the latter continue to suffer the disamenities.
>
> (Mishan, 1970, p. 234)

Similarly, in an article entitled 'slow is beautiful', Gleditsch (1975) noted, 'the severe environmental problems involved in an unlimited or uncontrolled further growth in aviation' (1975, p. 91), as well as the uneven social stratification of personal mobility, in which Gleditsch (1975, p. 91) hypothesized

> that topdogs will secure a disproportionately high share of the advantages and a disproportionately low share of the disadvantages of any new transportation system. . . . With resources such as education and income, topdogs are in a position to make use of new transportation technology – and avoid its cost.

The follow-up to the 1972 Stockholm was the World Commission on Environment and Development (WCED) established in 1983. Although works on sustainable development had been published before its release, the WCED (1987) report, *Our Common Future* (often referred to by the name of its chairwoman, Mrs. Brundtland), undoubtedly framed much of the sustainable development discourse up to the present day. As has been discussed elsewhere, the concept of sustainable development has been extremely successful, including in tourism (Hall, 2011a).

Dryzek (1997) suggests that the report was written in such a way as to ensure that it also received support from business interests. Although the WCED (1987, p. 44) noted the importance of 'consumption standards within the bounds of the ecological possible and to which all can reasonably aspire' are required as part of achieving greater equity, they nevertheless suggested that although ultimate ecological limits exist, reaching them could be delayed by technological innovation. Importantly for the present discussion, they also concluded that 'the international economy must speed up world growth while respecting the environmental constraints' (WCED, 1987, p. 89), primarily by encouraging qualitative economic growth that was less material/resource/energy (MRE) intensive and more equitable, i.e. more decarbonised and dematerialised – an approach that lies at the heart of much discussion of green growth to the present (Hall, 2015).

However, Hall (2015) argues that the WCED (1987) approach, which continues to dominate mainstream sustainability thinking in tourism with its focus on balance, failed to recognise several significant implications of their strategy. First, while dematerialisation may occur at a per-unit level, overall industrial expansion continues. Second, becoming more efficient leads to an increase in throughput (input plus output), what is otherwise known as the "Jevons' paradox" or "rebound effect" (Polimeni, Mayumi, Giampietro, & Alcott, 2008). Third, being 'part of an interdependent world economy' (WCED, 1987, p. 51) provided a rationale not only for further liberalisation of the global economy and the reduction of trade barriers by less developed countries (LDCs) but also for already wealthy countries to further pursue economic growth by increasing consumption so as to encourage economic growth in the LDCs. Indeed, this last point has become one of the cornerstones of so-called pro-poor tourism development. 'The alternative that poor countries could create their own markets' (Daly, 1991, p. 151), including with respect to tourism, is not one that has been greatly encouraged. However, the benefits of export-led growth as a means for poverty alleviation is moot (Perez-Carmona, 2013).

A further challenge to economic growth, and one that has become a growing focus in tourism, is climate change. The publication of the *Stern Review* (Stern, 2007) brought the relationship between economic growth and its environmental and social externalities back to the forefront of public policy and the attention of a wider business and general audience. However, as Jackson (2009, p. 11) noted: 'it's telling that it took an economist commissioned by a government treasury to alert the world to things climate scientists – most notably the Intergovernmental Panel on Climate Change (IPCC) – had been saying for years.' As Perez-Carmona (2013, p. 107) state, despite much of value in the report,

> When Stern published his review in 2006, the global economy already required almost 1.5 planets, yet a discussion on the causality's direction between economic growth and ecological obliteration . . . was completely absent in Stern's work. Economic growth was Stern's default assumption for the entire globe.

The emergence of the green growth paradigm following the Stern report and the global financial crisis only served to reinforce the lack of critique of economic growth and its ecological consequences. Green responses to the global financial crisis never really eventuated. Barbier (2010, p. 832) commented, 'most national recovery plans have missed this opportunity to invest in the planet while saving the economy'. As Barbier (2010) also observed, the G20 countries had not invested the recommended expenditure 1% of GDP on green initiatives (only China and South Korea had exceeded this target), nor had they removed resource-depleting energy, agriculture and fishing subsidies, advanced far on the taxing and trading of carbon emissions, or substantially aided the world's poor. The notion of green growth and a green economy has become firmly embedded in the discourse of sustainability, with the UNEP (2011a) Green Economy Report providing the new institutional orthodoxy of the significance of sustainable/green growth that has influenced tourism studies, if not the tourism industry (Hall, 2015).

According to the UNEP (2011a, p. 16), the green economy is 'one that results in improved human well-being and social equity, while reducing environmental risks and ecological scarcities'. Such a definition is as broad as that of the WCED's on sustainable development. However, like the concept of sustainable development, the acceptability – and potential weakness – of the green economy probably lies in its generality, so that it acts as a boundary object that can be positively interpreted from a wide range of different stakeholder perspectives. The UNEP also suggest, 'The concept of a "green economy" does not replace sustainable development, but there is now a growing recognition that achieving sustainability rests almost entirely on getting the economy right' (UNEP, 2011b, p. 2). So what does getting the economy right mean? (Hall, 2015). First, there is a continued commitment to growth, albeit "sustainable" and "green". Second, the UNEP (2011a) maintain that while there are a variety of causes for several concurrent crises that have unfolded since 2000: climate, biodiversity, fuel, food, water, and the global financial system, 'at a fundamental level they all share a common feature: the gross misallocation of capital' (UNEP, 2011a, p. 14). However, the trajectories of many socio-technical systems with limited environmental focus were set well before 2000, and the lifespan of such infrastructure can extend for several decades. Third, for how long can improvements in MRE efficiency and dematerialising an economy be sustained? Unfortunately, for those who advocate green growth strategies in isolation from changes in consumer behaviour, efficiency does not equal savings.

Nevertheless, the dominant approach in tourism is on technological efficiencies in reducing emissions and energy consumption.

> Tourism in a green economy refers to tourism activities that can be maintained, or sustained, indefinitely in their social, economic, cultural, and environmental contexts: "sustainable tourism". Sustainable tourism . . . aspires to be more energy efficient and more "climate sound" (e.g. by using renewable energy); consume less water; minimise waste.
>
> (UNEP, 2011a, p. 416)

The UNEP (2011a, p. 438) proposed that in a "Business As Usual" scenario for the period 2011–2050, growth in tourism over that time simply increases in energy consumption (111%), greenhouse gas emissions (105%), water consumption (150%) and solid waste disposal (252%). However, critically for evaluating the green growth possibilities of tourism, the tourism-related drawdown of natural capital still increases even in the optimistic greener investment scenario (Hall, 2015):

> the tourism sector can grow steadily in the coming decades (exceeding the BAU scenario by 7 per cent in terms of the sector GDP) while saving

> significant amounts of resources and enhancing its sustainability. The green investment scenario is expected to undercut the corresponding BAU scenario by 18 per cent for water consumption, 44 per cent for energy supply and demand, 52 per cent for CO_2 emissions.
>
> (UNEP, 2011a, p. 438)

Such a situation lead Hall (2015, p. 349) to respond: '*It is a cruel joke to describe a situation in which absolute growth in emissions and other impacts continue to expand as a result of tourism growth exceeding efficiency gains as 'green growth' or a 'green economy'*. But this is what is being done. Are you laughing yet?' (emphasis in original).

So here we are again

Green growth and the green economy is only a marginal reform of a socio-economic system geared towards economic growth. It is not a major shift in the sustainable tourism paradigm (Hall, 2011a). The dominant focus in tourism on managerial and material efficiency still does not fundamentally question either visitor growth or economic growth (Hall, 2019). The degrowth perspective to "rightsize" the visitor economy in keeping with biophysical and socio-cultural limits does offer an alternative. Such alternative tourisms have been part of the discourse of tourism since the 1970s and have, in reality, been more of a niche area of tourism and tourism research, although the attractiveness of eco-, green and sustainable tourism branding mean that some elements of alternative tourism become co-opted in the mainstream, but with only a marginal effect on the overall trajectory of tourism (Saarinen, 2015). But at least it means that some consumers and businesses can perhaps feel they are trying, or possibly they may even believe the hype of green growth, as it means they do not have to change behaviours too much.

Degrowth is an alternative approach to tourism (Hall, 2009), the characteristics of which are still being worked through. The work in this volume suggests several interesting tensions and future directions. The research on lifestyle enterprises in Scandinavia in Chapters 2 to 4 illustrate the complexities involved in shifting social and economic practices that potentially decommodify some tourism practices as the result of exchange for non-monetary repayment, or a mix of monetary and non-monetary exchange. They therefore show how market logic and gift-economic exchanges can be both intertwined and collide. Significantly, they also demonstrate that degrowth may be encouraged and sustained by practices that are not visible at the level of individual production and consumption. Some of these findings are therefore in line with previous studies that emphasise the social and cultural embeddedness of economic practices in relation to degrowth (Bogadóttir & Olsen, 2017; Gezon, 2017; Paulson, 2017). The overall impact of such businesses at a regional level may be negligible. Furthermore, Eimermann et al. (Chapter 4) concluded that the number of holistic simplifiers is low in Sweden, and while individuals meet the overall criteria for voluntary simplicity, they do not necessarily behave accordingly by cutting levels of consumption and living a simpler life. Clearly, future research on lifestyle entrepreneurship, as with entrepreneurship in general, and degrowth is welcome, particularly with respect to how entrepreneurship can meet their business and personal goals while not necessarily growing larger.

The second section of the book teased out a number of issues at the destination level. Building on some of the insights gained by looking at entrepreneurship and business strategies, a clear issue that emerges is the nature of relationships between business as well as other destination actors. Such an observation is not to reify collaboration for its own sake, but it is to support the importance of positive relationships to shift destination trajectories to more sustainable paths. Central to this possibility is a destination management organization that is sensitive to local views on local change. However, this may also require appropriate political agency for community members. What is often missing at research at this level is better insights into how those seeking degrowth solutions manage to enact them (or not) and the connection between individual and destination or community actions.

Another significant theme that emerges at the destination level and that also frames individual capacities is the role of institutions and their appropriateness. This also includes the development of appropriate tourism planning practices (see Prideaux & Pabel, Chapter 8, this volume), but there is clearly a need to better connect the relationship between policies/plans and actions, and the reasons for the implementation gap. Many of these gaps appear to arise from issues of scale, i.e. the connection of individual actors to what occurs at a destination (e.g. Kulusjärvi, Chapter 5), as well as the connection from the destination to higher levels of governance, including the international (e.g. Amore & Adie, Chapter 6). As Bernstein (2002, p. 2) notes, 'the institutions that have developed in response to global environmental problems support particular kinds of values and goals, with important implications for the constraints and opportunities to combat the world's most serious environmental problems' – the same applies to tourism. A further important point to emerge from the section is that degrowth strategies do need to be tailored to place, reminding us that in some cases appropriate degrowth may be – and paradoxically to some – increasing the number of visitors so that a particular level of economic well-being can be reached (Carson & Carson, Chapter 7).

The final section discussed the relationship between degrowth and tourism policy. The interplay between degrowth and the various forms of sustainable tourism were a major component of the critiques of tourism policy in Chapters 9 and 14 (this volume). Clearly, some of what has previously been regarded as degrowth is also potentially being described as post-growth, but the core issues remain the same. It is also interesting to see in these chapters that degrowth is positioned as part of the sequence of critiques of tourism that may broadly be described as alternative tourisms (Hall, 2015; Saarinen, 2015). Such intellectual continuity in terms of tourism research is also to be seen in Chapters 10 and 11, which focus on the community as a means of progressing a degrowth tourism agenda. In contrast, Ballantine (Chapter 12) notes the importance of domestic tourism, especially during the COVID-19 pandemic, and uses this as a way to highlight the potential role of voluntary simplicity perspectives in the consumption of tourism (Hall, 2011b; Kannisto, 2018), as opposed to the production perspective of Chapter 4. Such a contrast highlights the importance of positionality of the subject in degrowth studies and how an intellectual framing can shift depending on context. The role of institutions remains important. As Alexander (2013, p. 13) suggests, until 'a government seriously embraces a post growth model of progress – either voluntarily or by force of ecological or financial necessity – will a top down politics of voluntary simplicity be taken seriously'.

Both Ballantine (Chapter 12) and Demiroglu and Turhan (Chapter 13) also highlight the potential significance of the consumer as a change agent, the latter being particularly significant as a change agent in the case of ski tourism in a number of destinations. Such insights are extremely important as they suggest that the possible loss of valued tourism resources as the result of environmental change may provide a basis for effective pressure on business interests by consumers. This linkage also potentially provides a means for more effective social and demarketing in a tourism context (Hall, 2014). Such insights also emphasise that choices with respect to sustainability are also a problem of reflexivity at both the individual and collective level – the willingness to change yourself in order to be able to co-evolve with other humans and the environment (Polimeni et al., 2008). As Hall (2015) suggests, this applies to the tourism academy as well as the tourism industry and the wider community. This therefore means explicitly addressing 'the moral and cultural issues raised by the predominant emphasis in economic thinking on individual preferences, self-interest and competitive growth' (Ekins, 1993, p. 286) in considering how tourism education and research promulgates growthism, overconsumption and industry orthodoxies as part of supposedly 'good [BAU] practice' (Hall, 2015). To paraphrase Daly (1991), steady-degrowing economics is concerned as much with 'moral growth' as it is with biophysical equilibrium. This is also something that applies as much to corporations, institutions and policy-makers as it does to individuals.

Degrowth confronts the way in which neoliberal rationalities are embedded in many sustainable tourism policy, business and behavioural practices in a way the SDGs do not. The managerial efficiency approach advocated by the UNWTO, the WTTC and others, including most of the tourism academy, is rooted in the political and economic context of capitalistic resource extraction by which, as Hall (2019) noted, success means failure, i.e. continued growth in tourism leads to grossly uneven development and run-down of natural capital. Destination marketers and resource managers, mediated by state agencies, corporate interests and economic rationality, are therefore "managing" bio-physical and cultural resources into oblivion. In some ways this can be construed as "Brundtland-as-usual", as Brundtland's ambiguity and its advocacy by many tourism stakeholders has allowed businesses and policy-makers to equate "sustainable growth" and "green growth" or, in the case of the UNWTO and others, "sustainable tourism" with sustainable development, and they are most certainly not the same. 'The holy grail of manageability espoused by the UNWTO and others, the belief that all problems can be solved by exerting greater effort and demanding greater efficiency within the status quo of continued tourism growth and consumption, necessitates challenge' (Hall, 2019, p. 1056). It is hoped that the various chapters in this book and their insights into the potential of degrowth are part of such a challenge.

References

Alexander, S. (2013). Voluntary simplicity and the social reconstruction of law: Degrowth from the grassroots up. *Environmental Values*, *22*(2), 287–308.

Ayres, R. U., & Kneese, A. V. (1969). Production, consumption, and externalities. *The American Economic Review*, *59*(3), 282–297.

Barbier, E. (2010, April 8). How is the global green new deal going? *Nature*, *464*, 832–833.

Beckerman, W. (1972). Economists, scientists, and environmental catastrophe. *Oxford Economic Papers*, *24*, 327–344.

Bernstein, S. (2002). Liberal environmentalism and global environmental governance. *Global Environmental Politics*, *2*(3), 1–16.

Bogadóttir, R., & Olsen, E. S. (2017). Making degrowth locally meaningful: The case of the Faroese grindadráp. *Journal of Political Ecology*, *24*(1), 504–518.

Boulding, K. E. (1966). The economics of the coming spaceship earth. In H. Jarrett (Ed.), *Environmental quality in a growing economy* (pp. 3–14). Baltimore: John Hopkins University and Resources for the Future.

Carson, R. (1962 [2002]). *Silent spring*. New York, NY: Houghton Mifflin Harcourt.

Daly, H. E. (1972). In defence of a steady-state economy. *American Journal of Agricultural Economics*, *54*, 945–954.

Daly, H. E. (1974). The economics of the steady state. *The American Economic Review*, *64*(2), 15–21.

Daly, H. E. (1991). *Steady-state economics* (2nd ed.). Washington, DC: Island Press.

Dryzek, J. S. (1997). *The politics of the Earth*. New York, NY: Oxford University Press.

Ehrlich, P. (1968). *The population bomb*. New York, NY: Ballantine Books.

Ekins, P. (1993). "Limits to growth" and "sustainable development": Grappling with ecological realities. *Ecological Economics*, *8*, 269–288.

Georgescu-Roegen, N. (1971). *The entropy law and the economic process*. Cambridge, MA: Harvard University Press.

Gezon, L. L. (2017). Beyond (anti)utilitarianism: Khat and alternatives to growth in northern Madagascar. *Journal of Political Ecology*, *24*(1), 582–594.

Gleditsch, N. P. (1975). Slow is beautiful. The stratification of personal mobility, with special reference to international aviation. *Acta Sociologica*, *18*(1), 76–94.

Hall, C. M. (2009). Degrowing tourism: Décroissance, sustainable consumption and steady-state tourism. *Anatolia*, *20*(1), 46–61.

Hall, C. M. (2011a). Policy learning and policy failure in sustainable tourism governance: From first and second to third order change? *Journal of Sustainable Tourism*, *9*, 649–671.

Hall, C. M. (2011b). Consumerism, tourism and voluntary simplicity: We all have to consume, but do we really have to travel so much to be happy? *Tourism Recreation Research*, *36*(3), 298–303.

Hall, C. M. (2014). *Tourism and social marketing*. Abingdon: Routledge.

Hall, C. M. (2015). Economic greenwash: On the absurdity of tourism and green growth. In V. Reddy & K. Wilkes (Eds.), *Tourism in the green economy* (pp. 339–358). London: Earthscan.

Hall, C. M. (2019). Constructing sustainable tourism development: The 2030 agenda and the managerial ecology of sustainable tourism. *Journal of Sustainable Tourism*, *27*(7), 1044–1060.

IPCC. (2018). *Global warming of 1.5°C. An IPCC Special Report on the impacts of global warming of 1.5°C above pre-industrial levels and related global greenhouse gas emission pathways, in the context of strengthening the global response to the threat of climate change, sustainable development, and efforts to eradicate poverty*. Geneva, Switzerland: World Meteorological Organization.

Jackson, T. (2009). *Prosperity without growth*. London: Earthscan.

Kannisto, P. (2018). Travelling like locals: Market resistance in long-term travel. *Tourism Management*, *67*, 297–306.

Meadows, D. H., Meadows, D. L., Randers, J., & Behrens, W. W. (1972). *Limits to growth: A report for the Club of Rome's project on the predicament of mankind*. New York, NY: Universe Books.

Mishan, E. J. (1967). *The costs of economic growth*. London: Staples Press.

Mishan, E. J. (1970). What is wrong with Roskill? *Journal of Transport Economics and Policy*, *4*, 221–234.

Newell, R., & Dale, A. (2020). COVID-19 and climate change: An integrated perspective. *Cities & Health*. Retrieved from https://doi.org/10.1080/23748834.2020.1778844

Odum, H. T. (1971). *Environment, power and society*. New York, NY: Wiley-Blackwell.

Paulson, S. (2017). Degrowth: Culture, power and change. *Journal of Political Ecology*, *24*(1), 425–448.

Perez-Carmona, A. (2013). Growth: A discussion of the margins of economic and ecological thought. In L. Meuleman (Ed.), *Transgovernance: Advancing sustainable governance*. Dortrecht: Springer.

Pigou, A. C. (1943). The classical stationary state. *The Economic Journal*, *53*, 343–351.

Polimeni, J. M., Mayumi, K., Giampietro, M., & Alcott, B. (2008). *The Jevons Paradox and the myth of resource efficiency improvements*. London: Earthscan.

Saarinen, J. (2015). Conflicting limits to growth in sustainable tourism. *Current Issues in Tourism*, *18*(10), 903–907.

Stern, N. (2007). *The economics of climate change: The stern review*. Cambridge: Cambridge University Press.

United National Environment Programme (UNEP). (2011a). *Towards a green economy: Pathways to sustainable development and poverty eradication*. Nairobi: UNEP.

United National Environment Programme (UNEP). (2011b). *Towards a green economy: Pathways to sustainable development and poverty eradication – A synthesis for policy makers*. Nairobi: UNEP.

World Commission on Environment and Development (WCED). (1987). *Our common future* ["The Brundtland report"]. Oxford: Oxford University Press.

Index